T0178284

Vapor Liquid Two Phase Flow and Phase Change

Sarit Kumar Das · Dhiman Chatterjee

Vapor Liquid Two Phase Flow and Phase Change

Ane Books
Pvt. Ltd.

Sarit Kumar Das
Department of Mechanical Engineering
Indian Institute of Technology Madras
Chennai, Tamil Nadu, India

Dhiman Chatterjee
Department of Mechanical Engineering
Indian Institute of Technology Madras
Chennai, Tamil Nadu, India

ISBN 978-3-031-20926-0   ISBN 978-3-031-20924-6   (eBook)
https://doi.org/10.1007/978-3-031-20924-6

Jointly published with ANE Books India.
In addition to this printed edition, there is a local printed edition of this work available via Ane Books in
South Asia (India, Pakistan, Sri Lanka, Bangladesh, Nepal and Bhutan) and Africa (all countries in the
African subcontinent).
ISBN of the Co-Publisher's edition: 978-9-385-46227-6

This Springer imprint is published by the registered company Springer Nature Switzerland AG
The registered company address is: Gewerbestrasse 11, 6330 Cham, Switzerland

Dedicated to

Prof. W. Roetzel

and

Prof. V. H. Arakeri

who have inspired us to understand two-phase flow

Foreword

Engineers were using phase change devices even before James Watt separated the steam engine's condenser from its working cylinder in 1764. Today, phase change underpins numerous technologies that ease modern life. Electrical power generation, food refrigeration and air conditioning are each enabled by boiling and condensation.

After so much time, could anything more be said on this topic? Indeed, yes! New tools and growing datasets have opened new insights and opportunities. Modern numerical methods can simulate complex bubble dynamics from fundamental equations. Surface engineering and microfabrication technology have unveiled boilers and condensers with previously unimaginable properties. And vast amounts of high-quality data now support ever-improving models of the two-phase flows that cool nuclear reactors.

Nonetheless, phase change embodies a complexity not found in other heat transfer problems. Essential phenomena continue to defy systematic modelling across configurations, despite more than 100 years of research. As a child, I often visited my father's boiling laboratory. I remember well his 1960's experiments on pool boiling—especially those conducted in a centrifuge! Yet, even today, new designs lead us to the same basic questions: what determines the critical heat flux and how can we control it? Until we have comprehensively resolved these puzzles, engineers will need to understand the fundamental physics of two-phase systems, to develop models and to apply the latest knowledge.

I first met Prof. Sarit Das at the ISHMT conference in Guwahati in January 2006. There, I was startled by an audience member who asked an incisive question about evaporation in my liquid jet experiments. It was Sarit. Since that day, he and I have been good friends. Two years later, I hosted Sarit for a one-year visit to MIT. In 2011, we again appointed him as a visiting professor at MIT. Sarit's extensive research has touched most areas of heat transfer. And his work's high quality resulted in his election to both the Indian National Academy of Sciences and the Indian National Academy of Engineering.

Here, Profs. Das and Dhiman Chatterjee give an up-to-date account of our best knowledge of the subject. Their work has a strong focus on the two-phase flows that are of the greatest industrial importance. Their approach uniquely covers pressure-driven processes alongside thermally driven processes. And they discuss recent concepts in microengineering for two-phase devices. I invite the reader to enjoy and learn from this important new book.

October 2020   John H. Lienhard V
                                                  Massachusetts Institute of Technology
                                                                     Cambridge, USA

Preface

Before beginning to write any academic text, the authors have to come across the central question, "Why are we writing this book, what's the need for this book for this area of study?" A book on phase change in the gas–liquid regime is even more under question because the developments in this particular area spread over more than a century. Reputed and time-tested texts such as those by Collier and Thome (Convective boiling and condensation, 1972), Tong and Tang (Boiling heat transfer and two-phase flow, 1997) and edited volumes like Multiphase flow handbook (Eds, Michaelidas, Crowe and Schwarzkopf, 2016) exist, which cover almost the entire spectrum of studies in this field. Naturally, the need for another text in this area is questionable. However, the authors in course of guiding the masters and doctoral students found that the above texts, although quite comprehensive, are pitched at a level that is in between basic phenomenological understanding and state-of-the-art research. This makes them overwhelming to beginners in the area of two-phase heat transfer. Personally, we have experienced even cases where young scholars after the first reading have decided not to venture into this field because they found it to be "too complex and difficult". There are also scholars who found that the area is "more empirical and observational than scientific". In our opinion, these impressions are due to the lack of an introductory text that will approach the subject in a simple yet scientific way where the physical explanation combined with mathematical analysis can give an insight to a beginner which can kindle interest in him/her.

Apart from the above novelty in approach, the book contains a number of unique features. Usually, liquid–vapour phase change texts are limited to boiling and condensation alone, stressing on the thermal aspects of these phenomena. However, the pressure-dominated phase change like cavitation is usually not included in such books. We feel that from nucleation to bubble dynamics, there are a lot of common grounds between boiling and cavitation and hence decided to make it a part of the present text (Chap. 5).

The other unique feature of this book is the inclusion of the materials which happen to be the focus of research in two-phase heat transfer today. Boiling in mini and microchannels (Chap. 16) is one such chapter. The other one is an elaborated chapter on numerical modelling of two-phase flow and heat transfer (Chap. 14).

It goes without saying that these are included for the readers interested in these areas only and can be left out by others. Chapter 15 is devoted to the equipment for boiling and condensation. We felt that it is necessary to familiarize the reader with the devices used in different practical applications as these topics may be useful for their professional career.

We take this opportunity to thank a large number of people who helped us to bring out this text. First, we express our gratitude to Prof. B. Premachandran of Indian Institute of Technology Delhi who helped us to completely reframe and even write parts of the chapter on numerical modelling of two-phase heat transfer (Chap. 14). Thanks are also due to Dr. S. V. Diwakar of JNCASR for contributing the portion on QUASI algorithm in the same chapter. We also thank Dr. M. Venkatesan for his contributions towards a portion on boiling in microchannels and rewriting parts of Chap. 16. Next, we must thank our students in general and Mr. Allwyn Blessing Johnson, in particular, who have contributed in various ways from conceptualization to execution of this book writing project. Special thanks are due to Ms. Anitha and Ms. Maninder Kaur who prepared the manuscript along with the figures with utmost dedication. We also thank our collaborators who have enhanced our own understanding of the subject. To mention a few of them are Prof. A. R. Balakrishnan (IIT Madras) and Prof. Kausik Sarkar (University of Delaware). We thank profoundly the publisher Ane Book for their patience in waiting long for the completion of this book and the technical support they provided us. Last but not the least is the sacrifices made by our families who never complained to find that this book writing activity is further shrinking the already scarcely available time for them.

Finally, the proof of the pudding is in the eating. It is the readers, scholars and students who will decide the worth of this book. If they feel that this book has helped them anyway to grasp this rather complex yet interesting subject, the authors will consider their efforts to have been rewarded.

Chennai, India   Sarit Kumar Das
                                                                         Dhiman Chatterjee

Contents

About the Authors

Prof. Sarit Kumar Das is currently an institute professor of IIT Madras, in the Mechanical Engineering Department. After obtaining B.E. (Mechanical) in 1984 and M.E. (Heat Power) in 1987 from Jadavpur University, he completed his Ph.D. in the area of heat transfer from Sambalpur University in 1994 and joined University of BW Hamburg, Germany, where he did his post-doctoral research in the area of plate and liquid metal heat exchangers. He has published four books and more than 300 research papers. His current research interests include heat transfer in nanofluids, microfluidics, biological heat transfer, nanoparticle mediated drug delivery in cancer cells, heat exchangers, boiling in mini/microchannels, fuel cells, jet instabilities, heat transfer in porous media and computational fluid dynamics.

Dr. Dhiman Chatterjee is currently a professor in the Department of Mechanical Engineering at Indian Institute of Technology (IIT) Madras, India. He received his B.E. (Hons) in Mechanical Engineering from Jadavpur University, Kolkata, and M.S. and Ph.D. from IISc, Bangalore, India. He worked as a post-doctoral fellow at the University of Delaware, USA, before joining IIT Madras in 2004. His area of research includes bubble dynamics and cavitation and turbomachines. He has published 75 papers in journals and international conferences. At IIT Madras, he is involved in teaching courses like turbomachines, cavitation, energy conversion system, and incompressible fluid flow.

Nomenclature

A	Total cross-sectional area of the tube
a	Coefficient of evaporation
A_g	Fraction of the cross-sectional area of the tube occupied by gas
A_l	Fraction of the cross-sectional area of the tube occupied by liquid
Ar	Archimedes number
b	Laplace constant
Bo	Bond number
C	Chisolm parameter
C	Colour function
C_0	Saturation concentration of the dissolved gas in the liquid at a given temperature
C_p	Liquid specific heat
C_1	Absolute velocity at the inlet of pump impeller
C_∞	Concentration of a dissolved gas in a liquid far away from a bubble
c_i	Molar concentration of i-th component
D_l	Lift-off diameter
D_b	Bubble departure diameter
D_h	Hydraulic diameter
e	Energy convected per unit mass
E_s	Suction elevation
F	Enhancement factor
F	Volume fraction
f	Departure frequency
F_{B1}	Primary Bjerknes force
F_b	Buoyant force
F_n	Froude number
f_{sa}	Continuum surface force
f_{sv}	Interfacial volume force
F_s	Surface tension force
G	Gibbs free energy, Gibbs function
G	Mass velocity, or mass flux

g	Acceleration due to gravity
g	Molar Gibbs function
G_b	Gibbs number
Gr	Grashof number
H	Enthalpy
h	Heat transfer coefficient
h	Specific enthalpy
h_{ev}	Evaporative heat transfer coefficient
h_{fg}	Enthalpy of vaporization
h_l	Specific enthalpy of liquid
j	Volume flux
Ja	Jacob number
k	Polytropic exponent
K_a	Kapitza resistance
K_B	Bulk modulus
K_g	Kutateladze number
K_l	Kutateladze number
k_B	Boltzmann constant
k_f	Liquid thermal conductivity
L'	Laplace number
Le	Lewis number
MW_i	Molecular weight of i-th component
n	Number of moles
n	Unit normal vector
N_c	Nucleation site density
Nu	Nusselt number
p	Local static pressure
$P_A(t)$	Acoustic pressure amplitude
Pe	Peclet number
Pr	Prandtl number of the liquid
q_c	Transient heat conduction
q_{ME}	Microlayer evaporative heat flux
q_{tot}	Heat flux in nucleate boiling
R	Bubble radius
R_g	Gas constant
R_{max}	Maximum radius of a bubble
Ra	Surface roughness
Re	Reynolds number
S	Entropy
S	Slip
S	Suppression factor
s	Specific entropy
T	Period of the acoustic pressure
T	Temperature
t	Time

t_b	Bubble growth period
T_f	Mean film temperature
t_w	Waiting period
U	Superficial velocity
U	Velocity
u	Phase velocity
V	Volume of the combined two phases
$V(t)$	Instantaneous bubble volume
V_{ref}	Characteristics velocity
W	Weight
W_1	Relative velocity at the inlet of pump impeller
We	Weber number
X	Martinelli parameter
x^*	Flowing mass quality, or quality
x_{th}^*	Thermodynamic quality
x_i	Mole fraction of i-th component of liquid
y_i	Mole fraction of i-th component of vapour
Z	Length along flow direction
Z_s	Suction pipe friction loss

Symbols

$< q(t) >_{t,4}$	Weighted average of a quantity $q(t) = \dfrac{\int_0^T q(t) R^4 dt}{\int_0^T R^4 dt}$
α	Liquid thermal diffusivity
δt	Time step
δ	Film thickness
$\dot{M}$	Two-phase mass flow rate
$\dot{q}$	Heat flux
$\dot{V}_g$	Gas volume flow rate
$\dot{V}$	Two-phase volume flow rate
$\frac{dp}{dz}$	Pressure gradient
γ	Inclination angle
γ	Surface liquid interaction parameter
κ	Curvature of the interface
λ	Wavelength
λ_{KH}	Wavelength of Kelvin–Helmholtz wave
λ_T	Wavelength of Taylor wave
μ	Chemical potential
μ	Dynamic viscosity
$\nabla P_A(t)$	Gradient of the acoustic pressure at the bubble location
∇_s	Surface gradient
ν	Specific volume

ω	Driving frequency of piezoelectric crystal
ω_n	Resonant frequency of a bubble
$\bar{\chi}$	Mean of any quantity, χ
$\bar{F}$	Volumetric drag force
ϕ	Level set function
ϕ^2	Two-phase multiplier
ρ	Density
Σ	Cavitation number
σ	Surface tension
τ_w	Wall shear stress
θ	Contact angle
ε	Void fraction
ε^*	Flowing volumetric quality
ζ	Mass fraction

Subscripts

0	Outside, ambient, initial
a	Acceleration
b	Boiling
c	Critical
$conv$	Convective
$crit$	Critical
f	Frictional
fg	Liquid–vapour
g	Gas
g	Gravity
h	Homogeneous
i	Inside
i	Interface
in	Inlet
l	Liquid
ll	Laminar, laminar
lt	Laminar, turbulent
m	Mean
nc	Natural convection
r	Radial direction, radial coordinate
sat	Saturated
sub	Subcooled
tl	Turbulent, laminar
tt	Turbulent, turbulent
v	Vapour
w	Wall

Abbreviations

CB	Convective boiling
CHF	Critical heat flux
FC	Forced convection
FDB	Fully developed boiling
$LMTD$	Log mean temperature difference
NB	Nucleate boiling
$NPSH_a$	Net positive suction head, available
$NPSH_r$	Net positive suction head, required
ONB	Onset of nucleate boiling
SCB	Subcooled boiling
SPL	Single-phase liquid
TP	Two-phase

List of Figures

List of Tables

Chapter 1
Introduction to Two-Phase Flow—Flow Patterns and Maps

The subjects of two-phase gas–liquid flow and phase change processes have become increasingly important in a wide variety of engineering and technological applications as well as natural and biological processes. Although in the present text our main aim is to understand phase change phenomena (like boiling, condensation or cavitation), it must be understood that two-phase flow may occur even without phase change such as in air–water or oil–gas flows. The applications of adiabatic and diabatic two-phase flows in mechanical systems include thermal power plants, internal combustion engines, pumps, hydraulic turbines, naval propellers, heat exchangers, etc. Gas–liquid flows also occur in chemical process plants, nuclear technology and petrochemical industries. In recent times, realizing the limitations of gas and liquid cooling technology, the electronic cooling technology has also turned towards flow evaporation techniques. In all these applications proper understanding of two-phase flows can help in better design of equipment. This forms the scope of the present chapter as well as the next chapter.

1.1 Definitions Related to Two-Phase Flow

To specify two-phase flow, some basic definitions are important to know. In all these definitions the subscript 'g' stands for gas phase and 'l' stands for liquid phase, respectively. Let us first refer to Fig. 1.1 to show an annular gas–liquid flow in a vertical tube where the liquid phase is near the wall and the gas phase is at the core of the tube.

At any cross section, gas fill is given by the fraction of cross-sectional area covered by the gas as

$$\varepsilon = \frac{A_g}{A}, \tag{1.1}$$

© The Author(s) 2023
S. K. Das and D. Chatterjee, *Vapor Liquid Two Phase Flow and Phase Change*, https://doi.org/10.1007/978-3-031-20924-6_1

Fig. 1.1 Annular two-phase
flow in a vertical channel.
(*Courtesy:* "Heat Transfer in
Condensation and Boiling"
by Karl Stephan, with
permission from Springer)

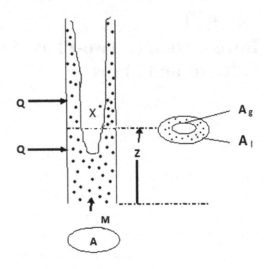

where A is the total cross-sectional area of the tube and A_g is the area of a particular
axial location which is filled by gas. Similarly, the liquid area fraction is given by

$$\frac{A_l}{A} = 1 - \frac{A_g}{A} = 1 - \varepsilon, \tag{1.2}$$

because $A = A_g + A_l$.

Let us now consider a small length of the tube Δz over which the fractions of
liquid and gas may be assumed to remain unchanged. Hence

$$\varepsilon = \frac{A_g \Delta z}{A \Delta z} = \frac{V_g}{V}, \tag{1.3}$$

where V is the volume of the combined two phases and V_g that of the gas phase only.
This ratio of volumes is called 'void fraction' and is also indicated by ε. Thus, the
amount of liquid by volume is given by

$$1 - \varepsilon = \frac{V_l}{V}. \tag{1.4}$$

This void fraction is also known as 'volumetric vapour content'. However, when the
fluids flow, one more quantity that is similar to the void fraction is often used. This is
called 'flowing volumetric quality', where, instead of static volumes, the flow rates
of fluids are used.

$$\varepsilon^* = \frac{\dot{V}_g}{\dot{V}}. \tag{1.5}$$

Here ε^* is the flowing volumetric quality, $\dot{V}$ stands for the two-phase volume flow rate and $\dot{V}_g$ for gas volume flow rate. In the same way, a 'flowing mass quality' (or just 'quality') is also defined, which is the ratio of the gas mass flow rate ($\dot{M}_g$) and the mass flow rate of the total two-phase fluid ($\dot{M}$)

$$x^* = \frac{\dot{M}_g}{\dot{M}}, \tag{1.6}$$

where $\dot{M} = \dot{M}_g + \dot{M}_l$.

Hence

$$1 - x^* = \frac{\dot{M}_l}{\dot{M}} \tag{1.7}$$

It has to be understood that this 'quality' is different from the thermodynamic quality (x_{th}^*) that we get from tables like steam table. Thermodynamic quality is the ratio of the mass of vapour and the vapour liquid mixture under the state of thermodynamic equilibrium and can be given by

$$x_{th}^* = \frac{M_g}{M} \tag{1.8}$$

The velocity of the two phases can thus be evaluated as

$$u_g = \frac{\dot{M}_g}{\rho_g A_g} = \frac{x^* \dot{M}}{\rho_g \varepsilon A} = \frac{x^* G}{\rho_g \varepsilon}, \tag{1.9}$$

$$u_l = \frac{\dot{M}_l}{\rho_l A_l} = \frac{(1 - x^*) \dot{M}}{\rho_l (1 - \varepsilon) A} = \frac{(1 - x^*) G}{\rho_l (1 - \varepsilon)}, \tag{1.10}$$

where G is called the 'mass velocity' or 'mass flux' and is defined as the mass flow rate per unit area having a unit of $kg/m^2 s$. Thus

$$G = \frac{\dot{M}}{A} \tag{1.11}$$

In general, the velocity of the liquid and the gas are different. This difference is designated by 'slip (S)', which is defined as the ratio of the two velocities

$$S = \frac{u_g}{u_l} = \left(\frac{x^*}{1 - x^*}\right)\left(\frac{1 - \varepsilon}{\varepsilon}\right)\left(\frac{\rho_l}{\rho_g}\right) \tag{1.12}$$

There may be special cases of two-phase flow where the liquid and gas can be assumed to flow with the same velocity ($u_g = u_l$) giving a slip value of unity ($S = 1$). Under this condition, void fraction and flowing volumetric quality become identical

$$\varepsilon^* = \frac{\dot{V}_g}{\dot{V}} = \frac{u_g A_g}{u_g A_g + u_l A_l} = \frac{A_g}{A} = \varepsilon. \tag{1.13}$$

The flowing mass quality and flowing volumetric quality are related to each other as

$$x^* = \frac{\dot{M}_g}{\dot{M}} = \frac{\dot{M}_g}{\dot{M}_g + \dot{M}_l} = \frac{\rho_g \dot{V}_g}{\rho_g \dot{V}_g + \rho_l \dot{V}_l} = \frac{\rho_g \varepsilon^* \dot{V}}{\rho_g \varepsilon^* \dot{V} + \rho_l(1 - \varepsilon^*)\dot{V}} \tag{1.14}$$

$$x^* = \frac{\varepsilon^*}{\varepsilon^* + (1 - \varepsilon^*)\frac{\rho_l}{\rho_g}}. \tag{1.15}$$

There is one more parameter often used for two-phase flow, particularly in adiabatic flows. It is superficial velocity. It is the velocity of one phase if it would have occupied the entire tube cross section through which the two-phase mixture is flowing. Thus, gas superficial velocity is given by

$$U_g = \frac{\dot{V}_g}{A} = \frac{\dot{M}_g}{\rho_g A} \tag{1.16}$$

Similarly, liquid superficial velocity is

$$U_l = \frac{\dot{V}_l}{A} = \frac{\dot{M}_l}{\rho_l A} \tag{1.17}$$

It may be noted that in both Eqs. 1.16 and 1.17 that the denominator is the total area of the tube. In calculating actual velocity of the gas and liquid phases, we have used gas and liquid areas, respectively, (please refer to Eqs. 1.9 and 1.10). Thus, the superficial velocities are fictitious (not actual) velocities and are defined for convenience. In flows with phase change, the flow rate of each phase (liquid or vapour) changes continuously because of mass transfer from one phase to another. Thus, the mass flow rate of each phase also changes. The liquid flow rate decreases and the vapour flow rate increases or *vice versa*. This causes the local quality of the mixture to change from the entry to the exit of the tube. This change of quality can be calculated using energy balance if the heat flux at the wall is known. Assuming that the liquid enters the tube at a subcooled state (at a temperature lower than the saturation temperature at the liquid pressure) with a specific enthalpy of h_l, and heated by a heating rate of q, neglecting kinetic and potential energy, the energy balance between entry and any location z, as shown in Fig. 1.2, gives

$$\dot{M}h_1 + q = \dot{M}[x^* h_g + (1 - x^*)h_l] \tag{1.18}$$

Now, if we assume that the vapour and liquid at that cross section are in thermodynamic equilibrium, then the enthalpies of the two phases at that cross section will

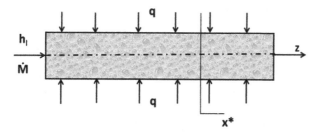

Fig. 1.2 Energy balance and associated phase change for constant wall heat flux

be the corresponding saturation enthalpies at local pressure $p(z)$ (as given in the steam table for water–steam flow). Thus, $h_g = h_g$ and $h_l = h_f$ (as used in phase equilibrium table, e.g., steam table for water–steam system)

Also, $h_{fg} = h_g - h_f$ is the latent heat of vaporization at the local pressure. Thus, we get

$$\dot{M}h_1 + q = \dot{M}[x_{th}^* h_g + (1 - x_{th}^*)h_f] = \dot{M}[x_{th}^* h_{fg} + h_f] \qquad (1.19)$$

Please note that we have used h_g to indicate saturation enthalpy of vapour, please distinguish it from the usually used gas enthalpy h_g. With this equation, we can calculate the local mean quality if we know the heating rate. The heating rate in this case can be the heat flux times the tube surface area between entry and position z measured along the axis.

$$q = \dot{q}\pi d_0 z, \qquad (1.20)$$

where $\dot{q}$ is the outer wall heat flux and d_0 is outer diameter of the tube.

The quality calculated from Eq. 1.19 is known as thermodynamic quality because it is calculated on the assumption of thermodynamic equilibrium. This quality differs from the actually observed quality, (i.e., if photographically liquid and vapour flow rates are determined). This happens more at the beginning and the end of the boiling process as shown in Fig. 1.3 due to lack of thermodynamic equilibrium.

At the beginning, bubbles are formed ($x_{real} > 0$) at the wall when the liquid core is still at a subcooled temperature, but the liquid near wall is hot enough to form bubbles. The average enthalpy here is still below the saturation enthalpy h_f and hence the thermodynamic quality calculated from Eq. 1.19 is negative. After the liquid dry out towards the end of the boiling process, superheated vapour flows near the wall, while still liquid droplets remain near the tube core. In this case, the enthalpy is higher than h_g and $x_{th}^* > 1$ thermodynamically, while due to the presence of liquid droplets $x*_{real} < 1$.

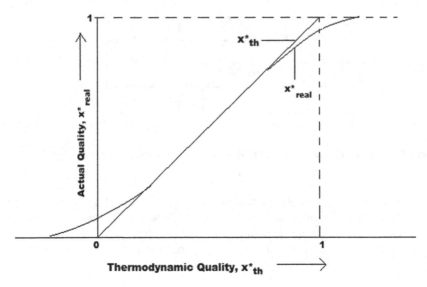

Fig. 1.3 Thermodynamic and actual quality for boiling inside a tube. (*Courtesy:* "Heat Transfer in Condensation and Boiling" by Karl Stephan, with permission from Springer)

1.2 Two-Phase Flow Patterns and Maps

Gas–liquids flows in ducts or channels occur in various patterns. The patterns essentially indicate various stages in which one phase is mixed with the other phase. The various forms in which the two phases coexist are primarily determined by the forms of the interface. The two phases can be different or the same fluid may be present in gaseous and liquid states. For example, air–water flow is an example of different fluids, while water–water vapour flow is a common example of same fluid flow. The major classification of flow patterns is made on the basis of heat transfer and the consequent phase change. When no heat transfer takes place to or from the system, it is called an adiabatic flow pattern. When heat is transferred to/from the system, which causes a change of phase, it is called flow boiling or flow condensation depending on whether liquid is transformed to vapour or *vice versa*. Another category of liquid-to-vapour phase change occurs in cavitation due to the reduction in pressure in the liquid. This will be dealt with separately. In the case of flow boiling and condensation, the flow pattern changes continuously within the conduit which is in sharp contrast with the adiabatic flow where the flow pattern remains more or less unchanged along the flow path. In the following sections, we will describe the adiabatic and phase change flow patterns. In reality, 'flow pattern' is a simplified term commonly used in the literature on two-phase flow. The more accurate term is 'flow regime' which indicates a region with a particular mode of influence of a type of force. These forces are:

1. Inertial forces of the liquid and gas
2. Buoyancy forces due to density difference of liquid and gas

3. Surface tension force
4. Viscous force

Depending on the magnitude of these forces relative to each other, the flow regime changes. However, flow pattern generally indicates morphological structures of the two phases which are visually observed such as bubbly flow, slug flow, annular flow, etc. Near the transition boundary of various flow regimes (such as between homogeneous and separated flows), we often observe hybrid flow patterns such as slug–annular, drop–annular flow, etc. Now, it is important that one understands the flow regimes and patterns not only in terms of their geometrical forms, but also in terms of the interaction of various forces in a two-phase flow. Based on these factors, the flow patterns can be analyzed for the four major cases stated below:

1. Adiabatic vertical flow
2. Adiabatic horizontal flow
3. Non-adiabatic vertical flow
4. Non-adiabatic horizontal flow

We will discuss adiabatic and non-adiabatic flows in the following sections.

1.2.1   Adiabatic Flow Regimes and Patterns

Adiabatic flow patterns are observed in tubes or channels without heat transfer. Since no change of phase takes place, the flow pattern more or less remains unchanged here unless disruption or instabilities are introduced by factors such as friction, impurities, etc. Two limiting cases namely, vertical and horizontal flow directions are given below as Hewitt (1982) has shown that the flow pattern are quite different in the two configurations.

1.2.1.1   Vertical Upward Flow in Tubes

Various flow patterns for upward vertical flow in unheated tubes are shown in Fig. 1.4. The patterns are: (a) bubbly flow, (b) plug or slug flow, (c) churn flow, (d) whispy annular flow, (e) annular flow and (f) drop or spray flow.

Bubbly flow (Fig. 1.4a) has liquid as continuous phase with large number of small bubbles nearly uniformly dispersed in it. The bubbles are spherical. This flow pattern is observed when the gas component is small compared to the liquid. This type of flow has a uniform mixture of gas and liquid, and hence bubbly flow can be treated as a homogeneous flow. The plug flow (Fig. 1.4b) (or sometimes called the slug flow) contains larger slugs of gas formed by the coalescence of the bubbles. The slugs are usually of bullet shape as shown in the figure. The slugs cover almost the entire cross section of the tubes. Between two slugs, smaller bubbles remain dispersed in the liquid. Churn flow (Fig. 1.4c) occurs when the mass flux of two-phase flow

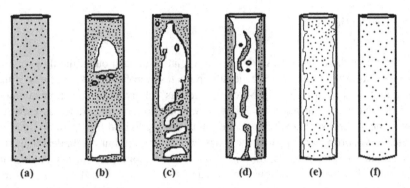

Fig. 1.4 Flow patterns in vertical tube with upward flow. **a** bubbly flow, **b** plug or slug flow, **c** churn flow, **d** whispy annular flow, **e** annular flow, **f** drop or spray flow. (*Courtesy:* "Heat Transfer in Condensation and Boiling" by Karl Stephan, with permission from Springer)

increases. This results in the disintegration of the slug giving unstable flow. The flow is often oscillatory with liquid near the wall continuously pulsating up and down. Tubes of large diameter and flow under high pressure produce such types of flow patterns. Whispy annular flow (Fig. 1.4d) is an intermediate flow pattern between the churn and annular flows. Here the liquid layer is thick at the wall and discontinuous inside the gas core. The liquid layers and small bubbles coexist along with some liquid droplets. Annular flow (Fig. 1.4e) is a more frequently observed flow pattern. Here, the liquid sticks to the wall in the form of an annular film. However, a large number of small liquid droplets remain distributed in the gas core region of the tube. With a further increase in gas velocity, the liquid disintegrates and forms a large number of small liquid droplets that remain suspended in the gas (Fig. 1.4f). This is just the inverse case of bubbly flow where the gas phase is continuous and the liquid remains in the small spherical well-dispersed state. Needless to say that this happens at a small liquid fraction. This pattern can also be treated like a single homogeneous fluid flow similar to that in a bubbly flow. It can also be mentioned here that channels of non-circular cross section also exhibit similar flow patterns. However, they critically depend on symmetry and aspect ratio of the cross section.

1.2.1.2 Horizontal Adiabatic Flow Patterns

Horizontal tubes show significantly different flow patterns compared to vertical tubes due to the influence of gravity forces. Since the gas density is lower than the liquid density, buoyancy force acts on the gas phase making it drift towards the upper part of the tube. The effect of gravity (and hence buoyancy) becomes more dominant at lower inertia force, i.e., when the flow velocity is small. Figure 1.5 shows various flow patterns in a horizontal tube.

It can be seen that, like the vertical upflow, here also bubbly flow (Fig. 1.5a) occurs. However, here the bubbles instead of being uniformly dispersed in the liquid get

Fig. 1.5 Flow patterns in horizontal tube without heat transfer. **a** bubbly flow **b** plug flow **c** stratified flow **d** wave flow **e** slug flow **f** annular flow and **g** spray or drop flow. (*Courtesy:* "Heat Transfer in Condensation and Boiling" by Karl Stephan, with permission from Springer)

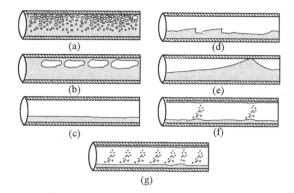

drifted towards the upper part of the tube. Subsequently, coalescence of the bubbles gives rise to plug flow (Fig. 1.5b) which remains in the upper part of the tube with the gas phase touching the wall. A new pattern, known as stratified flow (Fig. 1.5c) is observed when the gas and liquid velocities are small. The gas–liquid interface remains undisturbed only at extremely small gas velocities. With an increase of gas velocity, the interface becomes wavy giving rise to wavy flow (Fig. 1.5d). The waves after getting amplified make the liquid touch the upper wall giving rise to slug flow (Fig. 1.5e). As the velocity is further increased, annular flow (Fig. 1.5f) results in. However, unlike the vertical flow, the liquid layer in the horizontal annular flow is not of uniform thickness. The upper wall has thinner and the lower wall thicker liquid layer thickness. Finally, at high gas velocity and low liquid fraction, we get drop flow which is similar to vertical drop flow. It has to be mentioned that different authors use various nomenclature for flow patterns, and hence about 100 different names for flow patterns are available but they are essentially the same as mentioned above or are minor variations of those.

1.2.2 Non-adiabatic Flow Regimes and Patterns

Non-adiabatic flow patterns pertain to two-phase flow with heat transfer leading to phase change. Both boiling and condensation give such patterns but the flow patterns for these two processes have some differences. Hence, we will restrict our discussion to flow boiling here. The patterns for condensation of flowing vapour will be discussed in appropriate place. One important difference between adiabatic and non-adiabatic flow pattern is that, in adiabatic flow unless the flow conditions (like volume fraction, flow velocity) are changed, the pattern remains more or less unchanged throughout the tube. On the other hand, in flow boiling the flow regime continuously changes along the tube because liquid changes into vapour, resulting in a change in vapour volume fraction and velocity. Non-adiabatic flow can also be studied for vertical and horizontal configurations due to their difference owing to the gravity effect.

1.2.2.1   Non-adiabatic Vertical Flow in Tubes

When evaporation of liquid takes place in a tube during the upflow of fluid with heat being supplied from the wall, various flow patterns occur. Decreasing the amount of liquid and increasing the amount of vapour give rise to various shapes of vapour bubbles, slugs or core. Typically, boiling flow regimes are shown in Fig. 1.6. It can be seen that liquid enters the tube in a subcooled state. As the vapour starts forming near the wall, we get bubbly flow regime. It can be seen that more bubbles are observed near the tube wall because the bubbles condense near the liquid core which remains at a lower temperature than the liquid near the wall. Further up, the core liquid gets heated to saturation temperature and the bubbles remain more or less uniformly distributed over cross section.

Subsequently, the bubbles coalesce to form slugs giving rise to plug (or slug) flow regime. Further evaporation forms continuous vapour core. Towards the beginning it remains irregular in shape giving rise to semi-annular flow. This regime is associated with some amount of vapour bubbles. Further up, a symmetrical annular liquid layer

Fig. 1.6 Flow patterns in non-adiabatic vertically upward flow. (*Courtesy:* "Heat Transfer in Condensation and Boiling" by Karl Stephan, with permission from Springer)

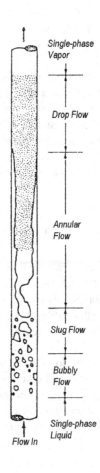

Fig. 1.7 Vertical up flow
boiling with mixed flow
patterns

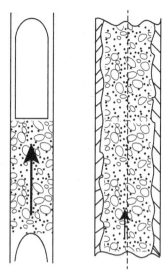

flows near the wall while the vapour flows at the core. In this region, bubbles decrease
gradually and finally disappear. The vapour at the core is also not pure vapour, but
it contains fine liquid droplets. The evaporation in this regime takes place from the
liquid vapour interface. This makes the liquid layer thinner and thinner. Finally the
liquid film completely disappears, causing what is known as 'dry out'. This makes
the vapour to come in contact with the tube wall and and this results in an increase
of the temperature of the wall. However, still finely divided liquid droplets remain
suspended in the vapour. This regime is known as drop flow. Finally, all the droplets
evaporate and we get what is known as single phase vapour flow. It must be kept in
mind that it is not always that such sharp changes from one pattern to other can be
observed. Quite often we see intermediate flow patterns such as bubble–slug, slug–
annular or drop–annular flow. Figure 1.7 shows such intermediate flow patterns.

1.2.2.2 Non-adiabatic Flow Patterns in Horizontal Tubes

Although flow in horizontal tubes gives essentially similar flow patterns as those
in vertical tubes, the gravity effect brings about important change in vapour phase
morphology. Like adiabatic horizontal tube flow, here also the bubbles and slugs
show a tendency to move towards the upper part of the tube as shown in Fig. 1.8.

This also results in the movement of the vapour core towards the upper wall
which can cause intermittent 'dry out' of the upper wall in the early flow section.
The eventual final dry out also occurs early for the upper wall and lately for the lower
wall. However, it has to be understood that the actual occurrence of the flow regimes
depends on factors such as fluid inertia (mass flux), fluid properties (like density,

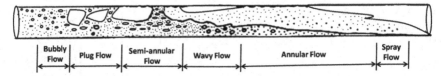

| Bubbly Flow | Plug Flow | Semi-annular Flow | Wavy Flow | Annular Flow | Spray Flow |

Fig. 1.8 Flow patterns in a horizontal tube flow with evaporation. (*Courtesy:* "Heat Transfer in Condensation and Boiling" by Karl Stephan, with permission from Springer)

viscosity, surface tension, latent heat, specific heat, thermal conductivity) and pipe geometry (circular, oval, square). It is interesting to note that the difference between vertical and horizontal flow patterns are more prominent when inertia (mass flux) and surface tension (tube diameter) effects are less, allowing gravity effect to dominate. As the mass flux is increased, the inertia effects starts dominating and the difference between vertical and horizontal flow reduces. Similarly, as the tube diameter reduces, surface tension starts dominating and the difference between vertical and horizontal flow reduces. This will be discussed in Chap. 16.

1.2.3   Flow Pattern Maps

The various flow patterns described in the previous sections were qualitative in description. Since the flow patterns are results of interplay of various forces such as the inertia, gravity as well as interfacial forces, the quantitative determination of the boundaries between various flow patterns is possible. This is usually done on a 2D plot with various areas depicting various regimes and the boundary lines contain same axes for all the transition regimes while the complex plots use different axes for different regimes. For horizontal tube flow, Baker (1954) was the first one to construct a flow pattern map. The two major parameters in the axes are given by:

$$G_g = \text{mass flux of gas} = \frac{\text{Mass flow rate of gas}}{\text{Cross-sectional area of tube}}$$

$$G_l = \text{mass flux of liquid} = \frac{\text{Mass flow rate of liquid}}{\text{Cross-sectional area of tube}}.$$

This 'mass flux' is also described as 'mass velocity' by some investigators as defined earlier. The original map of Baker was modified by Scott (1963). Two factors are included to the axes, λ and ψ primarily to accommodate the properties of various fluids as against water–air system for which the map was originally constructed. These factors are given by:

$$\lambda = \left(\frac{\rho_g \rho_l}{\rho_{air} \rho_{water}} \right)^{1/2}, \tag{1.21}$$

$$\psi = \frac{\sigma_{water}}{\sigma} \left[\frac{\mu_l}{\mu_{water}} \left(\frac{\rho_{water}}{\rho_l} \right)^2 \right]^{1/2}. \tag{1.22}$$

Here, ρ, μ, and σ are density, viscosity and surface tension, respectively. It can be easily seen when the liquid is water (l = water) and gas is air (g = air), these parameters reduce to unity ($\lambda = \psi = 1.0$). These factors thus can be seen as correction factor on gas and liquid mass flux in the form (G_g/λ) and ($G_l\psi$) which are taken as the axes of Baker's map as shown in Fig. 1.9.

For utilizing this map, standard values of properties of air and water to be used are the following:

$\rho_{air} = 1.23 \, \text{kg/m}^3$
$\rho_{water} = 1000 \, \text{kg/m}^3$
$\mu_{water} = 10^{-3} \, \text{Pa--s}$
$\sigma_{water} = 0.070 \, \text{N/m}$

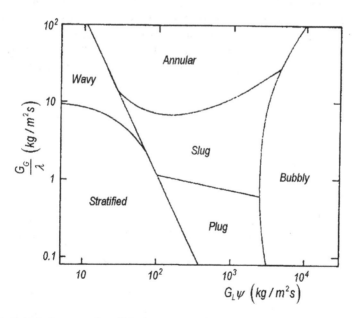

Fig. 1.9 Baker's flow map for adiabatic two-phase flow in horizontal tubes. (*Courtesy:* "Heat Transfer in Condensation and Boiling" by Karl Stephan, with permission from Springer)

However, a more general flow pattern map covering wider range of parameters like fluid properties, tube diameters and fluid mass flux was proposed by Hewitt and Roberts (1969). Instead of mass velocity, they found that superficial momentum flux is a physically more consistent parameter. To define superficial momentum flux, we must recall the definition of superficial velocity (which is used frequently in two-phase flow) as given previously by Eqs. 1.16 and 1.17. The momentum flux based on superficial velocity is given by

Gas momentum flux $= \rho_g U_g^2$
Liquid momentum flux $= \rho_l U_l^2$

The momentum flux can also be expressed in terms of mass flux as

$$\rho_g U_g^2 = \frac{G_g^2}{\rho_g} \tag{1.23}$$

$$\rho_l U_l^2 = \frac{G_l^2}{\rho_l}. \tag{1.24}$$

It can be noted that in Hewitt–Robert map (Fig. 1.10), unlike the simple regimes of wavy, annular, stratified, plug, slug and bubbly flow of Baker's map, mixed regimes such as 'churn flow' and 'whispy annular' flow are also depicted which refer to elongated bubbles and elongated droplets, respectively. It should also be remembered that the maps show sharp transition between various flow regimes which in reality does not take place. The real transition takes place over a wider band triggered by instability which are difficult to predict accurately because they depend critically on experimental conditions. It must be kept in mind that Hewitt–Robert map was

Fig. 1.10 Hewitt–Robert map applicable to vertical upflow in circular tubes. (*Courtesy:* "Heat Transfer in Condensation and Boiling" by Karl Stephan, with permission from Springer)

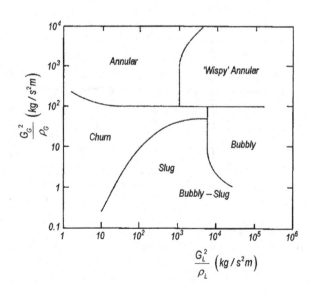

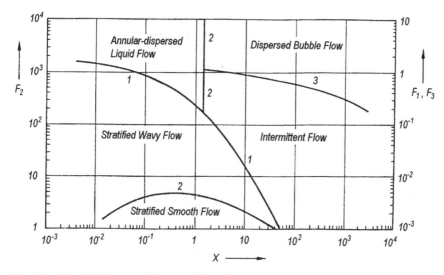

Fig. 1.11 Taitel–Dukler map. (*Courtesy:* "Heat Transfer in Condensation and Boiling" by Karl Stephan, with permission from Springer)

constructed primarily for vertical upflow in circular tubes and it also included data for steam–water system where phase change takes place.

However, both the flow maps are somewhat simplified picture of the reality. The flow regimes in reality which are followed not only by fluids of wider ranges, but tubes of various inclinations were also presented by Taitel and Dukler (1976). Taitel–Dukler map is complex but agrees with a much wider range of experimental data and is universally accepted as a reasonably accurate flow map for conventional tubes (tubes of diameter more than 3 mm). The map is shown in Fig. 1.11.

Taitel–Dukler map has some unique features. Firstly, as abscissa (*x* axis) it uses a well known quantity called the Martinelli parameter, given by Lockhart and Martinelli (1949):

$$X = \left[\frac{\left(\frac{dp}{dz}\right)_l}{\left(\frac{dp}{dz}\right)_g} \right]^{1/2} , \qquad (1.25)$$

where $\left(\frac{dp}{dz}\right)_l$ is the pressure gradient had the liquid of the two-phase flow alone were flowing through the entire tube and $\left(\frac{dp}{dz}\right)_g$ is the pressure gradient had the gas of the two-phase flow alone were flowing through the entire tube. This essentially reduce to pressure gradient due to superficial liquid and superficial gas velocities, respectively. The ordinate (*y* axis) for Taitel–Dukler map is a bit complicated. It does not use the same quantity in the *y*-axis for transition between various regimes. Instead, it uses three different quantities.

1. For transition between annular dispersed flow and stratified wavy flow, they used
   the parameter F_1 which is actually known as the Froude number (F_n)

$$F_1 = \frac{G_g}{[\rho_g(\rho_l - \rho_g)dg \cos \gamma]^{1/2}} = F_n \qquad (1.26)$$

where d is the tube diameter and γ is the angle of inclination of the tube. It is
interesting to note that the same parameter (F_1) can also be used for transition
from intermittent to stratified wavy flow. Froude number being a ratio between
inertia and gravitational force, it clearly indicates how increase of inertia (or mass
flux) can make an orderly stratified flow change into a dispersed flow pattern.

2. The transition between annular and dispersed bubbly or intermittent flow is governed by a parameter F_2 which is a combination of Froude number and ratio of
   inertia and viscous force (somewhat like liquid superficial Reynolds number).

$$F_2 = F_n \left[\frac{G_l d}{\mu_l} \right]^{1/2} \qquad (1.27)$$

It is interesting to note that F_2 is also the criterion for transition from stratified
flow to wavy flow. While stratification of liquid depends on gravity, the creation of
waves depends on the interfacial shear which is dominated by liquid viscosity. The
effect of this parameter can be explained physically from the above arguments.

3. Finally, the transition between bubbly flow and intermittent flow is dependent on
   the parameter F_3, which is a ratio between liquid pressure gradient and gravitational force.

$$F_3 = \left[\frac{\left| \left(\frac{dp}{dz} \right)_l \right|}{g(\rho_l - \rho_g) \cos \gamma} \right]^{1/2} \qquad (1.28)$$

It may be noted that the sign of modulus ensures that the numerator is always
positive and we get a real quantity for F_3. This means intermittent, plug or slug,
flow can get dispersed into bubbles of smaller sizes when the liquid experiences
large pressure forces.

It must be mentioned here that the transition lines in reality are not as sharp as
shown in the figure even for Taitel–Dukler map. They are rather broader bands. This
map is not just based on the experimental observations, but some simple analyses
are also done to find the transition. For example, in the case of stratified flow, the
liquid level in the tube is adjusted by equating the pressure drop of the gas to the
liquid phase. As shown in Fig. 1.12, it is postulated that intermittent flow will occur
for $h_l/d > 0.5$ because the waves in that case will touch the top wall.

It can be seen from the above flow maps that in conventional large diameter tubes,
inertia, gravity, fluid friction and interfacial drag are considered as dominant forces.

Fig. 1.12 Structure of
stratified flow

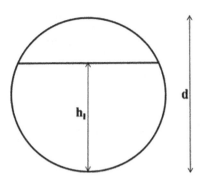

Surface forces such as surface tension are not considered to be important. We will
see later in case of mini- and microchannels that, with a decrease in diameter, the
scenario changes completely (Chap. 16).

Examples

Problem 1: Consider water flow in horizontal pipeline with a mass flow rate of
0.25 kg/s at 150 °C and quality of 0.12. The diameter of the pipeline is 1 cm. The
designer wants to know which model to use in this situation for further calculations
of pressure drop and heat transfer. Can you help him based on the Taitel and Dukler
Map?

Solution: Taking the properties at $T = 150$°C and steam as gas and water as liq-
uid present, $\rho_l = 918\,\text{kg/m}^3$; $\rho_g = 0.531\,\text{kg/m}^3$; $\mu_w = 1.84 \times 10^{-4}\,\text{Ns/m}^2$; $\mu_g = 13.925 \times 10^{-6}\,\text{Ns/m}^2$

$$\text{Re}_l = \frac{4\dot{m}(1 - x^*)}{\pi d \mu_l} = \frac{4 \times 0.25 \times (1 - 0.12)}{\pi \times 0.01 \times 1.84 \times 10^{-4}} = 1.52 \times 10^5$$

$$\text{Re}_g = \frac{4\dot{m}x^*}{\pi d \mu_g} = \frac{4 \times 0.25 \times (1 - 0.12)}{\pi \times 0.01 \times 13.926 \times 10^{-6}} = 2.74 \times 10^5$$

Hence, the flow is turbulent and so the friction factors are obtained as

$$C_{fl} = \frac{0.079}{\text{Re}_l^{1/4}} = 4 \times 10^{-3}$$

$$C_{fg} = \frac{0.079}{\text{Re}_g^{1/4}} = 3.45 \times 10^{-3}$$

$$G_l^2 = \left(\frac{4\times\dot{m}\times(1-x^*)}{\pi d^2}\right)^2 = 78.463 \times 10^5 \, \text{kg}^2\text{m}^{-4}$$

$$G_g^2 = \left(\frac{4\times\dot{m}\times x^*}{\pi d^2}\right)^2 = 1.45 \times 10^5 \, \text{kg}^2\text{m}^{-4}$$

$$\left(\frac{dp}{dz}\right)_l = C_{fl} \times \frac{G_l^2}{2d\rho_l} = 1709.4335 \, \text{Nm}^{-3}$$

$$\left(\frac{dp}{dz}\right)_g = C_{fg} \times \frac{G_g^2}{2d\rho_g} = 47105.41 \, \text{Nm}^{-3}$$

So, the Martinelli parameter is

$$X = \sqrt{\frac{\left(\frac{dp}{dz}\right)_l}{\left(\frac{dp}{dz}\right)_g}} = 0.19$$

$$F_1 = \frac{G_g}{\rho_g(\rho_l - \rho_g)dg} = 7.967$$

Since x^* is 0.12, from Taitel and Dukler's map, we find that the flow regime in the pipe is dispersed bubble flow. Hence, homogeneous model will work well.

Problem 2: Consider a cross flow shell and tube heat exchanger. As a thought experiment, assume that rod bundles have super hydrophobic surfaces. To achieve maximum efficiency of heat transfer, we wish to obtain spray flow conditions in this heat exchanger. The liquid phase is water at ambient temperature. Determine the mass flow rate of air to be supplied to obtain spray flow conditions. Take mass flow rate of water as 0.1 kg/s and ambient temperature as 20 °C. Pressure is 0.0233 bar and diameter of the tube is 10 cm.

Solution: Properties of air and water are:
$\rho_w = 1000 \, \text{kg/m}^3$;
$\rho_a = 0.0278 \, \text{kg/m}^3$;
$\mu_w = 1.006 \times 10^{-3} \, \text{Ns/m}^2$;
$\sigma_w = 0.00728 \, \text{N/m}$
We use the Grant's map for flow over rod bundles:

Average cross-sectional area $A' = (\varepsilon A + (1 - \varepsilon)A)/2 = A/2$, where ε is the void fraction and A is the cross-sectional area.

$$U_{sw} = \frac{1}{\rho_w} \times \frac{2\times\dot{m}_w}{\frac{\pi d^2}{4}} = 0.0254 \, \text{ms}^{-1}$$

$$U_{sa} = \frac{1}{\rho_a} \times \frac{2\times\dot{m}_a}{\frac{\pi d^2}{4}} = 9159.996\dot{m}_a \, \text{ms}^{-1}$$

$$U_{sw} \times \frac{(\rho_w\mu_w)^{1/3}}{\sigma_w} = 3.497$$

$$U_{sa} \times \sqrt{(\rho_a\rho_w)} = 49.296\dot{m}_a \, \text{ms}^{-1}$$

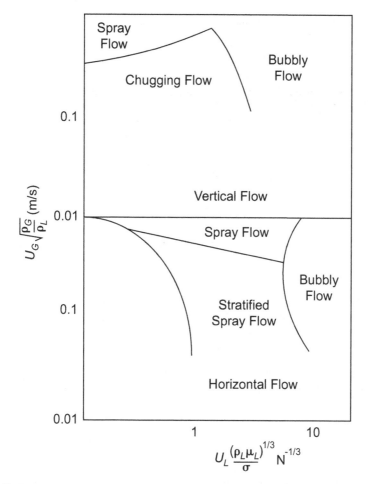

Fig. 1.13 Two-phase flow patterns for cross flow over tube banks

Thus, from Grant's Map, we find that as long as $49.296\dot{m}_a \simeq 0.01$, the flow is in spray flow regime, i.e., $\dot{m}_a$ must be approximately equal to 2.07×10^{-4} kg/s (Fig. 1.13).

Problem 3: Consider flow of water through a 3.6 cm diameter tube at a system pressure of 5 bar. Let the quality of vapour be 0.015 and take the mass flow rate to be 400 kg/m²s. Compare the flow patterns obtained in the above circumstances for a horizontal tube and a vertical tube. Can you explain the difference between the flow patterns?

Solution: Properties of steam and water at these conditions are:
$\rho_w = 815\,\mathrm{kg/m^3}$;
$\rho_g = 0.436\,\mathrm{kg/m^3}$;
$\rho_a = 7.44\,\mathrm{kg/m^3}$

(a) **Considering the tube to be horizontal**: Let's use the Baker's map to determine the flow regime

$$\psi = \frac{\sigma_w}{\sigma_l}\sqrt{\frac{\mu_l}{\mu_w}\left(\frac{\rho_w}{\rho_l}\right)^2} = 1$$

since the liquid is water in this case.

$$\lambda = \frac{\sigma_l}{\sigma_w}\sqrt{\frac{\rho_g}{\rho_{air}}} = 0.24$$

$$G_g = x^* G = 6\,\mathrm{kgm^{-2}}$$

$$G_l = (1 - x^*)G = 394\,\mathrm{kgm^{-2}}$$

$$\frac{G_g}{\lambda} = 24.78$$

$$G_l \psi = 394$$

From Baker's map, the flow regime is annular.

(b) **Considering the tube to be vertical**: Let's use the Hewitt's Map to determine the flow regime

$$\left(\frac{G_g^2}{\rho_g}\right) = 82.568\,\mathrm{kgm^{-1}}$$
$$\left(\frac{G_l^2}{\rho_l}\right) = 190.47\,\mathrm{kgm^{-1}}$$

From Hewitt and Roberts map, the flow regime is churn flow. This is obvious since gravity and pressure differential will destabilize the flow causing the churn flow pattern.

Exercises

1. Consider a process heating application where hot water is to be supplied. 0.8 kg/s of saturated water flows at 100°C through a horizontal pipe of 20 mm diameter and 4 m length, which is uniformly heated with a power of 100 W. The designer has used homogeneous model for calculation of pressure drop for all purposes ahead of this pipe. Verify if the designer is right is assuming this to be assuming this to be a homogeneous flow. Suddenly the valve in the water pipeline malfunctions and the flow rate drops to 0.5 kg/s. See if the designer's calculations are still valid?

2. Consider an Electrolux refrigerator using NH_3-H_2 system. Close to the evaporator some hydrogen leaks in the pipeline carrying ammonia at the rate 0.08
   kg/s. Comment on the effect on the performance of the evaporator based on the
   flow regime present as ammonia reaches the evaporator. Take pipe diameter as
   $d = 36$ mm, system pressure as 20 bar and temperature as $-15\,°C$ and mass flow
   rate of ammonia as 0.4 kg/s.

3. Water–air mixture flows in a 2.5 cm diameter pipe, vertically upwards at 80 bar
   and 250 °C. Mass flow rate of water is 0.13 kg/s and that of air is 0.02 kg/s. At
   a certain point, the air in the pipeline is completely evacuated and the engineer
   says that since only water must be present, the pressure can be lowered to 39 bar.
   Check till what point he is allowed to do so, if the flow regime in the pipeline
   has to be maintained as before.

4. At low volumetric fluxes of liquid phase, the critical gas velocity at flow reversal
   point is given by the relation $J_g^* = 0.9$, i.e., volumetric flux of gas reaches the
   value of $0.9\,m^3/s$. This is also the limit for churn flow. Verify the above statement
   from Hewitt's map for water at 100 °C and specify the limits for the volumetric
   flux of liquid. Take tube diameter as 56 mm.

5. Consider annular flow in horizontal helical coils. It has been observed that two-
   flow patterns are obtained: one where the liquid phase is forced to the outside
   of the coil and another where the liquid phase travels on the inside surfaces of
   the coil. Develop a crude flow map using explanations from the momentum flux
   considerations for the liquid and the gas phase. Also, illustrate its applicability
   using an example.

6. Flow maps for liquid metal-gas two-phase flows are extremely important in
   Nuclear applications. Consider a nuclear reactor system with NaK-78 and nitro-
   gen flow in the tube of diameter 22 mm. Mass flow rate is 0.4 kg/s and the quality
   is 0.1. Comment on the heat transfer capability under these circumstances. Take
   temperature as 700 °C.

7. In a process heat transfer application 0.01 kg/s of water at 100 °C is required.
   The quality is 0.03 and water flows through a horizontal pipe. Under these
   conditions, the flow regime is observed to be intermittent which is not desirable.
   A malfunction in the tube locators lowers the tube by 5 °C. Check from Barnea's
   map if it will affect the system seriously. Verify from Taitel and Dukler's Map.
   Take the diameter of the pipe as 1 in.

8. In a pressurized water heater reactor (PWR) cooling system, light water flows
   through an 1 in. diameter pipe at 155 bar and 315 °C with a mass flow rate of
   0.0996 kg/s and quality of 0.909. Obstructions are intended to be used to enhance
   the flow heat transfer. However, whether to use central obstructions or peripheral
   obstructions is to be decided so as to keep the design calculations for the earlier
   flow regime intact since it affects the whole system. Hence, determine the type
   of obstruction to be used.

9. In a condenser, 0.5 kg/s of water flows through the tube at 100 °C with quality of 0.125 and the void fraction is 0.18. Find the flow regime in the tube and comment on the condenser performance based on this. Use an appropriate flow map. Take tube diameter as 36 mm.

10. In hydrogen liquefier, hydrogen is compressed to 100 bar at 300 K. As a result, the hydrogen undergoes a phase change and is liquefied. After this, heat needs to be taken away. At a certain point, the quality is 0.08. Determine if this is the right place to take away heat. Take mass flow rate in the tube as 0.01 kg/s and diameter of tube as 1 in.

References

Baker O (1954) Simultaneous flow of oil and gas. Oil Gas J 53:185

Hewitt GF (1982) Flow regimes. Handbook of multiphase systems. McGraw Hill, New York

Hewitt GF, Roberts DN (1969) Studies of two-phase flow patterns by simultaneous flash and X-ray photography, AERE-M2159. UK, AEA Report

Lockhart RW, Martinelli RC (1949) Proposed correlation data for isothermal two phase, two-component flow in pipes. Chem Eng Prog 45:39–48

Scott DS (1963) Properties of concurrent gas liquid flow. Adv Chem Eng 4:199–277

Teitel Y, Dukler AE (1976) A model for predicting flow regime transitions in horizontal and near horizontal gas-liquid flow. AICHE J 22:47–55

Chapter 2
Two-Phase Flow—Pressure Drop and Flow Friction

Fluid mechanics of two-phase flow is significantly different from single-phase (only gas or only liquid) flow. Quite often these two-phase flow configurations are associated with a large number of bubbles or droplets, unsteady interface oscillations and phase separation (like liquid stratification or gas slug formation) which are very difficult to predict. For this reason, the analysis of two-phase flow first started as a purely empirical science based on large number of experimental observations. This is because, the discontinuous phases (like bubbles, slugs or droplets) are difficult to analyse with continuum assumptions with uniform properties. Hence, it is difficult to solve governing equations such as continuity, momentum or energy equations to yield flow and thermal variables (velocity, temperature, pressure). However, over the last half a century the scenario has changed significantly. There are several reasons for the tremendous progress in the research and the consequent enhanced understanding of the two-phase flow along with phase change (boiling and condensation) phenomena. Firstly, the development of nuclear reactors and thermal power plants in the 1950 and 60 s saw a huge investment in two-phase research. The development of electronics and consequent inventions of extremely accurate instrumentation accelerated this research. Finally development of high precision computing facilities has aided more theoretical understanding through multiphase flow simulation which was considered to be impossible even just two decades back. However, the first effort to analyse two-phase flow was more by using single-phase technique and introducing various correction factors to take care of few two-phase characteristics. Although this is an ad-hoc approach, many useful correlations can be developed by this approach and most of the industrial practices today are based on these correlations. This also helps to understand the basic features of two-phase flow and identify the influencing parameters.

In this chapter, we shall see how fluid mechanics of two-phase flow can be developed using a treatment similar to that of a single-phase flow. An important require-

© The Author(s) 2023
S. K. Das and D. Chatterjee, *Vapor Liquid Two Phase Flow
and Phase Change*, https://doi.org/10.1007/978-3-031-20924-6_2

ment of two-phase flow analysis is to estimate the pressure drop during flow which is important to calculate the pumping power required to force a two-phase mixture inside a conduit having a given orientation (vertical, horizontal or inclined).

2.1 Momentum Balance in an Inclined Tube and Pressure Drop

Let us consider the simplest case of flow inside a tube. We consider the tube to be inclined to horizontal plane at an angle of γ. This means that $\gamma = 0°$ gives a horizontal tube and $\gamma = 90°$ gives a vertical tube. At this point, we just consider a fluid is flowing up the tube (we do not specify here whether the fluid is single phase one like pure gas or pure liquid, or it is a two-phase fluid consisting of a mixture of gas and liquid). The balance of forces on a small element of this tube of length dz is shown in Fig. 2.1.

Let us assume that the flow is steady (does not change with time) and one-dimensional (balance of forces need to be carried out along the length of the tube). This means that we get forces acting at the cross section and wall uniformly irrespective of the location of gas or liquid phase. The forces acting on this fluid element are—the pressure forces at the inlet and outlet of the cross section, shear force at the wall acting opposite to the fluid flow direction and the gravity force in the form of fluid element's weight. The resultant of these forces will be equal to the inertia force which gives acceleration to the fluid element and is equal to the rate of change of momentum along the z-direction according to Newton's second law of motion. This can be written as

$$\frac{d}{dz}\left(\dot{M}\,u\right)dz = \left[p - \left(p + \frac{p}{dz}\right)\right]A - \tau_w F dz - \rho g A \sin\gamma dz \qquad (2.1)$$

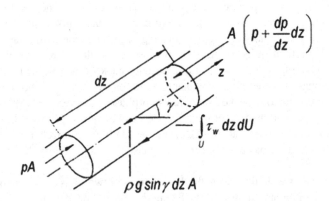

Fig. 2.1 Flow inside an inclined tube and the force balance (*Courtesy* "Heat Transfer in Condensation and Boiling" by Karl Stephan, with permission from Springer)

where u is the average fluid velocity at any cross section with area A (please note that although the mass flow rate and cross-sectional area remains constant, the average velocity may change due to change of phase), τ_w is the wall shear stress and F is the tube circumference based on tube inner diameter. The density used here is the average density and for a two-phase fluid, it has to be calculated based on void fraction.

Thus Eq. (2.1) can be finally cast in the form

$$-\frac{dp}{dz} = \tau_w \frac{F}{A} + \frac{1}{A}\frac{d}{dz}\left(\dot{M}u\right) + \rho g \sin\gamma \qquad (2.2)$$

Please note that the left-hand side is deliberately kept $-dp/dz$ to keep this term positive. This is because the pressure change in the flow direction is negative and hence dp/dz is negative. Thus the total pressure gradient in a tube flow is the sum of three pressure gradients, namely frictional, inertial and geodetic pressure gradients. Apart from these, change in tube cross section contributes to expansion and contraction losses in case the diameter of the tube changes along the length. Now let us consider various cases of single and two-phase flows.

2.1.1 Single-Phase Flow

In single-phase flow, properties like density is simply the fluid density. For incompressible flows, that is, low velocity liquid and gas flows, this can be taken as constant. F/A is equal to $4/d$ in case of circular tube with diameter d. The mass flow rate can be found when the entire cross section is filled by a single fluid, as $\dot{M} = GA = M(\pi/4)d^2$ and the fluid velocity can be taken from mass velocity as $u = G/\rho$. Hence, Eq. (2.2) takes the form:

$$-\frac{dp}{dz} = \frac{4\tau_w}{d} + G^2\frac{d}{dz}\left(\frac{1}{\rho}\right) + \rho g \sin\gamma \qquad (2.3)$$

The velocity being constant in a tube of constant cross section for incompressible flow, the entire inertial term (the second term in the right hand side of Eq. (2.3)) vanishes in this case. However, gravitational and frictional pressure drop remains. The frictional pressure drop can be predicted by well-known pipe flow correlation for friction factor, used in Hagen–Poiseuille equation.

$$c_f = \frac{16}{Re} \text{ for laminar flow}$$

$$c_f = \frac{0.0791}{Re^{1/4}} \text{ for turbulent flow}$$

and

$$\tau_w = c_f \frac{1}{2} \rho u^2 \tag{2.4}$$

where Reynolds number, Re is given by

$$Re = \frac{\rho u d}{\mu}$$

Thus the pressure gradient due to friction (the first term in the right-hand side of Eq. (2.3), $(dp/dz)_f$) reduces to

$$-\left(\frac{dp}{dz}\right)_f = \frac{4c_f \rho u^2}{4d} = \frac{2c_f G^2}{\rho d} \tag{2.5}$$

2.1.2 Two-Phase Flow

For two-phase flow, still Eq. (2.3) holds true because it has been developed on the basis of a general force balance. Let us make an assumption that the density of the two-phase mixture can be calculated by

$$\rho = \varepsilon \rho_g + (1 - \varepsilon)\rho_l \tag{2.6}$$

Similarly, the flow momentum can also be thought as a sum of the momentum of two phases along the tube as

$$\dot{M}u = \dot{M}_g u_g + \dot{M}_l u_l$$

Substituting values of u_g and u_l from Eqs. (1.9) and (1.10) of the previous chapter

$$\dot{M}u = \frac{\dot{M}^2}{A}\left[\frac{(x^*)^2}{\varepsilon \rho_g} + \frac{(1-x^*)^2}{(1-\varepsilon)\rho_l}\right] \tag{2.7}$$

$$\dot{M}u = G^2 A\left[\frac{(x^*)^2}{\varepsilon \rho_g} + \frac{(1-x^*)^2}{(1-\varepsilon)\rho_l}\right] \tag{2.8}$$

Inserting this into Eq. (2.8) gives

$$-\frac{dp}{dz} = \frac{4\tau_w}{d} + G^2 \frac{d}{dz}\left[\frac{x*^2}{\varepsilon \rho_g} + \frac{(1-x*)^2}{(1-\varepsilon)\rho_l}\right] + \left[\varepsilon \rho_g + (1-\varepsilon)\rho_l\right] g \sin \gamma \tag{2.9}$$

Hence for the consideration of two-phase flow also we get the same three components of pressure gradients.

(a) Frictional pressure gradient:

$$\left(-\frac{dp}{dz}\right)_f = \frac{4\tau_w}{d}$$

(2.10a)

(b) Acceleration pressure gradient:

$$\left(-\frac{dp}{dz}\right)_a = G^2 \frac{d}{dz}\left[\frac{x^{*2}}{\varepsilon \rho_g} + \frac{(1-x^*)^2}{(1-\varepsilon)\rho_l}\right]$$

(2.10b)

(c) Geodetic or gravitational pressure gradient:

$$\left(-\frac{dp}{dz}\right)_g = \left[\varepsilon \rho_g + (1-\varepsilon)\rho_l\right] g \sin\gamma$$

(2.10c)

The main source of pressure drop in two-phase flow in a tube or channel is the frictional pressure drop due to the shear that the fluid imparts at the wall. The interfacial interaction may also contribute to it. One needs to understand the various sources of pressure drop indicated by the three above components of pressure gradient. While fluid friction is the source of the pressure gradient due to shear stress, the acceleration pressure drop is due to change of phase. As liquid turns into vapour, both the liquid and gas undergo a change of momentum. This is because, channel cross section being unchanged, the sudden change in density (at usual pressures) cause a sudden change in momentum of both the phases due to acceleration. Now, the question is how can liquid change into vapour? There are two mechanisms. Firstly, due to frictional pressure loss, the two phase pressure of the gas–liquid mixture reduces. Hence, though the flow temperature is lower than the boiling point, the enthalpy is maintained by flashing a part of the liquid into vapour. This will take place even in the case of adiabatic flow (say, air-water flow or steam water flow in an insulated tube). The second way liquid can turn into vapour is by heat transfer as in the case of flow boiling where the latent heat supplied causes the change of phase. In both the cases the change of phase causes acceleration. Usually if frictional pressure drop is less than 20% of the system pressure, acceleration pressure drop can be neglected. For continuously changing void fraction, the acceleration pressure drop is given by

$$(p_{out} - p_{in}) = G^2\left[\frac{x_{out}^{*~2}}{\varepsilon_{out}\rho_g} - \frac{x_{in}^{*~2}}{\varepsilon_{in}\rho_{g_{in}}} - \frac{(1-x_{out}^*)^2}{(1-\varepsilon_{out})\rho_{lout}} - \frac{1-x_{in}^*}{(1-\varepsilon_{in})\rho_{l_{in}}}\right]$$

(2.11)

For evaluation of this equation, one needs an accurate equation for void fraction (ε) in terms of the flowing quality (x^*). A special case of no slip between the liquid and gas ($S = 1$, the significance of this will be discussed in the next section) can be used under some special condition. Putting $S = 1$ in Eq. (1.12) of the previous chapter, one gets this correlation as

$$\varepsilon = \frac{1}{\dfrac{\rho_g}{\rho_l}\left(\dfrac{1-x^*}{x^*}\right) + 1} \tag{2.12}$$

One has to remember that this equation is applicable only for very special case of $S = 1$ which will be discussed later.

2.2 Frictional Pressure Drop and Two-Phase Multiplier

The major part of two-phase pressure drop is the frictional pressure drop. Although we assume this pressure drop to be due to wall shear, in reality the frictional pressure drop also has contributions from momentum transfer between phases. Since in two-phase flow, the position of the gas–liquid interface is often irregular and indeterminate, the interfacial momentum transfer due to interfacial forces (like surface tension) and interfacial phenomena (like waves, liquid break up, bubble collapse) are also difficult to estimate accurately. The other factor which adds complication to these phenomena is the fact that there may be various regimes of the two phases of the fluids. For example, both liquid and gas can be in the turbulent state, or both can be in the laminar state. Even one can be in the laminar state and the other can be turbulent. These combinations make pressure drop and momentum interaction less predictable. Thus, although the pressure drop correlation, Eq. 2.9, has identified the various components of pressure loss (inertia, gravity or friction) it does not help in any way to estimate frictional pressure loss which is often the single largest component of pressure drop. The above difficulty in estimating interfacial momentum transfer prohibits any consistent method of deriving frictional pressure drop in two-phase flow theoretically. The approach usually used to overcome this deficiency is the method of analogy. We can treat the two-phase flow similarly as the single-phase flow and then see what correction needs to be done to take care of the deviation of two phase from single phase. One problem is that the frictional pressure gradient in two-phase flow is not a linear function of flowing quality as can be seen in Fig. 2.2. It shows a maximum in between $0.7 < e* < 0.9$. The maximum is more distinct when the density difference between liquid and gas flow is large (or in other words, when the system pressure is much lower compared to the critical pressure of the fluids). The method of analogy is applied to determine this pressure drop by using a single-phase pressure drop as suggested by Lockhart and Martinelli (1949). In this method we assume that the pressure drop of two-phase flow can thought as the pressure drop of only the liquid phase multiplied by a factor known as the two-phase multiplier, ϕ^2. Thus,

$$\left(\frac{dp}{dz}\right)_f = \phi_l^2 \left(\frac{dp}{dz}\right)_l \tag{2.13}$$

Fig. 2.2 Typical friction pressure drop of two-phase flow (*Courtesy* "Heat Transfer in Condensation and Boiling" by Karl Stephan, with permission from Springer)

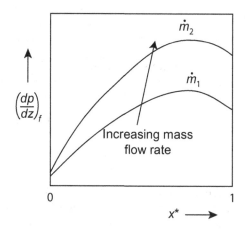

Similarly, the frictional pressure drop of the two-phase flow can also be related to the gas phase pressure drop, as

$$\left(\frac{dp}{dz}\right)_f = \phi_g^2 \left(\frac{dp}{dz}\right)_g \tag{2.14}$$

Here, we must remember that the liquid phase pressure drop $(dp/dz)_l$ and gas phase pressure drop $(dp/dz)_g$ are calculated based on superficial velocity of liquid, U_l and that of gas U_g respectively (these have been defined in the last chapter). In other words, we can say that $(dp/dz)_l$ is the pressure gradient if only the liquid would have flown through the entire tube (with no gas) and $(dp/dz)_g$ is the pressure gradient if only the gas would have flown through the entire tube (with no liquid). In these cases the assumptions is that, when only liquid or only gas flows through the entire tube, they are flowing with the same mass flow rate as their respective mass flow rates are in the two- phase flow. Thus, these gradients can be easily calculated using single-phase pressure gradient estimation technique, using Hagen–Poiseuille equation

$$-\left(\frac{dp}{dz}\right)_l = c_{fl} \frac{L \rho_l U_l^2}{2d} \tag{2.15}$$

where c_{fl}= liquid friction factor, U_l = liquid superficial velocity, L = length of the tube and d = diameter of the tube. This can be further reduced to the following form using Eqs. (1.23) and (1.24),

$$-\left(\frac{dp}{dz}\right)_l = c_{fl} \frac{LG_l^2}{2d \rho_l} = c_{fl} \frac{LG^2(1 - x^*)^2}{2d \rho_l} \tag{2.16}$$

Similarly, the gas phase pressure drop is given by

$$\left(\frac{dp}{dz}\right)_g = c_{f_g}\frac{LG_g^2}{2d\rho_g} = c_{f_g}\frac{LG^2(x^*)^2}{2d\rho_g} \tag{2.17}$$

The friction factors for the two phases c_{f_l} and c_{f_g} can be calculated using standard correlation involving single-phase Reynolds number, as

$$c_{f_l} = \frac{c_1}{Re_l^{n_1}} \tag{2.18a}$$

$$c_{f_g} = \frac{c_2}{Re_g^{n_2}} \tag{2.18b}$$

Here c_1, c_2 and n_1, n_2 depend on the flow regime, that is laminar or turbulent, and the shape of the cross section. For example, for circular tube $n = 1$ and $c = 16$ for laminar flow and $n = 0.25$ and $c = 0.0796$ for turbulent flow (Blasius equation). The use of the two sets of values of c and n (c_1, c_2 and n_1, n_2) clearly indicates that in a two-phase flow the two phases (liquid and gas) may be in different regimes. They can be laminar for liquid and turbulent for gas or vice-versa (they can also be in the same regime like laminar-laminar or turbulent-turbulent). It is important here to define the Reynolds number for the two phases, as

$$Re_l = \frac{\rho_l U_l d}{\mu_l} = \frac{G_l d}{\mu_l} = \frac{G(1-x^*)d}{\mu_l} \tag{2.19a}$$

$$Re_g = \frac{\rho_g U_g d}{\mu_g} = \frac{G_g d}{\mu_g} = \frac{Gx^* d}{\mu_g} \tag{2.19b}$$

There is an alternative way of defining two-phase multiplier. Instead of defining the individual liquid and gas phase pressure drops on the basis of superficial velocities of each phases, here it is assumed that the entire mass of two-phase flow (i.e., liquid and gas) is flowing as liquid (or as gas). This definition is convenient in case of flow boiling because here the vapour phase is formed at the expense of liquid phase and hence the superficial velocity of both the liquid and vapour changes. By adopting this definition here the single-phase pressure drop (as liquid or vapour) remains unchanged because the combined mass of liquid and vapour remains unchanged at every cross section even when boiling takes place. We call it $(dp/dz)_{lo}$ or $(dp/dz)_{go}$

$$\left(\frac{dp}{dz}\right)_f = \phi_{lo}^2 \left(\frac{dp}{dz}\right)_{lo} \tag{2.20a}$$

$$\left(\frac{dp}{dz}\right)_f = \phi_{go}^2 \left(\frac{dp}{dz}\right)_{go} \tag{2.20b}$$

where,

$$\left(\frac{dp}{dz}\right)_{lo} = c_{f_{lo}} \frac{LG^2}{2d\rho_l} \tag{2.21a}$$

$$\left(\frac{dp}{dz}\right)_{go} = c_{f_{go}} \frac{LG^2}{2d\rho_g} \tag{2.21b}$$

Frictional factors $c_{f_{lo}}$ and $c_{f_{go}}$ can be obtained in a similar manner as described in the previous case, namely, Eq. 2.18. However, the definition of Reynolds number changes here as

$$Re_l = \frac{G_l d}{\mu_l} \tag{2.22a}$$

$$Re_g = \frac{G_g d}{\mu_g} \tag{2.22b}$$

It has to be kept in mind here that this second approach is difficult to be applied to different fluids as liquid and gas flow (such as air–water flow) because we have to assume that the entire mass is flowing as only liquid or vapour. In the air-water flow, can we assume the total mixture flowing as air or as water? What will be its property then? While in case of boiling the fluid is same (say, water and water vapour) and the assumption of the total mass flowing as water or water-vapour is consistent. Thus, the task of evaluating frictional pressure drop reduces to estimating the two-phase multiplier. This is not a very straight forward task and the value of it depends on the flow pattern discussed in the previous chapter. In the following sections, we will introduce the various models for evaluation of ϕ and discuss their applicability to different flow patterns.

2.2.1 Homogeneous Flow

The simplest type of two-phase flow which can be analysed in a straight forward way using the fluid mechanics of single-phase flow is the homogeneous flow. The homogeneous model considers the two-phase flow as a homogeneous fluid flow where one phase is uniformly dispersed into other. In other words, in homogeneous flow we assume the flowing fluid as a single-phase homogeneous *pseudo-fluid* whose properties are neither that of liquid nor of gas. The properties of the homogeneous pseudo-fluid (hypothetical fluid) are something in between the properties of liquid and gas. This assumption, although is a convenient one for analysis, suffers from the drawback of being too simplistic. Hence, two queries arise before we proceed further to describe the homogeneous model.

1. What are the appropriate properties (like density, viscosity, thermal conductivity, etc.) for this assumed pseudo-fluid? How can they be calculated?
2. What are the flow patterns and conditions where the homogeneous model can be applied?

To evaluate the properties we need to answer the second question first. The homogeneous flow model assumes that the two phases are indistinguishably mixed and the flow patterns are such that one phase is finely dispersed in the other. This is realized when $\varepsilon \to 1.0$ and $\varepsilon \to 0$. In the first case ($\varepsilon \to 0$) we have bubbly flow where large number of small gas bubbles are dispersed in liquid giving a homogeneous phase distribution. In the second case ($\varepsilon \to 1$) we have spray flow or mist flow in which large number of small liquid droplets are dispersed into the gas phase. In these cases since the two phases are very well mixed together, we can assume that the slip between the two phases does not exist, giving $S = 1$. Under this condition Eq. 2.12 is applicable and is reproduced here

$$\varepsilon = \frac{1}{\dfrac{\rho_g}{\rho_l}\dfrac{(1-x^*)}{x^*}+1} \tag{2.23}$$

The density of the homogeneous mixture can be assumed as a weighted average of density in the form

$$\rho_h = \varepsilon\rho_g + (1-\varepsilon)\rho_l \tag{2.24}$$

Substitution of Eq. (2.23) in Eq. (2.24) gives

$$\frac{1}{\rho_h} = \frac{x^*}{\rho_g} + \frac{(1-x^*)}{\rho_l} \tag{2.25}$$

where, ρ_h is the density of the homogeneous fluid. For other properties like viscosity and thermal conductivity people have suggested various options based on their individual results of experiments. Isbin et al. (1959) suggested a harmonic mean of viscosity weighted by flowing quality similar to that of density for evaluation of effective homogeneous viscosity as

$$\frac{1}{\mu_h} = \frac{x^*}{\mu_g} + \frac{(1-x^*)}{\mu_l} \tag{2.26}$$

This was also suggested by McAdams et al. (1942). However, Cicchiti et al. (1960) proposed a simple weighted mean

$$\mu_h = \varepsilon\mu_g + (1-\varepsilon)\mu_l \tag{2.27}$$

while Duckler et al. (1964) suggested

$$\mu_h = \frac{j_l}{j}\mu_f + \frac{j_g}{j}\mu_l \tag{2.28}$$

where j stands for the volume flux given by

$$j_g = \varepsilon v_g \qquad (2.29a)$$
$$j_l = (1 - \varepsilon)v_l \qquad (2.29b)$$
$$j = j_g + j_l. \qquad (2.29c)$$

When we consider the homogeneous fluid as a single-phase liquid, the frictional pressure drop is given by

$$-\left(\frac{dp}{dz}\right)_f = C_f \frac{L\rho_h U_h^2}{2d} = C_f \frac{LG^2}{2d\rho_h} \qquad (2.30)$$

where, U_h is the homogeneous fluid viscosity, G its mass velocity and its density is given by Eq. (2.25). Now if we consider only the liquid flowing through the tube, the pressure drop based on the superficial velocity is given by

$$-\left(\frac{dp}{dz}\right)_l = C_{fl} \frac{LG^2(1-x^*)^2}{2d\rho_l} \qquad (2.31)$$

If we take a ratio of Eqs. (2.31) and (2.30), we get the two-phase multiplier for homogeneous flow as

$$\phi_l = \frac{\left(\frac{dp}{dz}\right)_f}{\left(\frac{dp}{dz}\right)_l} = \frac{C_f}{C_{fl}}\left[\frac{1}{(1-x^*)^2}\right]\frac{\rho_l}{\rho_h} = \frac{C_f}{C_{f,l}}\frac{\left[1 + x^*\left(\frac{\rho_l}{\rho_g} - 1\right)\right]}{(1-x^*)^2} \qquad (2.32)$$

To evaluate C_f and $C_{f,l}$ we take resort to Blasius equation.
  For homogeneous fluid:

$$C_f = \frac{C}{Re_h^{n_1}}, \quad \text{where,} \quad Re_h = \frac{Gd}{\mu_h} \qquad (2.33)$$

for liquid:

$$C_{fl} = \frac{C_1}{Re_l^{n_2}}, \quad \text{where,} \quad Re_l = \frac{G(1-x^*)d}{\mu_l} \qquad (2.34)$$

Please note that liquid and the homogeneous fluid may be in two different regimes—one can be laminar and the other can be turbulent. When both are in the same regime (laminar for both or turbulent for both), $c_1 = c_2$ and $n_1 = n_2 = n$. Under this condition, using Eqs. (2.33) and (2.34) in Eq. (2.32), one can get an expression for two-phase multiplier in homogeneous model as

$$\phi_l^2 = \frac{\left(\dfrac{dp}{dz}\right)_f}{\left(-\dfrac{dp}{dz}\right)_l} = (1 - x^*)^{n-2} \frac{1 + x^* \left(\dfrac{\rho_l}{\rho_g} - 1\right)}{\left[1 + x^* \left(\dfrac{\mu_l}{\mu_g} - 1\right)\right]^n} \qquad (2.35)$$

It is found that the homogeneous model gives a good prediction of void fraction and frictional pressure drop when $\rho_l/\rho_g < 10$ or if $G > 2000$ kg/m^{2}s (Whally 1996). For the steam water system $\rho_l/\rho_g < 10$ only occurs for pressure above 120 bar. This means, the homogeneous model can be uniformly used for $p_{system} > 120$ bar. For steam–water system, this is because as we go to higher pressures the saturated liquid and saturated vapour states come closer in property plane, eventually merging at critical point. However as $x^* \to 0$ and $x^* \to 1$ (if we have bubbly or drop flow), the homogeneous model can also be used if the $p_{system} < 120$ bar because more complex separated model (to be discussed in the next section) may not improve prediction substantially. The energy equation for homogeneous flow, as given by Barcozy (1966), is

$$\rho_h \left(\frac{\partial e}{\partial t} + U \frac{\partial e}{\partial z}\right) = \frac{\dot{q} \cdot F}{A} + \dot{q}_v + \frac{\partial p}{\partial t} \qquad (2.36)$$

This is the most general transient form of the energy variation where time (t) is also accounted. In this equation F is the tube circumference (πd) and A its cross sectional area ($\pi d^2/4$). $\dot{q}$ is the heat transfer from periphery (wall heat flux) and $\dot{q}_v$ is the volumetric heat generation in the fluid. Here e stands for the energy convected per unit mass and is given by

$$e = h + \frac{U^2}{2} + gz \sin \gamma \qquad (2.37)$$

where h is the specific enthalpy of the homogeneous fluid, U is its superficial velocity and γ is the inclination of the tube. In the above derivations, we have used one-dimensional pipe flow as an example as it is very common and easy to comprehend. However one must understand that homogeneous flow is not limited to one-dimensional flow alone. It can be used for two and three dimensional situations as well. The only thing to be remembered is that in homogeneous flow we convert the two-phase fluid into a pseudo-fluid and then solve usual fluid mechanics and heat transfer equations (continuity, momentum and energy equations). In case of two or three dimensional cases we can use the standard Navier-Stokes equations with homogeneous fluid properties. In this connection one more word of caution—we are solving for the single-phase homogeneous flow to predict overall pressure drop. Based on our solution for local pressure, temperature if one tries to find out where vapour and where liquid exists (using equilibrium data like steam table), that will be absolutely erroneous. Once we have assumed the pseudo-fluid, vapour and liquid do not exist in our model and we should not try to extract them from the solution.

2.2.2 Separated Flow—Lockhart–Martinelli Model

The method prescribed by Lockhart and Martinelli (1949) is the most widely used model for separated flow. Separated flow is the flow pattern where homogeneous flow assumption is not valid because liquid and gas are not in a dispersed state but are completely separated. Annular and stratified flows are typical examples. However, sometimes the slug flow regime also behaves like separated flow. This model works on the assumption that the total two-phase pressure can be seen as a combination of liquid pressure drop, gas pressure drop and an interaction (between liquid and gas) pressure drop. The model was developed on experimental data for two-phase flow in horizontal tubes with air-water and air-oil flow. Although the model is not very accurate (50% uncertainity) it does a reasonably good job for diameters below 100 mm and low pressures and low mass fluxes for which the experiments were conducted. However due to fact that it came as one of the very first two phase correlations and did reasonably well against applications of that time (1950–1960s), it became very popular. The simplicity of the model is also a reason behind its popularity. In fact a large number of subsequent models were based on this model.

This model defines a model parameter known as Martinelli parameter. We used it in the Taitel–Dukler flow map in the previous chapter. It is given by

$$X^2 = \frac{\left(\dfrac{dp}{dz}\right)_l}{\left(\dfrac{dp}{dz}\right)_g} \tag{2.38}$$

From the equation of two-phase multipliers

$$X^2 = \frac{\phi_g^2}{\phi_l^2} \tag{2.39}$$

As stated earlier, the value of Martinelli parameter depends on the flow regimes of the two fluids. Four combinations are possible and are given in Table 2.1.

Table 2.1 Martenelli parameter

Flow regime	Marteneli parameter
Gas laminar, liquid laminar (ll)	X_{ll}
Gas laminar, liquid turbulent (lt)	X_{lt}
Gas turbulent, liquid laminar (tl)	X_{tl}
Gas turbulent, liquid turbulent (tt)	X_{tt}

In a majority of applications, we come across flows where both phases are in turbulent flow regime and $X_{t,t}$ is to be used. Now the pressure gradients in Eq. (2.38) can be evaluated as

$$-\left(\frac{dp}{dz}\right)_g = C_{fg}\frac{LG^2(x^*)^2}{2d\rho_g} \tag{2.40}$$

$$-\left(\frac{dp}{dz}\right)_l = C_{fl}\frac{LG^2(1-x^*)^2}{2d\rho_l} \tag{2.41}$$

Hence Eq. (2.38) reduces to

$$X^2 = \frac{C_{fl}}{C_{fg}}\left(\frac{1-x^*}{x^*}\right)^2\frac{\rho_g}{\rho_l} \tag{2.42}$$

$$C_{fl} = \frac{C_1}{Re_1^{n_1}} = \frac{C_1}{\left[\frac{G(1-x^*)d}{\mu_l}\right]^{n_1}} \tag{2.43}$$

$$C_{fg} = \frac{C_2}{Re_g^{n_2}} = \frac{C_2}{\left[\frac{Gx^*d}{\mu_g}\right]^{n_2}} \tag{2.44}$$

Now, let us consider the most common case where both gas and liquid are turbulent, $n_1 = n_2 = n$ and $c_1 = c_2$, giving

$$X_{tt}^2 = \left(\frac{1-x^*}{x^*}\right)^{2-n}\left(\frac{\mu_l}{\mu_g}\right)^n\frac{\rho_g}{\rho_l} \tag{2.45}$$

If we assume usual values of n such that $n(0.2 < n < 0.25)$, then

$$X_{tt} = \left(\frac{1-x^*}{x^*}\right)^{0.9}\left(\frac{\mu_l}{\mu_g}\right)^{0.1}\left(\frac{\rho_g}{\rho_l}\right)^{0.5} \tag{2.46}$$

Lockhart and Martinelli proposed that X and ϕ are correlated. However it depends on whether it is ll, lt, tl or tt regime indicating laminar or turbulent flow of the phases. The plots for ϕ_g and ϕ_l are different and the set of eight curves showing ϕ_g and ϕ_l for various flow regimes are shown in Fig. 2.3.

In Fig. 2.3, two-phase multiplier are plotted for various flow regimes according to Lockhart–Martinelli model.

Chisolm (1967) showed that these curves can be represented by the following equations:

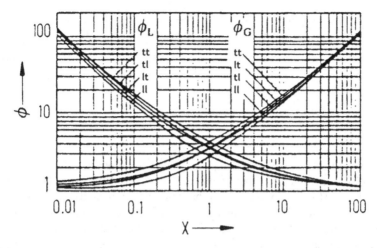

Fig. 2.3 Two-phase multiplier in various flow regimes according to Lockhart–Martinelli model (*Courtesy* "Heat Transfer in Condensation and Boiling" by Karl Stephan, with permission from Springer)

$$\phi_l = 1 + \frac{C}{X} + \frac{1}{X^2} \tag{2.47a}$$

$$\phi_g = 1 + CX + X^2 \tag{2.47b}$$

The data of Lockhart and Martinelli (1949) yields the following values of the constant C (sometimes called the Chisolm parameter).

Due to inherent instability of two-phase flow the transition Reynold numbers are some what lower than single-phase flow. Usually, $Re > 2000$ gives turbulent flow and $Re < 1000$ gives laminar flow. The range in between is for transition for which it is difficult to suggest the value of C. If we substitute the definition of Martinelli parameter

$$\left[X^2 = \frac{\left(\dfrac{dp}{dz} \right)_l}{\left(\dfrac{dp}{dz} \right)_g} \right]$$

to any of the above equations, we get

$$\left(-\frac{dp}{dz} \right)_f = \left(-\frac{dp}{dz} \right)_l + C \left[\left(-\frac{dp}{dz} \right)_l \left(-\frac{dp}{dz} \right)_g \right]^{1/2} + \left(-\frac{dp}{dz} \right)_g \tag{2.48}$$

One gets a very insightful understanding from this equation. This equation states that the total frictional pressure drop of the two-phase flow is the sum of the liquid flow pressure drop, the gas/vapour flow pressure drop and the pressure due to interaction

Table 2.2 Value of Chisolm parameter

Gas/vapour flow	Liquid flow	Flow regime	C
Laminar	Laminar	ll	5
Laminar	Turbulent	lt	10
Turbulent	Laminar	tl	12
Turbulent	Turbulent	tt	20

term is the geometric mean of the liquid and gas phase pressure drop, multiplied by the Chisolm parameter actually indicating the level of interaction between the liquid and gas. That is why when both the flows are laminar (ll) the value of C is the minimum (Table 2.2). The value increases to 10 or 12 when one of the fluids is laminar (lt or tl). Finally, for the case of both gas and liquid being turbulent (tt) it is the highest amongst the four at 20. However, these are not the real extreme cases and are mostly representative of regimes like annular or slug flow. The most intense interaction can be expected when the liquid-gas interface is really large as in the case of homogeneous flow. It can be shown that, in this case if the friction factor is constant, the Chisolm parameter is given by

$$C = \left(\frac{\rho_g}{\rho_l}\right)^{1/2} + \left(\frac{\rho_l}{\rho_g}\right)^{1/2} \tag{2.49}$$

This is somewhat the upper limit of C. For air-water flow at atmospheric pressure, $C = 28.6$. The other limit is the case of fully stratified flow where the interaction between the two phases is minimum. Under such cases for turbulent-turbulent case (tt), the value of C is 3.66.

For laminar flow (ll), this value is even lower at 2. The Lockhart–Martinelli model was based on data for flow in horizontal tube at low (near atmospheric) pressure and hence it cannot be applied to conditions varying too much from that. The Lockhart–Martinelli relation particularly works well for annular flow where the relation between two-phase multiplier and void fraction can be given by

$$\phi_l^2 = (1 - \varepsilon)^{-2} \tag{2.50}$$

Substituting Eq. (2.45) into it, the void fraction can also be calculated from the value of pressure drop through Martinelli parameter. This is shown in Fig. 2.4.

2.2.3 Martinelli–Nelson Method for Diabatic Separated Flow

The Lockhart–Martinelli model is primarily for low pressure flows like air-water flow. In such flows the void fraction remains more or less unchanged (if we consider liquid flashing to vapour due to pressure drop to be negligible). In case of diabatic

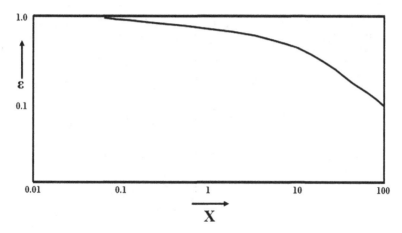

Fig. 2.4 Correlation between void fraction Martinelli parameter

flow with phase change, particularly in flow boiling the void fraction changes due
to evaporation. Here, a change of liquid to vapour prohibits finding the value of
Martinelli parameter accurately as its value changes along the tube. Due to this
limitation, Martinelli and Nelson (1948) extended this to the critical point using
steam water data in evaporation condition. They assumed the regime to be always
tt (turbulent-turbulent) which always gives a conservative estimate of pressure drop
(i.e., higher than or equal to actual pressure drop) when phase change takes place.
They first examined the data near critical pressure where the density and viscosity of
the two phases (liquid and vapour) are close to each other. Assuming thermodynamic
equilibrium at all the points (which means, thermodynamic quality = actual quality)
they established correlation between ϕ and x_{tt}. This happens to give a Chisolm
parameter of $C = 1.36$ (Chisolm 1963). The near atmospheric conditions are known
from the original Lockhart–Martinelli model ($C = 20$). The curves for intermediate
pressures between critical and atmospheric pressure was found by trial and error
using the data of Davidson et al. (1943). This correlation was then converted to
correlations based on pressure from that based on x^*. The definition for ϕ is given by
Eqs. (2.20) and (2.21) and the advantage of using ϕ_{lo} under the condition of phase
change is also described there. From Eqs. (2.41) and (2.21), we get

$$\frac{\left(\dfrac{dp}{dz}\right)_l}{\left(\dfrac{dp}{dz}\right)_{lo}} = \frac{\phi_l^2}{\phi_{le}^2} = \frac{C_{fl}}{C_{fl_e}}(1 - x^*)^2 \tag{2.51}$$

Fig. 2.5 Martinelli–Nelson
correlation for two-phase
multiplier at higher pressures

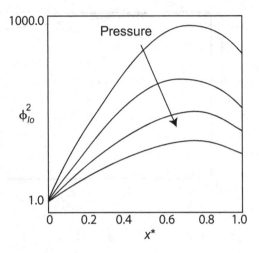

Assuming Blasius correlation to be valid for both of these

$$C_{fl} = \frac{C_1}{Re_l^{n_1}} = \frac{C_1}{\left[\dfrac{G(1-x^*)d}{\mu_l}\right]^{n_1}} \qquad (2.52)$$

$$C_{fg} = \frac{C_2}{Re_g^{n_2}} = \frac{C_2}{\left[\dfrac{Gx^*d}{\mu_g}\right]^{n_2}} \qquad (2.53)$$

$$\phi_l^2 = \phi_{lo}^2 (1-x^*)^{2-n} \qquad (2.54)$$

The variation of two-phase multiplier ϕ_{lo}^2 against quality is given in Fig. 2.5 for steam–water system at various pressure between atmospheric (1.01 bar) and critical (221.2 bar) pressures.

For evaporation in flow boiling where Martinelli–Nelson method is relevant, The changing void fraction has to be accounted for in pressure drop calculation. For circular tube, by integrating Eq. (2.2) over the length of the tube, we get

$$\Delta p = \int_0^L \tau_w dz + G\left[x^*v_g + (1-x^*v_l)\right]_0^L + g\cos\gamma \int_0^L \left[\varepsilon\rho_g + (1-\varepsilon)\rho_l\right]dz \qquad (2.55)$$

If at entry to the tube we assume $x^* = 0$, using two-phase multiplier ϕ_{lo}, the above equation reduces to

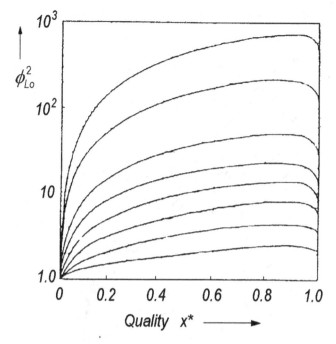

Fig. 2.6 Two-phase multiplier for changing void fraction (*Courtesy* "Heat Transfer in Condensation and Boiling" by Karl Stephan, with permission from Springer)

$$\Delta p = \frac{2c_{le}G^2L}{d\,p_l}\left(\frac{1}{x^*}\int_0^{x^*}\phi_{lo}dx^*\right)+G^2\left\{\frac{x^{*2}}{\varepsilon\rho_g}+\frac{1}{\rho_l\left[\frac{(1-x^*)^2}{1-\varepsilon}-1\right]}\right\}$$

$$+g\cos\gamma\int_0^L\left[\varepsilon\rho_g+(1-\varepsilon)\,\rho_l\right]dz$$

(2.56)

The integral in the first term can be evaluated by the $\phi_{lo}-x^*$ correlation of Martinelli, which is plotted in Fig. 2.6

$$\overline{\phi_{lo}^2}=\frac{1}{x^*}\int_0^{x^*}\phi_{lo}^2dx^*$$

(2.57)

The second term can be written as

$$\gamma_2=\frac{1}{\rho_l}\left[\frac{x^{*2}}{\varepsilon}\frac{\rho_l}{\rho_g}+\frac{(1-x^*)^2}{1-\varepsilon}-1\right]$$

(2.58)

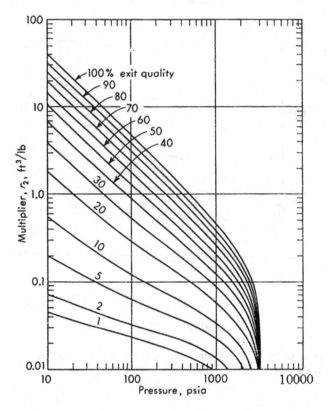

Fig. 2.7 Coefficient for acceleration term for changing void fraction (*Courtesy* "Prediction of pressure drop forced circulation boiling of water" by R.C. Martinelli and D.B. Nelson, Trans. ASME, 1948, with permission from The American Society of Mechanical Engineers)

using the above two curves, the pressure drop by Martinelli–Nelson method can be evaluated. This is plotted in Fig. 2.7. However, the method is not very accurate for $G > 1500 \, \text{kg/m}^2\text{s}$.

2.2.4 Barcozy and Chisolm Model

In the above method the fluids were taken to be steam–water and mass flux did not appear as a parameter. Hence, the applicability of Martinelli–Nelson method was limited to $G < 1500 \, \text{kg/m}^2\text{s}$. To take care of various fluids and high mass flux, Barcozy (1966) determined frictional pressure drop from two curves. The curve shown in Fig. 2.8 is the one in which the two-phase multiplier is plotted against the following parameter for turbulent-turbulent (tt) regime.

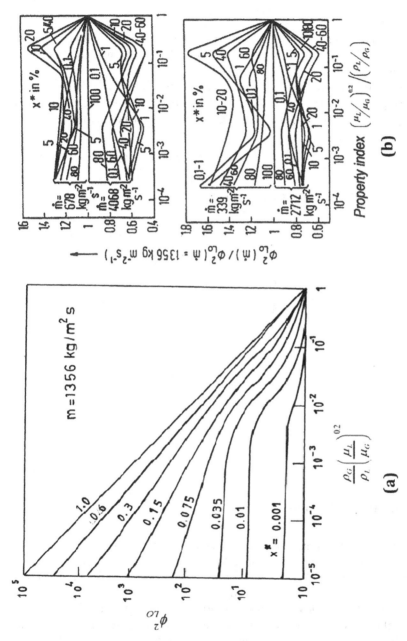

Fig. 2.8 Pressure drop multiplier of Baroczy various values of mass quality, x^* (*Courtesy* "Heat Transfer in Condensation and Boiling" by Karl Stephan, with permission from Springer)

$$Y^2 = \frac{\phi^2_{go}}{\phi^2_{lo}} = \frac{\rho_g}{\rho_l}\left(\frac{\mu_l}{\mu_l}\right)^{0.2} \tag{2.59}$$

where, $c_{fl} = \frac{c}{Re_l^{0.2}}$

It has to be remembered that Fig. 2.8 is plotted for a reference mass flux of $G = 1356\,kg/m^2s$ (actually it is taken because in FPS units, this corresponds to $1 \times 10^6 lb/hrft^2$).

The pressure drop is given by the two-phase multiplier at $G = 1356\,kg/m^2$-s (as given in Fig. 2.8a) and a correction factor (ω) corrected to the real mass flux. Barcozy had suggested a complicated plot for omega which is presented here in Fig. 2.8b.

$$-\left(\frac{dp}{dz}\right) = \frac{2c_{fl}G^2}{\rho_l d}\phi^2_{lo}(\text{at } G = 1356)\Omega \tag{2.60}$$

Chisolm (1973) presented a simple correlation for these curves, as

$$\phi^2_{lo} = 1 + \left(Y^2 - 1\right)\left[Bx^{*\frac{2-n}{2}}(1 - x^*)^{\frac{2-n}{2}} + x^{*2-n}\right] \tag{2.61}$$

$$B = \frac{55}{G^{0.5}} \quad for \quad 0 < Y \leqslant 9.5$$

$$B = \frac{520}{YG^{0.5}} \quad for \quad 9.5 < y \leqslant 28$$

$$B = \frac{1500}{Y^2G^{0.5}} \quad for \quad Y \geqslant 28$$

where

$$Y^2 = \frac{\left(\frac{dp}{dz}\right)_{go}}{\left(\frac{dp}{dz}\right)_{lo}}$$

2.3   Drift Flux Model

The drift flux model is an elegant model which gives a deeper understanding of two-phase flow phenomenon in certain flow patterns. It gives a very good insight into phenomena such as flooding which is difficult to understand from conventional homogeneous or separated model. We will discuss flooding when discussing in-tube condensation. Wallis (1969) developed the drift flux model. The main focus of drift flux model is the relative motion of the two fluids and not the absolute motion of the phases. It is important in bubbly, slug and spray (or drop) flow regimes. The model is not for regimes such as annular flow, but there are efforts to adopt it to these regimes as well. The model works very well for particulate flow and foam drainage. As stated earlier the entire concept revolves around relative movement of the two phases and works well when the relative motion is independent of flow rate for each phase. The model requires a new set of variables which are given below.

Let us recall the definition of actual phase velocity u_g and u_l for the gas and the liquid respectively as well as that of superficial velocities U_g and U_l. From these definitions it follows that

$$u_g = \frac{U_g}{\varepsilon} \tag{2.62a}$$

$$u_g = \frac{U_l}{1 - \varepsilon} \tag{2.62b}$$

We define slip velocity us as the velocity of gas relative to the liquid,

$$u_s = u_g - u_l \tag{2.63a}$$

$$u_s = \frac{U_g}{\varepsilon} - \frac{U_l}{1 - \varepsilon} \tag{2.63b}$$

Now we define drift flux of gas relative to the liquid, as

$$j_{gl} = u_s \varepsilon (1 - \varepsilon) \tag{2.64}$$

From Eq. (2.63), we can also write drift flux, as

$$j_{gl} = U_g (1 - \varepsilon) - U_l \varepsilon \tag{2.65}$$

Now, we define the drift velocity of gas relative to mean fluid (u_{gj}) and liquid relative of mean fluid (u_{lj}), as

$$u_{gj} = u_g - j \tag{2.66a}$$
$$u_{lj} = u_l - j \tag{2.66b}$$

The j here stands for the mean velocity, given by

$$j = \frac{liquid \; volume \; flow \; rate + gas \; volume \; flow \; rate}{total \; tube \; cross \; sectional \; area}$$

$$j = U_g + U_l \tag{2.67}$$

Physically, drift flux is the volumetric flux of a phase relative to a transverse surface moving with the average velocity j. Thus, correlation between drift flux and drift velocity can be given by

$$j_{gl} = \varepsilon u_{gj} \tag{2.68}$$

Similarly, the liquid drift flux can be given by

$$j_{lg} = (1 - \varepsilon)u_{lj} \qquad (2.69)$$

Using Eqs. (2.62), (2.66) and (2.67), it can be shown from Eqs. (2.68) and (2.69)

$$j_{gl} = -j_{lg} \qquad (2.70)$$

In other words, the liquid and gas drift fluxes as equal and opposite in direction. Conventionally, the upward direction is taken as positive and downward negative. The drift flux is zero when there is no slip between the two phases (homogeneous flow is such as case).

2.3.1 One-dimensional Drift Flux Model

Let us consider one-dimensional flow without wall shear, as shown in Fig. 2.9. The force balance in this case in the control volume gives the following equations for liquid and gas phases,

$$\frac{dp}{dz} + \rho_l g - \frac{\overline{F}}{1 - \varepsilon} = 0 \qquad (2.71)$$

$$\frac{dp}{dz} + \rho_g g - \frac{\overline{F}}{\varepsilon} = 0 \qquad (2.72)$$

This is particularly a case in which gravity force is balanced by the pressure gradient given by (dp/dz) in the equations. The volumetric drag force, $\overline{F}$ appearing in the equation is a function of relative motion, void fraction, geometry of the interface as well as the properties of the phases. From the above equations, this force can be found as

Fig. 2.9 One-dimensional wall shear free two-phase flow

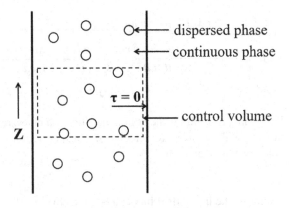

$$\overline{F} = \varepsilon(1 - \varepsilon)(\rho_l - \rho_g)g. \tag{2.73}$$

2.3.2 Application to Bubbly Flow

The application of the drift flux model to bubbly flow is very straight forward. For this we use an equation to relate the slip to the single bubble rise velocity, u_b as

$$u_s = u_b(1 - \varepsilon) \tag{2.74}$$

Giving $j_{gl} = u_s\varepsilon(1 - \varepsilon)$

$$j_{gl} = u_b\varepsilon(1 - \varepsilon)^2 \tag{2.75}$$

This can be plotted against void fraction as shown in Fig. 2.10 as the curved line giving the drift flux to be zero in both the limits $\varepsilon \to 0$ and $\varepsilon \to 1$.

Again drift flux can also be written as

$$j_{gl} = U_g(1 - \varepsilon) - U_l\varepsilon \tag{2.76}$$

This is plotted in Fig. 2.10 as three straight lines. Thus the points at which both Eqs. (2.75) and (2.76) are satisfied is the solution. This can be found out as the intersection of the two types of curves as shown in Fig. 2.10. The figure clearly shows that in a vertical tube for co-current flows (both gas and liquid are in the same direction, upward or downward), there is always a solution. For gas moving downward and liquid upward, there is no solution. For gas moving upward and liquid moving downward, three possibilities are there: (a) either there are two solutions or (c) none or (b) as a limiting case, the two solutions merge giving what is known as flooding point. This is the limit of counter-current flow. The physical phenomenon related to this is entrainment of liquid by gas and flow reversal of liquid during flooding which we will discuss along with condensation.

Fig. 2.10 Solution for void fraction (*Courtesy* "Boiling Condensation and Gas-Liquid flow" by P.B. Whalley, with permission from Oxford University Press)

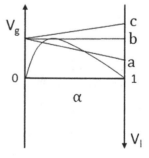

Similar analysis can be carried out for plug or slug flow where plug rising velocity can be related to drift velocity as (Whally 1996)

$$u_{gj} = u_p = 0.35(gd)^{0.5} \qquad (2.77)$$

2.3.3 Two-dimensional Drift Flux Model

Zuber and Findlay (1965) extended the original drift flux model to the case where the void fraction variation across the tube direction is considered. From definition

$$\varepsilon = \frac{U_g}{u_g} = \frac{U_g}{j + u_{gj}} \qquad (2.78)$$

$$\varepsilon = \frac{\frac{U_g}{j}}{1 + u_b(1 - \varepsilon)^2 j} \qquad (2.79)$$

using

$$u_{gj} = \frac{j_{gl}}{\varepsilon} = u_b(1 - \varepsilon)^2$$

If we consider void fraction variation across the tube, we can define

$$c_o = \frac{\overline{\varepsilon j}}{(\varepsilon)(j)} \neq 1$$

where the over score indicates average. Equation (2.79) then gets transformed to

$$\varepsilon = \frac{\frac{U_g}{j}}{c_o + \frac{u_{gj}}{j}} \qquad (2.80)$$

Zuber and Findlay suggested $c_o = 1.13$ for vertical upflow and for bubbly flow

$$u_{gj} = 1.4\left[\frac{\sigma g(\rho_l - \rho_g)}{\rho_l^2}\right]^{\frac{1}{4}} \qquad (2.81)$$

where σ = surface tension of the liquid.
This gives a reasonably good prediction at $\varepsilon < 0.3$.

Examples

Problem 1: A mixture of air and water at atmospheric pressure and temperature (300 K) flows through a copper tube of inner diameter 36 mm. If the mass flow rates of water and air are 0.3 and 0.1 kg/s respectively, estimate the frictional pressure drop per metre length of the tube.

Solution: At 300 K, properties of water are $\rho_l = 997 \text{kg/m}^3$, $\mu_l = 855 \times 10^{-6} \text{ Ns/m}^2$

$$\frac{\dot{m}_L}{\pi \frac{d^2}{4}} = \frac{0.3}{\frac{\pi(0.036)^2}{4}} = 294.73 \text{ kg/m}^2 s$$

$$Re_L = \frac{G_L d}{\mu_L} = \frac{294.73 \times 0.036}{855 \times 10^{-6}} = 12410$$

friction factor,

$$\varsigma_L = \frac{0.079}{Re_L^{1/4}} = \frac{0.079}{(12410)^{1/4}} = 0.00748$$

$$\left(\frac{dp}{dz}\right)_L = -\frac{\varsigma G_l^2}{2d\rho_l} = \frac{-0.00748 \times (294.72)^2}{2 \times 0.036 \times 997} = 9.06 \text{ N/m}^2$$

At 300 K, properties of air are $\rho_l = 1.1614 \text{ kg/m}^3$, $\mu_l = 184.6 \times 10^{-6} \text{ Ns/m}^2$

$$G_G = \frac{\dot{m}_G}{\frac{\pi d^2}{4}} = \frac{0.1}{\pi \frac{(0.036)^2}{4}} = 98.24 \text{ kg/m}^2$$

$$Re_G = \frac{G_G d}{\mu_G} = \frac{98.24 \times 0.036}{184.6 \times 10^{-7}} = 191584$$

friction factor,

$$\varsigma_G = \frac{0.079}{Re_G^{1/4}} = \frac{0.079}{(191584)^{1/4}} = 0.00378$$

$$\left(\frac{dp}{dz}\right)_G = -\frac{\varsigma_G G_G^2}{2d \rho_G} = -\frac{0.00378 \times (98.24)^2}{2 \times 0.036 \times 1.1614} = 436.27 \text{ N/m}^2\text{m}$$

Since, both Re_L and Re_G are greater than 2300, liquid and gas phases are turbulent. Lockhart and Martinelli parameter, $X_{tt}^2 = \frac{(dp/dz)_L}{(dp/dz)_G} = \frac{9.06}{436.27} = 0.0207$, $X_{tt} = \sqrt{0.0207} = 0.144$. Two-phase multiplier,

$$\phi_L^2 = 1 + \frac{C}{X_{tt}} + \frac{1}{X_{tt}^2}$$

The valve of C is equal to 20 since both phases are turbulent

$$\phi_L^2 = 1 + \frac{20}{0.144} + \frac{1}{0.144^2} = 188.11$$

$$\phi_G^2 = 1 + CX_{tt} + X_{tt}^2 = 1 + (20 \times 0.144) + (0.144)^2 = 3.90$$

Frictional pressure drop of two-phase flow is given by

$$\left(\frac{dp}{dz}\right)_{friction} = \left(\frac{dp}{dz}\right)_G \phi_G^2 = 436.27 \times 3.9 = 1701.45\,\text{N/m}^2\text{m}$$

Also,

$$\left(\frac{dp}{dz}\right)_{friction} = \left(\frac{dp}{dz}\right)_L \phi_L^2 = 9.06 \times 188.11 = 1704.28\,\text{N/m}^2\text{m}$$

It may be noted that two-phase frictional pressure drop calculated assuming water alone and that calculated assuming air alone are almost equal when multiplied by their respective two-phase multipliers.

Problem 2: Calculate the frictional pressure drop during the condensation of R134a flowing through a 1.5 mm diameter tube at a mass flux of 250 kg/m²s at a quality of 0.6 and pressure of 20 bar.

Solution: Given:

$Mass\ Flux = 250\,\text{kg/m}^2\text{s}$

$Diameter\ of\ Tube = 0.0015\,\text{mm}$

$Pressure = 20\,\text{bar}$

$Refrigerant\ property\ T_{sat} = 340.63\,\text{K},$

$\rho_g = 107.411\,\text{kg/m}^g$

$\mu_g = 10.45 \times 10^{-5}\,\text{kg/m.s}$

$\sigma = 0.0316\,\text{N/m}$

$\rho_l = 1010\,\text{kg/m}^3$

$\mu_l = 1.24 \times 10^{-4}\,\text{kg/m.s}$

Liquid Reynolds number

$$R_{el} = GD(1-x)/\mu_l = 250 \times 0.0015 \times (1-0.6)/1.24 \times 10^{-4} = 1209$$

vapour Reynolds number

$$R_g = GD(1-x)/\mu_g = 250 \times 0.0015 \times (0.6)/1.45 \times 10^{-5} = 15517$$

Single-phase friction factor is calculated by the following Churchill (1977) correlation

$$f = 8\left[\left(\frac{8}{Re}\right)^{12} + \left\{\left[2.457x \ln\left(\frac{1}{(\frac{7}{Re})^{0.9}+0.27\varepsilon/D}\right)\right]^{16} + \left(\frac{37530}{Re}\right)^{16}\right\}^{-1.5}\right]^{\frac{1}{12}}$$

$$f_l = 0.052936$$

$$f_g = 0.029342$$

$$\left(\frac{dP}{dz}\right)_l = \frac{f_l G^2(1-x)^2}{2D\rho l} = \frac{0.052936 \times 250^2(1-0.6)^2}{2\times0.0015\times1010} = 174.71$$
$$\left(\frac{dP}{dz}\right)_g = \frac{f_g G^2(x)^2}{2D\rho g} = \frac{0.029342 \times 250^2(0.6)^2}{2\times0.0015\times107.411} = 2048.81$$

The Martinelli parameter is given by

$$X = \left[\frac{\left(\frac{dP}{dz}\right)_l}{\left(\frac{dP}{dz}\right)_g}\right]^{\frac{1}{2}} = \left[\frac{174.71}{2048.81}\right]^{\frac{1}{2}} = 0.292$$

Two-phase multiplier is given by

$$\phi^2 = 1 + \frac{C}{X} + \frac{1}{X^2} = 1 + \frac{12}{0.292} + \frac{1}{0.292^2} = 53.82$$

Since liquid flow is in laminar region and vapour in turbulent region, the value of constant is being taken as 12. Therefore two phase pressure gradient is given by

$$\frac{\Delta P}{L} = \phi^2 \left(\frac{dP}{dz}\right)_l = 53.82 \times 174.71 = 9402.8922 \text{ Pa/m}$$

The frictional pressure drop per length is given by Lockhart and Martinelli correlation is 9402.8922 Pa/m.

Problem 3: Water at 0.13 kg/s flows through a vertical tube with a power of 100 kW applied to the tube. The tube has 10 mm inner diameter and is 3 m long heated uniformly over its length. The water enters the tube at 86 bar and 250 °C. Calculate the frictional pressure drop using

(a) the homogenous model
(b) the Martinelli–Nelson model.

Solution:

$\dot{m} = 0.13\,\text{kg/s}$

$\Delta h = \frac{power}{Mass\ flow\ rate} = \frac{100 \times 10^3}{0.13} = 769.23 \times 10^3\,\text{J/kg} = 769.23\text{kJ/kg}$

Water enters at 86 bar, 250 °C

$T_{sat}(86\,\text{bar}) = 300\,°\text{C}$

Enthalpy of the sub-cooled water, hi, entering the tube is equal to the enthalpy at the inlet temperature.

$h_i = h_f(250\,°\text{C}) = 1085.8\,\text{kJ/kg}$

$h_f(86\,\text{bar}) = 1345\,\text{kJ/kg}$

The length of the tube lSC required to preheat the water to the saturation temperature is given by

$$\frac{l_{sc}}{L} = \frac{1345 - 1085.8}{769.23} \quad \left(\because \frac{h_f - h_i}{\Delta h} = \frac{l_{sc}}{L} \right)$$

Length of the tube, $L = 3\,\text{m}$.

Therefore $l_{sc} = 0.337 X L = 0.337 X 3 = 1.011\text{m}$

The outlet mass quality x_o is given by

$h_o = h_i + \Delta h$

$(h_f + x h_{fg})_o = h_i + \Delta h$

$1345 + x_o X 1404 = 1085.8 + 769.23$

$\therefore x_o = 0.363$

Mass velocity or mass flux

$G = \frac{\dot{m}}{\frac{\pi d^2}{4}} = \frac{0.13}{\frac{\pi (0.01)^2}{4}} = 1655\,\text{kg/m}^2\text{s}$

At 250 °C (T_{fi}), $\mu_{fi} = 1.09 \times 10^{-4}\,\text{Ns/m}^2$

$v_{fi} = 1.25 \times 10^{-3} m^{-3}./\text{kg}$

$$Re_{fi} = \frac{Gd}{\mu_{ji}} = \frac{1655 \times 0.01}{1.09 \times 10^{-4}} = 1.518 \times 10^5$$

Friction factor, $f_{fo} = 0.079(Re_{fi})^{-1/4} = 0.079(1.518 \times 10^5)^{-1/4} = 0.004$

At $300\,°C(T_{sat})$, $\mu_f = 0.907 \times 10^{-4}\,Ns/m^2$

$v_f = 1.404 \times 10^{-3}\,m^3/kg$

$$Re_f = \frac{Gd}{\mu_f} = \frac{1655 \times 0.01}{0.907 \times 10^{-4}} = 1.825 \times 10^5$$

$f_{fo} = 0.079(Re_f)^{-1/4} = 0.079(1.825 \times 105)^{-1/4} = 0.00382$

$\bar{f}_{fo} = \frac{0.004 + 0.00382}{2} = 0.00391$

$\bar{v}_f = \frac{1.25 \times 10^{-3} + 1.404 \times 10^{-3}}{2} = 1.327 \times 10^{-3}$

Frictional pressure drop in preheating section, taking average property values

$$\Delta p_F = \frac{2\bar{f}_{fo}G^2\bar{v}_f l_{sc}}{d} = \frac{2 \times 0.00391 \times 1655^2 \times 1.327 \times 10^{-3} \times 1.011}{0.01}$$

$$= 2873.6\,N/m^2$$

$$= 2.874\,kN/m^2$$

Frictional pressure drop in two phase region

(a) Homogenous model:

$\frac{1}{\mu} = \frac{x}{\mu_g} + \frac{1-x}{\mu_f}$

$\mu_g(86\,bar) = 19.73 \times 10^{-6}\,Ns/m^2$

$\mu_f(86\,bar) = 90.7 \times 10^{-6}\,Ns/m^2$

$\frac{1}{\mu} = \frac{0.363}{19.73 \times 10^{-6}} + \frac{1-0.363}{90.7 \times 10^{-6}} = 39.33 \times 10^{-6}\,Ns/m^2$

$f_{TP} = 0.079\left(\frac{Gd}{\mu}\right)^{-1/4} = 0.079\left(\frac{1655 \times 0.01}{39.33 \times 10^{-6}}\right)^{-1/4}$

$= 0.0031\,(at\ two\ phase\ outlet\ or\ at\ x = x_o)$

At two phase inlet or at $T_{sat} = 300\,°C$, $f_{TP} = 0.00382$

$\bar{f}_{TP} = \frac{0.0031 + 0.00382}{2} = 0.00346$

$v_{fg}(86\,\text{bar}) = 20.236 \times 10^{-3}\,\text{m}^3/\text{kg}$

$\Delta P_F = \frac{2\bar{f}_{TP}G^2v_f(L-l_{sc})}{d}\left[1 + \frac{v_{fg}}{v_f}\frac{x_o}{2}\right]$

$\frac{2\times0.00346\times1655^2\times1.404\times10^{-3}\times(3-1.011)}{0.01}\left[1 + \frac{20.24\times10^{-3}}{1.404\times10^{-3}}\times\frac{0.363}{2}\right]$

$= 19142\,\text{N}/\text{m}^2$

$= 19.142\,\text{kN}/\text{m}^2$

$\Delta P_F, Total = \Delta P_{F,preheating} + \Delta P_{F,two-phase} = 2.874 + 19.142$

$= 22.016\,\text{kN}/\text{m}^2$

Martinelli–Nelson Method

$\Delta P_F = \frac{2f_{fo}G^2v_f(L-l_{sc})}{d}\left[\frac{1}{x_o}\int_0^{x_o}\phi_{fo}^2 dx\right]$

$\frac{2\times0.00382\times1655^2\times1.404\times10^{-3}\times(3-1.011)}{0.01}\left[\frac{1}{0.363}\int_0^{0.363}\phi_{fo}^2 dx\right]$

The value of $\left[\frac{1}{x_o}\int_0^{x_o}\phi_{fo}^2 dx\right]$ for $p = 86\,\text{bar}$ and $x_o = 0.363$ using Fig. 2.6,

$\Delta P_F = \frac{2\times0.00382\times1655^2\times1.404\times10^{-3}\times(3-1.011)}{0.01} \times 5.2$

$= 30387.47\text{N}/\text{m}^2$

$= 30.39\text{kN}/\text{m}^2$

$\Delta P_{F,total} = \Delta P_{F,preheating} + \Delta P_{F,two-phase} = 2.874 + 30.39$

$= 33.264\,\text{kN}/\text{m}^2$

Problem 4: A simple device has been proposed to cool a chip that is in a remote location where it is difficult to keep a large heat sink. An evaporator in the form of a thick copper slab with drilled horizontal holes is placed with thermal contact on the chip and refrigerant R134a flows through it. The chip produces 2.5 KW heat. The slab is 5 cm × 5 cm × 2 cm copper and there are 9 smooth bores of 0.5 cm apart and 0.15 cm dia each placed 1.5 cm above the chip surface. The saturated refrigerant at 0 °C enters these tubes at $x = 0.2$. The refrigerant flow rate is 18 g/s. Find the pressure drop across the tubes. Find the approximate average temperature of the chip surface assuming the pipes as a plane held at the temperature of saturation. The chip produces 900 W. Actual pressure drop is found to be less than 1 Pa.

Solution: The flow rate across each tube is 2 g/s. Given diameter 1.5 mm.

The chip produces W. So the change in enthalpy across each tube is

$$\Delta h = \frac{Q}{m} = \frac{kJ}{kg}$$

Assuming that the pressure drop is not as drastic as to change the temperature of the flowing refrigerant more than a few degrees, we can take the properties of the refrigerant.

$\mu_l = 0.2728 \times 10^{-3}$ Pa.s and $\mu_g = 10.8 \times 10^{-6}$ Pa.s

$Initial\ Enthalpy = 0.8 X h_l + 0.2 X h_g = 240$ kJ/kg

$Leaving\ enthalpy = 240 + 900/18 = 290$ kJ/kg

The approximate exiting quality is given by $x = 0.455$.

As the quality is not changing very much, me will calculate approximately at mean thermodynamic quality $x^* = 0.327$. The Reynolds number for the liquid,

$$\left(\frac{dp}{dz}\right)_l = -9.28 \times 10^{-3} \times \frac{\left((1 - 0.327) \times \dot{m}\right)^2}{2 d \rho_l} = 4.33 \times 10^{-9} \text{pa}$$

The Reynolds number for gas is

$$Re = \frac{0.327 \times 4 \dot{m}}{\pi d^2} \times \frac{d}{\mu_g} = 51401$$

Both the phases are turbulent.

Friction factor

$$f = 0.079(Re)^{\frac{1}{4}}$$

Friction factors are

$$f_l = 9.82 \times 10^{-3}$$

$$f_g = 5.25 \times 10^{-3}$$

The pressure drop for liquid is

$$\left(\frac{dp}{dz}\right)_l = -9.28 \times 10^{-3} \times \frac{\left((1-0.327) X \dot{m}\right)^2}{2 d \rho_l} = -4.33 \times 10^{-9} \text{pa}$$

That for gas is

$$\left(\frac{dp}{dz}\right)_g = -5.25 \times 10^{-3} \times \frac{\left(0.327\,\dot{m}\right)^2}{2d\rho_g} = -5.18 \times 10^{-5}\,\text{pa}$$

The Lockhart–Martinelli parameter is given by

$$X^2 = \frac{f_l(1-x^*)\rho_g}{f_g x^* \rho_l} = 0.05$$

Since both the phases are turbulent, $C = 20$

$$\phi_l^2 = 1 + \frac{C}{X} + \frac{1}{X^2}$$

$$\phi_g^2 = 1 + CX + X^2$$

$$\phi_l^2 = 801$$

$$\phi_g^2 = 2.0024$$

Therefore the pressure drop is $3.5\,\mu$ Pa. which is very negligible. The actual pressure drop is larger than this as micro-effects come into picture as we are very near to 1.5 mm dia. Let the temperature of the chip be T. The overall heat transfer takes place between a temperature of $0\,°\text{C}$ and $T\,°\text{C}$ over a length $= 1.7\,5\,\text{cm}$ and a area of $5\,\text{cm} \times 5\,\text{cm}$.

$$\Delta T = 0.0175 \times \frac{900}{0.05 \times 0.05 \times 393} = 16\,°\text{C}$$

Hence the chip is at $16\,°\text{C}$.

Problem 5: A vertical tubular cross section is to be installed in an experimental setup high pressure water loop. The tube is 10 mm and 4 m long heated uniformly over its length. Estimate the pressure drop across the test section if water is sent as saturated liquid at 80.037 bar at 0.16 kg/s. The pressure drop has to be kept lower than 150 kN/m². What changes would you make in mass flow rate if the pressure drop exceeds the above limit? The power can be assumed to be as 100 kW.

Solution: $-h = \text{power/mass flow rate} = \frac{100 \times 10^{-3}}{0.16} = 625$ kJ/kg

Water enters as saturated liquid at 80.037 bar ($T_{sat} = 295\,°\text{C}$)
$h_f = 1317.3$ kJ/kg

$$x \times h_{fg} = \Delta h$$

$$625 = x \times 1442.5$$

$x = 0.4333$

$f_{tp} = 3.456 \times 10^{-3}$

$\bar{f}_{tp} = \frac{(2.85+3.456)\times10^{-3}}{2} = 3.15 \times 10^{-3}$

Mass velocity $= G = \frac{\dot{m}\times4}{\pi\times d^2} = \frac{0.16\times4}{\pi\times(0.01)^2} = 2037.2\,\text{kg/sm}^2$

$\nu_f = 1.384 \times 10^{-3}\ \text{m}^3/\text{kg}$

$\mu_f = 0.977 \times 10^{-4}\ \text{Ns/m}^2$

$\mu_g = 0.209 \times 10^{-4}\ \text{Ns/m}^2$

(a) Homogeneous model:

$\frac{1}{\mu} = \frac{x}{\mu_g} + \frac{(1-x)}{\mu_f} = \frac{0.4333}{0.209\times10^{-4}} + \frac{(1-0.4333)}{0.977\times10^{-4}}$

$\mu = 0.3774 \times 10^{-4}\,\text{Ns/m}^2$

$f = 0.079\left[\frac{Gd}{\mu}\right]^{-0.25} = 0.079\left[\frac{2037.2\times0.01}{0.3774\times10^{-4}}\right]^{-0.25} = 2.92 \times 10^{-3}$

$f_{sat} = 0.079\left[\frac{2037.2\times0.01}{0.977\times10^{-4}}\right]^{-0.25} = 3.7 \times 10^{-3}$

$\bar{f} = f(f_{sat} + f) \div 2$

$\bar{f} = 3.31 \times 10^{-3}$

$\nu_{fg} = 22.129 \times 10^{-3}\,\text{m}^3/\text{kg}$

$\Delta P_f = \frac{2\bar{f}_{tp}G^2\nu_f L}{d}\left[1 + \frac{\nu_{fg}x_0}{\nu_f 2}\right] = \frac{2\times3.31\times10^{-3}\times2037.2^2\times1.384\times10^{-3}\times4}{0.01}$ [1]

$\left[1 + \frac{22.129\times10^{-3}\times0.4333}{1.384\times10^{-3}\times2}\right] = 67.9\,\text{kN/m}^2$

(b) Martinelli–Nelson method:

$\Delta P_f = \frac{2f_0 G^2\nu_f L}{d}\left[\frac{1}{x_0}\int_0^{x_0}\phi^2{}_{fo}dx\right]$

$= \frac{2\times3.7\times10^{-3}\times2037.2^2\times1.384\times10^{-3}\times4\times7}{0.01} = 119\,\text{kN/m}^2$

$119\,\text{kN/m}^2 < 150\,\text{kN/m}^2$

The prescribed pressure drop. Therefore, it is safe.

Exercises

1. Pure saturated steam at $180\,^\circ C$ with the mass flux of $400\,kg/m^2s$ condenses in a horizontal tube with $5\,mm$ inner diameter and quality is 0.8. The inner surface temperature maintained at $150\,^\circ C$. Find the frictional pressure drop.

2. A mixture of air and water at atmospheric pressure and temperature $(300\,K)$ flows through a copper tube of inner diameter $36\,mm$. If the mass flow rates of water and air are 0.3 and 0.1 kg/s respectively, estimate the frictional pressure drop per unit length of the tube. (Properties of water at $300\,K$ are $\rho_l = 997 kg/m^3$, $\mu_g = 855 \times 10^{-6}\,Ns/m^2$; and properties of air at $300\,K$ are $\rho_g = 1.1614\,kg/m^3$, $\mu_g = 184.6 \times 10^{-7}\,Ns/m^2$).

3. Water at $0.13\,kg/s$ flows through a vertical tube with a power of $100\,kW$ applied to the tube. The tube has $10\,mm$ inner diameter and is heated uniformly over its length. The water enters the tube at 86 bar and $250\,^\circ C$. Calculate the frictional drop using (a) homogeneous model and (b) Martinelli–Nelson model.

4. A vertical tubular test section of $12.5\,mm$ ID and $4\,m$ long is heated uniformly over its length. The water enters the section at 60 bar and $200\,^\circ C$ with a flow rate of $0.25\,kg/s$ and with a power of $80\,kW$. Calculate the pressure drop over the test section using, (1) homogeneous model (2) Martinelli–Nelson model (3) Thom correlation (4) Baroczy correlation.

5. A mixture of air and water enters a tube of inner diameter $36\,mm$. The liquid and gas pressure drops are 10 and 450, respectively. Find the frictional pressure drop per metre length of the tube and also the mass flow rates of the fluids.

6. In a steam generator, water at $30\,^\circ C$, 1 atm enters a pipe of ID $20\,mm$ First the water is uniformly heated with $150\,KW/m^2$ for a length of $3\,m$ in an economizer. Then it enters the boiler and exits as saturated vapour. The length of the boiler tube is $1m$. Then it is superheated till $150\,^\circ C$ by supplying const. heat input of $100\,KW/m^2$. Calculate the total pressure drop across the steam generator. Assume all components work ideal.

7. Water at $64.2\,bar$ as saturated liquid enters into a vertical tube of dia $8\,mm$ and length $3\,m$ and leaves as saturated steam. Calculate the mass flow rate required to have a constant power supply to the tube as $80\,kW$. Calculate the pressure drop across the pipe using Martinell–Nelson method. Does homogeneous method appropriate in calculating the pressure drop? Explain.

8. A R134a refrigerator working on standard VCR cycle operates at $40\,^\circ C$ condensing temperature. The evaporator temperature may be taken as $-20\,^\circ C$. Take capacity of the system as $10\,kW$. Neglecting the pressure drop across the condenser, calculate the pressure drop across the evaporator if saturated liquid enters the throttle valve at condenser pressure and the refrigerant at outlet of the compressor can be assumed to be as saturated vapour. Assume tube did $10\,mm$ and tube length $0.5\,m$.

9. A mixture of heptane and octane 25% mass fraction of octane flows in a horizontal $2.21\,cm$ bore tube at mass flow flux of $407\,kg/m^2s$ and a pressure of $310\,kPa$. If the quality is 15%, calculate the pressure gradient in the tube using the Lockhart and Martinelli correlations. The acceleration pressure drop can be neglected

and gravitational pressure drop can be neglected for the horizontal tube under following conditions, The fluid undergoes following sequence of flow,

   (i) A sudden contraction from 4 cm to a 2.21 cm dia pipe.
   (ii) A straight 5 m long 2.21 cm dia horizontal pipe.
  (iii) A 90° bend with centerline radius (R_c) of 20 cm.
  (iv) A straight 5 m long 2.21 cm dia horizontal pipe.
   (v) A sudden expansion into a 4 cm dia pipe.

10. Calculate the pressure drop around a 30 cm radius 90° bend in a horizontal 75 mm bore pipe for the flow of a steam–water mixture. The system pressure is 15 bar, the mass quality is 15% and the mass flow rate is 1.4 kg/s.

References

Barcozy CJ (1966) A systematic correlation for two-phase pressure drop. Chem Eng Prog Symp Ser 62:232-249

Chisolm D (1963) The pressure gradient due to friction during the flow boiling of water. NEL, Report No, p 78

Churchill SW (1977) A comprehensive correlating equation for laminar, assisting, forced and freeconvection. AIChE Journal. Jan 23(1):10-6

Cicchitti A, Lombardi C, Silvestri M, Soldaini G, Zavatarlli R (1960) Two phase cooling experiments pressure drop, heat transfer and burnout measurement. Energia Nucl 7:407–425

Davidson WF et al (1943) Studies of heat transmission through boiler tubing at pressures from 500 to 3300 pounds. Trans ASME 65:553–591

Dukler AE, Wicks M, Cleveland RG (1964) Pressure drop and hold-up in two phase flow, Part A: a comparison of existing correlations, Part B: an approach through similarity analysis. AICh E J 10:38–51

Isbin HS, Moen RH, Wickey RO, Mosher DR, Larson HC (1959) Two-phase steam-water pressure drops. In: Chemical engineering progress symposium series-nuclear engineering, vol 55, pp 75–84

Lockhart RW, Martinelli RC (1949) Proposed correlation data for isothermal two phase, two component flow in pipes. Chem Eng Prog 45:39–48

Martinelli RC, Nelson DB (1948) Prediction of pressure drop during forced circulation boiling of water. Trans ASME 70:695

McAdams WH, Woods WK, Heroman LC (1942) Vaporization inside horizontal tubes-II- benzene oil mixtures. Trans ASME C4:193–200

Wallis GB (1969) One dimensional two phase flow. McGraw-Hill, New York

Whally PB (1996) Two phase flow and heat transfer. University of Oxford, Department of Engineering Science

Zuber N, Findlay JA (1965) Average volumetric concentration in two phase flow systems. J Heat Trans Tans ASME 87:453–68

Chapter 3
Thermodynamics of Phase Change

Matter in the universe remains in three phases usually—solid, liquid and gaseous.[1] The change from one phase to other involves the interaction of energy, thermal energy to be precise. The interaction of energy is also related to temperature. Classically, interaction of thermal energy and temperature with other forms of energy (such as mechanical, electrical and chemical energy) and properties (such as pressure, volume and entropy) is the subject matter of thermodynamics. Hence, the change of phase between solid, liquid and vapour (or gas) is primarily a thermodynamic phenomenon. However, only a thermodynamics viewpoint does not give the real picture of the phase change process. This is because in almost all occasions, phase change is associated with transport phenomena. This can be in the form of fluid flow (momentum transport), heat transfer (thermal transport) or mass transfer (species transport).

Thus, although thermodynamics acts as the scientific basis, it alone cannot describe or explain the phenomena related to phase change or co-transport of multiple phases of matter. In simple terms, thermodynamics deals with the quantity of heat transfer from/to a system to its surroundings and its relation with other forms of energy. It does not take care of details of the heat transfer mechanism and most importantly 'the rate of heat transfer'. Thus, thermodynamics tells 'how much' heat transfer takes place at the expense of other forms of energy, but it does not tell 'in what way' and 'at what rate' it takes place. The science of 'heat transfer' answers these questions. However, the situation gets more complicated when, along with heat transfer, a change of phase takes place. The most important difference between single-phase and two-phase heat transfer lies in the perceptible change in temperature. The heat transfer rates are higher in two-phase cases involving a phase change compared to single-phase heat transfer because here heat transfer is primarily in the form of latent heat. The complexity arises in the assessment of 'in what way' and

---

[1] Often, the state of 'plasma' is defined as the 4th state of matter.

© The Author(s) 2023
S. K. Das and D. Chatterjee, *Vapor Liquid Two Phase Flow
and Phase Change*, https://doi.org/10.1007/978-3-031-20924-6_3

'at what rate' heat transfer takes place during a phase change because of one more considerations which is 'equilibrium'. Thermodynamics, due to its development over the last two centuries, can accurately describe multiphase (involving more than one of the three phases—solid, liquid and gaseous) systems when it is in equilibrium.

However, the phase change process is also associated with transport processes with finite time scales and the multiphase system from a thermodynamics or fluid mechanics point of view alone is not sufficient to understand the process. There is a need to combine and understand the interaction of these two aspects (thermodynamics and fluid mechanics) to have a complete knowledge of multiphase systems particularly associated with phase change. In addition, when two phases are present, additional surface forces act at the interface (such as surface tension) which makes it more complex.

Since thermodynamics acts as a basis for dictating the possibility of various multiphase interactions, including phase change, we will start with the basic aspects of thermodynamics associated with phase change in this chapter. Kindly note that we presume that the reader has knowledge of elementary undergraduate thermodynamics. One may refer to Borgnakke and Sonntag (2009) for a good introduction to the subject. Also since this text is confined to liquid/vapour interaction and phase change, we will restrict our discussion to liquid–vapour only and exclude other phase changes such as solid–liquid and solid–vapour change. This chapter presents a concise view of all the important topics related to phase change. Advanced readers may refer to Carey (1992), Das (2009) or Kandlikar et al. (1999). In particular, Chap. 1 of the book by Kandlikar et al. (1999) contains a good description of the relevant topics.

3.1   Pure Substance and State Principle

The change of phase for a substance depends primarily on what kind of substance it is. Is it a single component or a mixture of different materials with different chemical compositions? Are the various components miscible or immiscible? Thermodynamically, we define a pure substance and use it as a starting point for our analysis. Before defining it, we need to define something called a simple system. A simple system is a thermodynamic system in which effects of phenomena like surface forces such as surface tension, forces due to gravitation, electric and magnetic fields and shear stresses resulting from the deformation of solids and bulk motion are negligible. Although this appears to be a gross simplification, it actually represents a close approximation to many real systems. The most important consequences of simple system assumption are that it results in uniform pressure and temperature and the work done in a simple system can be written as $W = \int p \, dV$ where p is the pressure and V is the volume of the system. In the absence of a chemical reaction, a simple system reduces to a pure substance. The definition of a pure substance can be given as follows. "A pure substance is a simple system which is uniform in chemical composition, uniform in chemical aggregation and time-invariant in chemical aggregation". What it means is that the substance should be made of the same chemical

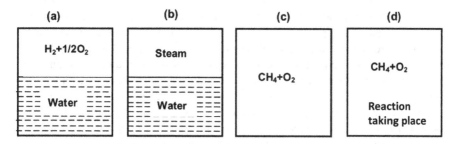

Fig. 3.1 Examples of pure and non-pure substances

matter combined chemically in the same way, and this chemical aggregation is not changing with time. Let us take an example as shown in Fig. 3.1.

Figure 3.1a shows a closed vessel containing water and a mixture of H_2 and O_2 gases in the same proportion as in water inside the closed vessel. This is not a pure substance because although everywhere chemical composition is the same H_2O or $H_2 + 1/2O_2$, they are not chemically combined in the same way (aggregation is different). In water, it exists as H_2O, whereas in the gas mixture they remain in separate identities as H_2 and O_2 molecules. In Fig. 3.1b, we have water and steam and both have the same chemical composition and aggregation (H_2O), and hence it is a pure substance. It is important to realize that even if water changes to steam or *vice versa*, water-steam still forms a pure substance as long as no chemical reaction takes place. Figure 3.1c shows a mixture of methane (CH_4) and oxygen (O_2) in the vessel. The chemical composition aggregation is uniform, and it is also invariant in time and may be considered as uniform. Hence, it is a pure substance. Finally in Fig. 3.1d we have a case with the same composition as in Fig. 3.1c. However, in this case combustion of methane according to chemical reaction $CH_4 + O_2 \rightarrow H_2O + CO_2$ takes place. Even if we assume a homogeneous (infinite speed) combustion, giving a uniform mixture of CH_4, O_2, H_2O and CO_2 throughout the vessel, it is not a pure substance because the proportion of these substances changes with time. As the combustion advances, the amount of CH_4 and O_2 decrease and that of H_2O, and CO_2 increase. The consequence of a simple system and pure substance assumption is a principle known as the 'state principle'. This can be stated as

Any two independent properties of a simple system are sufficient to identify its state if it is in a stable thermodynamic equilibrium. For each identifiable departure from simple system one additional independent property is required to specify the state.

For pure substance, this is known popularly as the 'two property rule' and is given as

The state of a pure substance of given mass can be specified in terms of two independent properties in the absence of other effects like gravity, bulk motion, electricity, magnetism, elastic deformation, surface tension.

The first observation from the above principle is that it talks about a stable equilibrium state. This is central to the present text because thermodynamic relations and calculations are only valid when there is a thermodynamically stable equilibrium. Thermodynamic equilibrium for a system is assumed to exist in a system when it is in thermal, mechanical and chemical equilibrium. In turn, mechanical equilibrium indicates no unbalanced force within the system and between the system and the surroundings. Similarly, thermal equilibrium indicates the absence of temperature difference and chemical equilibrium indicates the absence of chemical reaction. The second and the most important consequence of this principle is that this fixes the number of independent variables required to specify the state of a pure substance. This has important implications during phase change. If the system has only internal energy as the mode of stored energy then, for a given mass, two independent properties are enough to specify its state. Now, since internal energy can also be due to the interaction of two modes of energy interaction, namely heat and work (from the first law of thermodynamics) given by $du = Tds - pdv$, we can say that *"an independent property is required per mode of energy interaction"*. This also includes the additional changes due to each of the departures from simple systems (like electricity, gravity and magnetism). For example, if we have a pure substance with gravity, we will need three independent properties. With gravity and surface tension, we will require four independent properties to specify a state. Going back to the 'two property rule', we are aware that for specifying the state of a given mass of gas any two of pressure, volume or temperature are sufficient. This is because in this state all these properties are independent properties. For liquid, it can be pressure and temperature. However, the situation is different when phase change takes place. When we have two phases, liquid and vapour in equilibrium, pressure and temperature no longer remain independent properties, and we need to look for other properties which are independent. We will discuss this aspect in the next section. What the above discussion entails is that there are some independent properties and other properties bear a relationship with them. We will look at these relationships for pure substances during phase change.

3.2 Properties of Liquid–Vapour System in Thermodynamic Equilibrium

Let us confine our attention only to liquid–vapour systems which is the subject matter of this book. We can turn liquid into vapour in two ways: (a) by heating the liquid, and (b) by reducing the pressure of the liquid. Let us examine how the properties of the liquid substance behave under these two processes.

Constant Pressure Heating of Liquid

Let us consider a piston cylinder arrangement shown in Fig. 3.2a containing a liquid. We put a certain weight on the piston to keep the pressure of the liquid vapour system constant. Now we heat the content of the cylinder slowly by giving heat through its walls. The volume of the liquid will slowly increase from point 1 to point f as

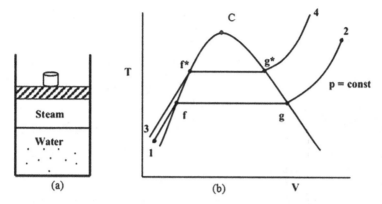

Fig. 3.2 Constant pressure heating of liquid

shown in Fig. 3.2b. The temperature of the liquid will also increase. At f (called saturated liquid), the liquid starts changing to vapour. As long as this change takes place (from f to g), both the pressure and temperature remain constant. This temperature is called the saturation temperature corresponding to the pressure. Conversely, this pressure is called saturation pressure at that temperature. For example, at a pressure of 1.01325 bar (1 atm), the saturation temperature for water is 100°C. Or at 100°C temperature saturation pressure is 1.01325 bar. At the point g (called the saturated vapour), the entire liquid gets converted to vapour. By adding more weight to the piston, we follow a different line as shown in Fig. 3.2b. Here, we get the state point 3 and on heating, the vapour formation starts to point f^*. The liquid turns completely to vapour at g^* at this pressure. If we find the locus of f and g points, we get the 'dome' shown in Fig. 3.2b with c as the critical point. At this point, f and g points merge indicating that what we discussed in the previous section, in the part 1-f or g-2, the two properties pressure and temperature could specify a state (i.e., uniquely identify a point), however in the region f-g, pressure and temperature cannot specify a state because both pressure (saturation pressure) and temperature (saturation temperature) remain constant and hence these two are dependent properties unable to specify a state. However, volume remains an independent property and hence either pressure and volume or temperature and volume can specify the state when phase change is taking place. Lastly, we remind you that this diagram is constructed assuming that at every point on the curve the system is in thermodynamic equilibrium, and hence the rules stated in the previous sections and the property relations (to be given later) are applicable.

Constant Temperature Pressure Reduction of Liquid

Now, let us use the other technique of reducing pressure of a liquid by a similar piston cylinder arrangement shown in Fig. 3.3a. Here, the temperature of the water-steam system is kept constant using an isothermal bath and the piston is pulled out slowly to

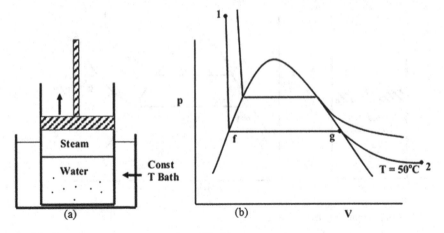

Fig. 3.3 Constant temperature expansion of liquid

change volume and pressure. First, the liquid pressure decreases (1-f), bringing it to a saturated liquid state (f). Then the pressure remains constant (saturated pressure) till it completely evaporates (at g) in the process f-g and finally the superheated vapour behaves like a gas (in g-2).

The p-V (Fig. 3.3) and T-V diagrams are in reality projections of a more general diagram representing all (p, V and T) the three properties. Thus, a 3D plot of p-v-T (v is specify volume, volume per unit mass) is a more complete representation of property variation during phase change.

One word of caution here. We should mention here that p-v-T is not the only property combination by which a state can be defined. We can also construct diagrams with other properties such as enthalpy (h), entropy (s), Gibbs free energy, etc. p-v-T is more popular because these are directly measurable properties, while the others need to be evaluated from various measured quantities.

3.3 Gibbs Phase Rule

We made some important observations in the last section. Firstly, we have seen that in the liquid vapour two-phase region, pressure and temperature cannot be varied independently. Once one is specified, the other gets automatically fixed. It was also observed that for each pure substance, we get a triple point which is a single point in the property plane and the values of pressure and temperature are fixed at this point for a given pure substance. J. W. Gibbs in 1876 proclaimed the following rule which generalizes these observations. This is called the 'Gibbs phase rule'. "The thermodynamic degrees of freedom (f) of a given multi-component system with π number of phases in equilibrium with each other" can be expressed by

$$f = C + 2 - \pi \tag{3.1}$$

where C is the number of chemically independent components. Kindly note that this is a more general case of a component mixture existing in π different phases. The thermodynamic degrees of freedom, f, is defined as the number of intensive properties of the system which can be independently specified. For a liquid–vapour system, usually pressure and temperature are chosen as intensive properties (i.e., the properties which are not proportional to the mass of the system). For a multi-component system (a system having many components, for example mixture of H_2 and N_2), the concentration of each component is also an intensive property.

Now, the first observation we can make is the fact that when $C = 1$ (that is only one chemical component is there in the system such as a gas), the system reduces to a pure substance. We will deal with multi-component systems in a later chapter, and let us concentrate on the Gibbs phase rule for pure substance here. Under this condition, the Gibbs phase rule reduces to ($C = 1$):

$$f = 3 - \pi \tag{3.2}$$

Now, if only a single phase (say, gas or liquid) exists, $\pi = 1$. Thus, Eq. (3.1) gives $f = 2$. This means that pressure and temperature can be varied arbitrarily in this case. Hence, as long it remains in that phase, any combination of pressure and temperature will fix its state and all other properties will automatically get fixed. For example at 1.01325 bar and 20 °C or 1.01325 bar and 200 °C, specific internal energy, specific enthalpy and specific entropy will have unique values. However, if the pure substance exists in two phases in equilibrium (say at 1.0325 bar and 100 °C for water), we have $\pi = 2$ and hence $f = 1$. Thus, only one of pressure and temperature can be arbitrarily fixed (either 100 °C or 1.01325 bar), and the other will automatically get fixed. This can be understood very well from Fig. 3.4 which plots the intensive properties p and T. Figure 3.4 curve (a) gives a plot of p-T for substances which expand upon melting (say, metals) and curve (b) gives a plot for substances which contract upon melting (water). It is clear that when phase change takes place, p and T vary in a correlated way while in solid, liquid or vapour region they can be varied independently.

Finally, when all three phases (solid–liquid–vapour) exist in equilibrium, $\pi = 3$. This gives $f = 0$. Thus, the thermodynamic degree of freedom being zero, none of the intensive properties (p and T) can be varied arbitrarily, and it gets fixed for a given substance (e.g., 0.01 °C and 611.3 Pa for water). The important point to be noted here is that the three phases need to be in thermodynamic equilibrium. Suppose water is boiling in a vessel, and we put some ice into it; we will have three phases, but they are not in equilibrium. They will be in one or two phases when they come to an equilibrium unless at the equilibrium state pressure is 611.3 Pa and temperature 0.01 °C. This is evident from Fig. 3.4 where the triple point appears as a definite point.

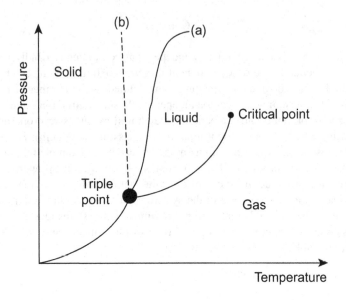

Fig. 3.4 Intensive properties of a pure substance for substances which expand on melting (curve a) and those contract on melting (curve b)

3.4 Phase Equilibrium and Associated Conditions

In the previous sections, we have seen that the assumptions of pure substance lead to the consequence of uniform pressure and temperature. It also leads to the consequence that the work can be represented by $\int p\,dV$. Hence, the first law of thermodynamics for a fixed mass of a pure substance can be written as

$$\delta Q = dU + \delta W \qquad (3.3a)$$

or as

$$T\,dS = dU + p\,dV \qquad (3.3b)$$

Now we introduce a well-known property of the thermodynamic system, Gibbs free energy (or Gibbs function), as

$$G = H - TS \qquad (3.4)$$

We know that enthalpy is defined as

$$H = U + pV \qquad (3.5)$$

Combining Eqs. (3.4) and (3.5), we get

$$G = U + pV - TS \qquad (3.6)$$

Taking differential,

$$dG = dU + pdV + Vdp - TdS - SdT, \qquad (3.7)$$

and combining Eqs. (3.7) and (3.3b) gives

$$dG = Vdp - SdT \qquad (3.8)$$

Now in a pure substance in phase equilibrium since pressure (p) and temperature (T) are constant, $dp = dT = 0$. This gives

$$dG|_{p,T} = 0 \qquad (3.9)$$

This means that for a pure substance in phase equilibrium, the Gibbs function will be minimum. This can be taken as the "condition for phase equilibrium". Since in the thermodynamic approach the equilibrium (under which the property diagrams presented in the previous section are valid) is assumed, the condition for phase equilibrium is important. The consequence of the condition for phase equilibrium is important and useful to understand. Gibbs free energy for a pure substance existing in two phases in equilibrium will be a function of the intensive properties, pressure and temperature, and also on the number of moles of each phase. Thus,

$$G = f(p, T, n_l, n_v) \qquad (3.10)$$

where n_l and n_v are the number of moles of the liquid and vapour, respectively (we are considering only liquid vapour equilibrium here). Thus, we get

$$dG|_{(p.T)} = \left(\frac{\partial G}{\partial p}\right)_{T,n_l,n_v} dp + \left(\frac{\partial G}{\partial T}\right)_{p,n_l,n_v} dT$$
$$+ \left(\frac{\partial G}{\partial n_l}\right)_{p,T,n_v} dn_l + \left(\frac{\partial G}{\partial n_v}\right)_{p,T,n_l} dn_v \qquad (3.11)$$

Since the total number of moles (n) will remain constant during a phase change process $(n = n_l + n_v)$, under the assumption of constant pressure and temperature (which holds if the liquid—vapour change takes place under equilibrium as we have seen in the last section), Eq. (3.11) transforms into

$$dG|_{(p.T)} = \left(\frac{\partial G}{\partial n_l}\right)_{p,T,n_v} dn_l + \left(\frac{\partial G}{\partial n_v}\right)_{p,T,n_l} dn_v \qquad (3.12)$$

Now if $n = n_l + n_v$ is constant, we get

$$dn_l = -dn_v \qquad (3.13)$$

Thus, Eq. (3.12) can be written as

$$dG|_{(p.T)} = \left(\left(\frac{\partial G}{\partial n_l} \right)_{p,T,n_v} - \left(\frac{\partial G}{\partial n_v} \right)_{p,T,n_l} \right) dn_l = 0 \qquad (3.14)$$

and defining molar Gibbs function,

$$g_l = \left(\frac{\partial G}{\partial n_l} \right)_{p,T,n_v}$$

and

$$g_v = \left(\frac{\partial G}{\partial n_l} \right)_{p,T,n_l}$$

we get

$$g_l = g_v \qquad (3.15)$$

Thus, for a pure substance to be in two-phase equilibrium between liquid and vapour, the necessary condition is that the molar Gibbs free energy of each phase is equal. Or, in other words, the molar Gibbs function remains invariant across the phases.

3.5   Clausius–Clapeyron Equation

It is very important to know the relation between various properties for a (liquid–vapour) two-phase system. Elementary thermodynamics tells us how to evaluate these properties from figures and charts (such as a steam table or the Mollier diagram for water and refrigeration tables). These tables usually relate properties such as specific internal energy (u), specific volume (v), specific enthalpy (h) and specific entropy (s) using experimental results or empirical correlations. However, the basic question remains about how the two fundamental intensive properties of pressure and temperature are related when liquid–vapour phase change takes place. Although the tables give the values of saturation temperature corresponding to a saturation pressure or *vice versa*, the basis of these correlations as depicted in Fig. 3.4 needs to be known. The well-known Clausius–Clapeyron equation provides this theoretical basis which is useful to understand the liquid–vapour phase change process. Let us recollect Eq. 3.15 which states that Gibbs free energy remains unchanged across phases. Differentiation of this equation gives

$$\left(\frac{\partial g_l}{\partial T} \right)_p dT + \left(\frac{\partial g_l}{\partial p} \right)_T dp = \left(\frac{\partial g_v}{\partial T} \right)_p dT + \left(\frac{\partial g_v}{\partial T} \right)_T dT \qquad (3.16)$$

From Eq. (3.8) in molar form, we get

$$\left(\frac{\partial g}{\partial p}\right)_T = v \quad \text{and} \quad \left(\frac{\partial g}{\partial T}\right)_p = -s \tag{3.17}$$

Combining Eqs. (3.16) and (3.17), we get

$$-s_l + v_l\left(\frac{dp}{dT}\right) = -s_v + v_v\left(\frac{dp}{dT}\right) \tag{3.18}$$

or,

$$\frac{dp}{dT} = \frac{s_v - s_l}{v_v - v_l} \tag{3.19}$$

Noting that

$$g_l = h_l - Ts_l = g_v = h_v - Ts_v \tag{3.20}$$

we get

$$s_v - s_l = \frac{h_v - h_l}{T} \tag{3.21}$$

Please note that temperature is the same for both phases as they are in equilibrium. Substituting Eq. (3.21) in (3.19) gives

$$\frac{dp_{sat}}{dT_{sat}} = \frac{h_{fg}}{v_{fg}T_{sat}} \tag{3.22}$$

Here, $h_{fg} = h_v - h_l$, enthalpy difference between vapour and liquid states and $v_{fg} = v_v - v_l$, specific volume change due to phase change. Temperature and pressure are written as T_{sat} and p_{sat} since during phase change the temperature and pressure remain constant at saturation value. It is also noted that since h_{fg}, hence v_{fg} both are molar quantities and the number of moles can be cancelled hence mass specific enthalpy (enthalpy per unit mass) and mass specific volume (volume per unit mass) can also be used for this equation.

Actually, Eq. (3.22) is known as the Clapeyron equation after Emile Clapeyron (1799–1864). If we use this equation between solid and liquid, we understand the difference between the curves (a) and (b) in Fig. 3.4. If the subscripts are changed from l and v to s and l (solid and liquid), the Clapeyron equation gives

$$\frac{dp}{dT} = \frac{h_l - h_s}{(v_l - v_s)T} \tag{3.23}$$

Now, $(h_l - h_s)$ is always positive (because heat has to be supplied for melting but $(v_l - v_s)$ is positive for the case when volume expands on melting (as in metals), and it is negative when volume contracts on melting (as in water). Hence, we get the right-hand side of Eq. (3.23) positive in one case (corresponding to Fig. 3.4, curve a) and negative in the other (for Fig. 3.4, curve b)). Rudolf Clausius (1822–1888) simplified the Clapeyron equation by using the approximation that liquid specific volume is much smaller than vapour specific volume ($v_l \ll v_v$). For water at atmospheric pressure, v_v is approximately 1000 times that of v_l. Also, vapour phase can be modelled as an ideal gas, giving $v_{fg} \approx v_v$ or v_g and $v_g = RT/p$ (usually for vapour v_g and for liquid v_f is used in thermodynamics).

Using this, the Clapeyron equation reduces to

$$\frac{dp}{dT} = \frac{ph_{fg}}{R_g T^2} \tag{3.24}$$

It should be noted that this gives the slope of the vapour pressure curve (the liquid–vapour part of Fig. 3.4). This equation is called Clausius–Clapeyron equation. Now the correlation between temperature and pressure during phase change can be obtained by integrating this correlation

$$\int_{p_1}^{p_2} \frac{dp}{p} = \frac{h_{fg}}{R} \int_{T_1}^{T_2} \frac{dT}{T^2} \tag{3.25}$$

giving

$$\ln \frac{p_2}{p_1} = -\frac{h_{fg}}{R} \left(\frac{1}{T_2} - \frac{1}{T_1} \right) \tag{3.26}$$

This is the correlation for two points on the vapour pressure curve. Thus, any point on the curve can be given by

$$\ln p = A - \frac{B}{T} \tag{3.27}$$

Usually, the constants A and B are obtained by fitting the best curve to experimental data. The Clausius–Clapeyron equation is also useful for evaluating the boiling process as we will see later.

So far, we have discussed thermodynamics of phase change and, in particular, we are more interested in liquid-to-vapour (or vice versa) phase change either under constant pressure heat addition (Fig. 3.2) or under constant temperature pressure reduction (Fig. 3.3). As shown in those figures, phase change begins as soon as point 'f' or point 'g' is reached along the isobaric or isothermal process. In course of this phase change, line segment f-g in Figs. 3.2 and 3.3 indicate that both vapour and liquid phases can coexist in equilibrium. This is generally true for liquids (or, vapours) having a copious supply of nuclei. For example, phase change can start exactly at point f (heating, evaporation and cavitation) or point g (cooling, condensation) of Fig. 3.2 or 3.3, only if already some nucleations (simplest examples being vapour bubbles in a liquid or tiny liquid droplets in a vapour). Thermodynamics of phase change in the absence of a nucleus is now taken up for discussion.

3.6 Thermodynamics of Phase Change of Pure Liquid

In the case of any liquid, like water, pressure, temperature and specific volume changes are depicted in Fig. 3.5. In the absence of a nucleus, the liquid-to-vapour phase change does not take place at saturation pressure (for constant temperature in case of cavitation) or saturation temperature (at constant pressure in case of boiling). Thus, phase changes do not proceed along f-g but will follow the extended isotherm line (called theoretical isotherm) and first vapour may appear at E, whose pressure is lower than the saturation pressure at point f. If the same liquid is heated at constant pressure, then vapour formation will not start at E' (on the saturated liquid line) but will begin at E.

In an analogous way, if pure vapour without the presence of any liquid droplet is cooled, the condensation does not start at point g but the vapour gets subcooled to point G along gB. We thus have fA and gB processes where the previous phase (liquid/vapour) continues to exist even after crossing the saturation condition. These regions are termed as 'metastable equilibrium' as these satisfy the condition of mechanical equilibrium of $(\partial p/\partial v)_T \leqslant 0$, but can undergo phase change with the slightest perturbation. Portion AB of theoretical isotherm does not satisfy this mechanical equilibrium condition. Hence, the metastable region can be realized in practice under certain specific conditions of lack of nuclei and absence of any other disturbance. The line joining the minima of different theoretical isotherms is called liquid spinodal while the locus of the maxima is vapour spinodal and the two spinodal meet at a critical point as shown in Fig. 3.5.

To understand the nature of the theoretical isotherm better, let us adopt a 'gas-like' approach to liquids. Before discussing features of the liquid phase, we may recall that $pv = RT$ is the ideal gas law and for constant temperature, we get Boyle's law, namely $pv = $ constant. This relation is valid for gas with low density and hence with negligible effects of intermolecular forces. Under standard conditions

Fig. 3.5 Phase diagram of water

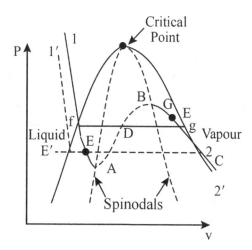

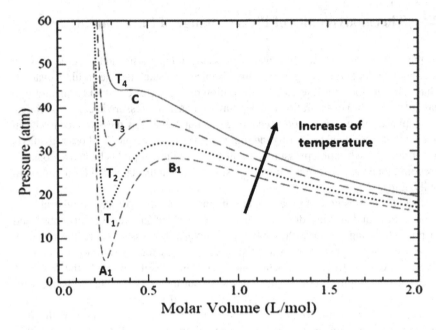

Fig. 3.6 Van der Waal isotherms

of temperature and pressure (STP), that is at $0\,^\circ$C and 1 atm pressure, a mole of gas contains 6.023×10^{23} molecules and occupies 22.4 L. Thus, the distance between two molecules is about 4 nm and hence is one order of magnitude more than the molecular diameter. Thus, a departure from 'ideal' gas laws under these conditions can be neglected. However, if the gas density is high, then molecular interactions can no longer be neglected. Van der Waals showed that in the presence of short-ranged repulsive force and long-range attractive forces, the equation of state for gas changes to

$$\left(p + \frac{a}{v^2}\right)(v - b) = RT \tag{3.28}$$

and for fixed T we can draw (P, v) isotherms as shown in Fig. 3.6.

Let us focus our attention on the curve at temperature T_1. It shows the minimum at A_1 and the maximum at B_1. Portion $A_1 B_1$ does not satisfy the mechanical equilibrium condition described earlier. At such temperature T_1 below critical temperature (T_c), there exist three real roots of specific volume for any pressure P. This curve has two asymptotes: (i) when specific volume (v) tends to become infinity (density tending to zero) and we get back ideal gas law, and (ii) when specific volume becomes very small $(v \rightarrow b)$, pressure tends to infinity. As temperatures increase, the minima and maxima come closer and at a critical point, we get point of inflection at C (Fig. 3.7).

Now, let us focus our attention on the region where $T < T_c$. If a vapour is compressed progressively, its pressure increases till it reaches B. However, it cannot follow the curve between B and A as it is mechanically unstable. Maxwell was the

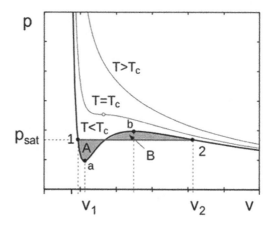

Fig. 3.7 Pressure specific volume curves for temperatures below, at and above the critical temperature. Maxwell's construction of equal area is highlighted by the shaded area on the curve for which $T < T_C$

first to observe this, and he indicated it as a transition of phase with the liquid phase having density (ρ_L) different from that of the vapour phase (ρ_V).

So he suggested that a horizontal line corresponding to constant pressure be drawn connecting two phases with different densities such that the areas of the theoretical isotherm above and below the horizontal line are the same (Fig. 3.7). This can be justified by the fact that at equilibrium, the two phases have the same chemical potential which gives

$$0 = \mu_L - \mu_V = \int_V^L d\mu = \int_V^L v \, dP \qquad (3.29)$$

Above the critical point, such an unstable region does not exist (look at the curve corresponding to T_4 in Fig. 3.6), and we do not encounter any phase change. The idea of a metastable state can now be better appreciated if we summarize it as a condition where mechanical stability is guaranteed, but it need not be thermodynamically stable. This is examined further later.

The drop in pressure between f and A (Fig. 3.5) is termed as the liquid tension indicating the pressure below the saturation pressure (at the liquid temperature) which a pure liquid can sustain before the formation of vapour. Likewise, the increase in temperature (T_E-$T_{E'}$, Fig. 3.5) is termed as liquid superheat and of course, the liquid tension and liquid superheat are related to the Clausius–Clapeyron equation discussed earlier. Similarly, temperature difference of T_E-$T_{E'}$ (Fig. 3.5) is termed as the extent of subcooling.

This tensile strength of a liquid can be explained following Frenkel (1955). Let us consider the interaction potential between two molecules of a liquid separated by a distance x as shown in Fig. 3.8. x_0 denotes the equilibrium separation between the molecules which is typically of the order of 1 Angstrom (10^{-10} m). For distances less than x_0, the molecules experience repulsion, while they face attraction for $x > x_0$. Let us say that the maximum attraction is felt at $x = x_1$, and so the molecule needs

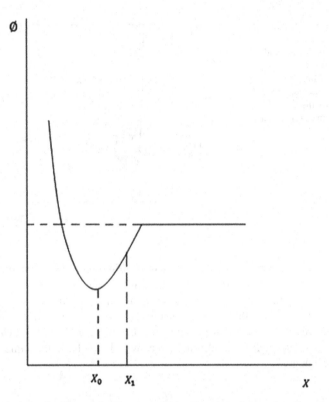

Fig. 3.8 Typical sketch of intermolecular potential as a function of separating distance

to overcome this attractive force in order to bring in rupture. Typically, $\frac{x_1}{x_0} \sim 1.1\text{--}1.2$ and hence this means the volume change $(\Delta V/V_0) \sim 1/3$. This requires a pressure p given by the relationship $p = -K_B(\Delta V/V_0)$, where K_B is the bulk modulus of the liquid and is of the order of $10^{10} - 10^{11}$ Pa. Thus, $p \sim -(10^9 - 10^{10})$ Pa. This limiting pressure, or equivalent liquid superheat, needed to produce vapour formation is hardly seen in real life.

With this, we come to an end of the discussion on the thermodynamics of phase change and show that the actual thermal thermodynamic path followed by a liquid or a vapour depends critically on the presence of nuclei. Hence, the role of nucleation is very important for a better understanding of phase change phenomena in cavitation, boiling and condensation and will be taken up in the next chapter.

Examples

Problem 1: Using the Clausius–Clapeyron equation, find out the ratio of (dP/dT) if the liquid is water at 40 bar. Liquid superheat is $105\,°C$. Find out the pressure of the vapour formed and the temperature of the liquid. Also find out the bubble radius.

Solution: Given:

$P = 40\ bar$

$T_{sat} = 250.40\,°C$

$fg = 0.04978 - 0.00125 = 0.04853\ \mathrm{m^3/kg}$

$h_{fg} = 1714.1\ \mathrm{kJ/kg}$

$\frac{dP}{dT_{sat}} = \frac{h_{fg}}{T_{sat}v_{fg}} = \frac{1714}{0.04853 \times 250.40} = 141.04\ \mathrm{kJ/m^3\,°C}$

Liquid superheat given is $105\,°C$.

$\Delta T = 105\,°C = T_l - T_{sat}$,

$T_l = 355.4\,°C$

$P_l = 40\ bar,\ \Delta P = 141.04 \times \Delta T = 14.8\ bar\ P_g - P_l = 14.8\ bar$

$P_g = 14.8 + 40 = 54.8\ bar$

we know $P_g - P_l = \frac{2\sigma}{r}$.

$r = \frac{2\sigma}{P_g - P} = \frac{2 \times 0.0030}{14.8 \times 10^5} = 4.05 \times 10^{-6}\ \mathrm{mm}.$

Exercises

1. Derive an expression for the vapour quality or flow quality equation used for diabatic two-phase flow using the Clausius–Clapeyron equation. δQ is the heat supplied to the fluid; t is the total mass flow rate of the two-phase flow. Consider the spontaneous flashing effect of two-phase flow which becomes significant when pressure drops.
2. Using the expression in question 1, find out the heat supplied to the fluid for diabatic two-phase flow if the fluid is at temperature $120\,°C$, bubble radius of 2 mm. Quality is 0.2, $\sigma = 0.0058\ \mathrm{N/m}$, $dP = 0.8\ bar$, $C_p = 0.2\ \mathrm{J/kg\,°C}$ and mass flow rate $= 0.1\ \mathrm{kg/s}$.
3. If the heat supplied is 1 kW, the temperature of the fluid is $150\,°C$ with a quality of 0.3, $\sigma = 0.0058\ \mathrm{N/m}$, $C_p = 0.4\ \mathrm{J/kg\,°C}$ and mass flow rate $= 0.1\ \mathrm{kg/s}$. Find out the flashing effect pressure of the fluid and the bubble radius.

4. Derive the slip for two-phase flow. Derive the condition when slip becomes a unity. Find out the slip for a tube of radius 1 mm with a mass flow rate of gas as 0.002 kg/s and velocity of the two-phase fluid is 2 m/s. Radius of gaseous fill is 0.2 mm with the fluid at 150 °C.

5. Find out the entropy generation for a two-phase flow fluid if the heat transfer coefficient at the tube wall is 20 W/m^2 K. The tube is of diameter 10 mm. Quality is 0.4 with a pressure drop of 2 N/m^3, mass flow rate of 2 kg/s. Superheat at the wall is 120 °C. Also find the temperature of the fluid and the bubble radius.

References

Borgnakke C, Sonntag RE (2009) Fundamentals of thermodynamics. Wiley
Carey VP (1992) Liquid–vapor phase-change phenomena. Hemisphere, Washington, DC
Das SK (2009) Fundamentals of heat and mass transfer. Alpha Science, Oxford
Frenkel J (1955) Kinetic theory of liquids. Dover Publication, Mineola
Kandlikar SG, Shoji M, Dhir VK (1999) Handbook of phase change: boiling and condensation. CRC Press, Boca Raton

Chapter 4
Nucleation and Bubble Dynamics

In Chap. 3, we have discussed the thermodynamics of phase change. In particular, we have presented two cases: one in which the two phases, liquid and vapour, are in thermodynamic equilibrium and coexist, and the second, in which a metastable situation may exist and the phase change does not take place at the specified saturation pressure or temperature. As mentioned in Sect. 3.6, such a metastable condition can happen when the liquid is *pure*. In other words, when the liquid is in thermodynamic equilibrium, we may expect the liquid to be *impure*. Thus, we should be able to define two types of nucleations leading to a change of phase in the bulk fluid. These are termed homogeneous and hetergeneous nucleations.

4.1 Homogeneous Nucleation

Formation of a second phase (e.g., vapour in a liquid or liquid in a vapour) requires a certain kind of nucleation for the phase change process to initiate. In most of the engineering processes, contact of a liquid with a solid surface (boundary wall) is unavoidable, and in certain processes like heat transfer, boiling usually starts from the surface heater. In such cases, it is easy to locate a tiny bubble or gas pocket to act as a nucleation agent. The immense importance of the role played by surface will be discussed in detail later. Role of nuclei in condensation of vapour is also important and the role of nucleation is also discussed later. In the remaining portion of this section, we will focus on the bulk of the liquid and deal with the formation of vapour from a pure liquid. Such a scenario is important in the phenomenon like cavitation where the appearance of bubbles in the bulk of the liquid is quite frequent.

If the liquid is 'pure' such that no 'dirt' or 'contamination' or 'foreign particle' exists, then the liquid may be expected to nucleate only due to thermal fluctuations of molecules which are ephemeral in time. For such a liquid, we should ask

© The Author(s) 2023
S. K. Das and D. Chatterjee, *Vapor Liquid Two Phase Flow
and Phase Change*, https://doi.org/10.1007/978-3-031-20924-6_4

Fig. 4.1 Schematic of
different forces acting on a
bubble to produce static
stability

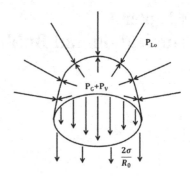

ourselves what should be the difference between the pressure when cavitating bubble
is first seen and the saturation pressure? Also, what is the experimentally achievable
highest pressure difference recorded? And most importantly, the value which is typically reached in experiments is the same as the theoretical difference predicted? The
answers to these questions are now explained.

But before that, let us consider a bubble inside a liquid (Fig. 4.1). We postpone
the question of how this gas and vapour bubble comes into existence inside the
liquid. Then, we can see from this figure that the internal pressure of the bubble,
$p_i = p_G + p_V$, tries to increase the size of the bubble, while the liquid pressure (p_L)
and surface tension effect try to counter this. Thus, for the bubble to be in equilibrium
with the liquid, there must be a balance of the forces arising due to liquid pressure
from the outside (acting inward), internal pressure of the bubble acting in a radially
outward direction and interfacial (surface) tension. Thus, for equilibrium, we get

$$p_i - p_0 = \frac{2\sigma}{R} \tag{4.1}$$

where p_i and p_0 are bubble interior and outside liquid pressures, respectively, and
σ is liquid–vapour surface tension. This pressure difference is called the Laplace
pressure difference.

If we consider that the Laplace pressure difference (given in Eq. 4.1) between
the interior and exterior (liquid) pressures due to the surface tension between a
bubble and the liquid exists down to a few intermolecular distance, then for a vapour
bubble ($p_i = p_v$) and ($p_i - p_0$) yields liquid tension as explained in Chap. 3. For a
bubble of size R_0, this relation yields $\Delta p = 2\sigma/R_0$. For vacancies created by thermal
fluctuations, R_0 is of the size 10^{-10} m and for a surface tension value of 0.07 N/m
(value for water is chosen as it is the most ubiquitous fluid), this yields a tensile
strength of 10^9 Pa similar to what was shown earlier.

This can also be viewed in terms of the energy required to create a surface of
vapour–liquid interface in an otherwise homogeneous liquid. The work done (W_{SC})
associated with this surface creation is given by $W_{SC} = 4\pi R^2\sigma$ and, in the process,
the liquid needs to be pushed away against the Laplace pressure difference discussed

before. This work is given by $W_p = -(4/3)\pi R^3 \triangle p$, and hence the resultant net energy required is

$$W_{net} = 4\pi R^2 \sigma - \frac{4}{3}\pi R^3 \triangle p = \frac{4}{3}\pi R^2 \sigma = \frac{16}{3}\frac{\pi \sigma^3}{(\triangle p)^2} \quad (4.2)$$

This energy is to be provided by thermal fluctuations of molecules and is related to $k_B T$, where k_B is the Boltzmann constant. Thus, we obtain Gibbs number (Gb), $Gb = W_{net}/k_B T$, and the total number of nucleation events occurring per unit time in a unit volume of liquid is given as

$$J = J_0 \exp{(-Gb)} \quad (4.3)$$

Combining Eqs. (4.2) and (4.3), together with definition of Gb, we get

$$\triangle p = \left[\frac{16\pi}{3} \frac{\sigma^3}{k_B T \ln\left(\frac{J_0}{J}\right)} \right]^{\frac{1}{2}} \quad (4.4)$$

This gives us some idea about the different terms on which the liquid tension depends. Surface tension decreases almost linearly with an increase in temperature and goes to zero at critical temperature. Thus, it is clear that at elevated temperatures, close to critical temperature, the change of surface tension is more strongly felt because of σ^3 term and for the same reason, at about the room temperature, the effect of temperature is not very strong. On the other hand, nucleation term has less importance due to its appearance within the logarithm term. Theoretical estimates thus talk about 1000's of bars of negative pressure (tension). But what do we observe in the experiments?

In experiments, we find that the highest experimentally recorded data for liquid tension obtained by Briggs in 1950 could reach only 277 bar. Thus, we can conclude that, for engineering applications, it is almost impossible to produce ultraclean liquid and hence experimental observations fall significantly short of theoretical limits. A good account of cavitation and liquid tension is given by Trevena (1984). The difficulty of producing ultraclean liquid free of contact with a solid surface was partially addressed by Apfel (1971), who carried out experiments with a droplet of test liquid inside an immiscible host liquid (Fig. 4.2). The two liquids are so chosen that the droplet density is less than that of external liquid and the test droplet rises due to buoyancy. The rising of the droplet is then arrested by an imposed acoustic field which levitates the droplet. The levitated droplet ruptures with a popping sound when the acoustic pressure amplitude is increased beyond a point. This yields liquid tensile and the results indicate agreement between experiment and theory at an elevated temperature. However, no counterpart experiment at room temperature exists. This brings out the possibility of another type of nucleation, namely, heterogeneous nucleation.

Fig. 4.2 Drop levitation
experiment for determination
of tensile strength and extent
of superheat permissible for
a liquid

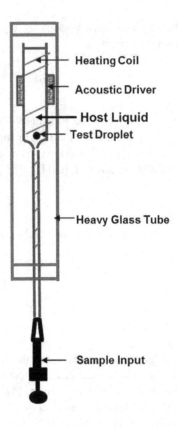

Heating Coil

Acoustic Driver

Host Liquid
Test Droplet

Heavy Glass Tube

Sample Input

4.2 Heterogeneous Nucleation

The very fact that in most of the applications, much lower tensile strengths are
recorded indicate that 'weak spots' must be present in the liquid. These weak spots
or nuclei give rise to favoured sites for nucleation of phase change and such a process
of nucleation is termed as heterogeneous nucleation. Now what could be the potential
sources of heterogeneous nucleation?

Miscible liquids and dissolved solids cannot provide nucleation sites. Thus, we
are left with either insoluble solids or free gas bubbles. We start with free gas bubbles.
If a free gas bubble has to satisfy the requirement of nucleation site, then it should be
able to remain in the liquid for a large time. But we know that free gas bubble will
rise due to buoyancy and may escape from the free surface. This is particularly true
if the bubble size is big and for a bubble with 1 mm diameter, the terminal velocity
(V_t) is in the range of 10 cm/s. This can be estimated by the relation for

$$V_t = \sqrt{\frac{4gd}{3C_d}\left(\frac{\rho_L - \rho_G}{\rho_L}\right)}$$

assuming the bubble as a rigid sphere with diameter d. ρ_L and ρ_G are liquid and gas densities, respectively, and C_d is drag coefficient. As this simple relation shows, the terminal velocity reduces with reduction in the size of the bubble. So, is it possible for a $10\,\mu m$ bubble to remain in water for a long enough time. For example, substituting the value of $20\,\mu m$ in above equation would give 0.03 cm/s which is slow enough. However, if the bubbles are small in size, then the bubble size may reduce further due to diffusion of gas going out of the bubble. This has been explained in detail by Epstein and Plesset (1950). Here we discuss the essential physics behind the dissolution mechanism and present the salient result.

We have already noted that the pressure inside the gas is more than the surrounding liquid because of surface tension and this pressure differential is given in Eq. 4.1. We also know from Henry's law that the amount of gas dissolved in the neighbourhood of a liquid surface in contact with a gas at a partial pressure P_G is given by $C = K(T)P_G$, where constant K is called Henry's constant and is a function of liquid temperature. Thus, we may consider two cases—saturated liquid in contact with a flat gas interface (Fig. 4.3a) and the same liquid in contact with a gas bubble (Fig. 4.3b). In the first case of flat interface, $P_G = P_L$, and hence the dissolved gas in the liquid can be at its saturated value (corresponding to P_L). In the second case, the same liquid (having pressure P_L same everywhere) has dissolved gas concentration corresponding to case (a) and proportional to P_L. However, in this case, $P_G > P_L$, and hence in the neighbourhood of the bubble, liquid is locally capable of dissolving more gas proportional to P_G and hence a local unsaturation is created. This leads to gradual dissolving away of gas from the bubble to the liquid. As the bubble size shrinks, surface tension (proportional to $1/R$) effect becomes more dominant and the rate of dissolution increases. Thus, Epstein and Plesset (1950) have shown that even in a saturated liquid a $20\,\mu m$ bubble dissolves in about 6 s. Thus, it is clear from these examples that a free gas bubble is an unlikely candidate for nucleation in liquid. Then the question about possible source of nucleation still remains unanswered. This has been addressed by two different models—due to Harvey et al. (1944) and due to Fox and Herzfeld (1954). Fox and Herzfeld assumed that the bubble may be covered by a layer of organic skin which prevents gaseous diffusion and cavitation onset can take place with the release of gas when the organic skin ruptures. Thus, this threshold is determined by the breaking strength of the organic film and the size of the gas bubble. Recent papers on interfacial rheology has renewed interest in this model. But a more popular model is provided by Harvey et al., and this is discussed in the next section.

Fig. 4.3 Contact of gas with liquid. **a** flat interface, and **b** curved interface in the form a bubble

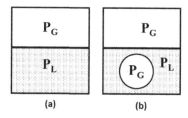

4.2.1   *Trapped Gas Pockets (Harvey's Model)*

Let us consider the presence of small-sized hydrophobic solid inside the liquid (Fig. 4.4a). If we look closely at the surface of this hydrophobic solid, we can see that the surface contains crevice (Fig. 4.4b–d). In that hydrophobic crevice, it is possible that a gas pocket is trapped. Now there can be three situations: liquid is saturated (Fig. 4.4b), unsaturated (Fig. 4.4c) and supersaturated (Fig. 4.4d). We know that for hydrophobic surface, static contact angle is greater than 90°. Thus, unlike the free bubble scenario, here the liquid is at a higher pressure and condition for equilibrium is $P_L = P_G + (2\sigma/R)$. Thus, the gas pocket can remain stable against dissolution. If the outside liquid pressure is increased (or the liquid is degassed and its gas content is reduced below saturation limit), gas will come out of pocket and go into solution. It will come to new equilibrium as shown by dotted line in Fig. 4.4c. In fact, this is one of the successful ways to explain the experimentally observed phenomenon of changing cavitation threshold by pressurizing. Opposite phenomenon is observed when liquid pressure is reduced (or, gas concentration is increased over saturation level), gas pocket size increases and reaches a new equilibrium. Thus, it is established that a gas pocket can remain in liquid for an extended period of time. When this solid passes through sufficiently low pressure region in flowing system or the liquid containing it is subjected to low pressure achieved by the passage of acoustic wave, this gas pocket will come out of the crevice and provide necessary free bubble which is the simplest source of nucleation site.

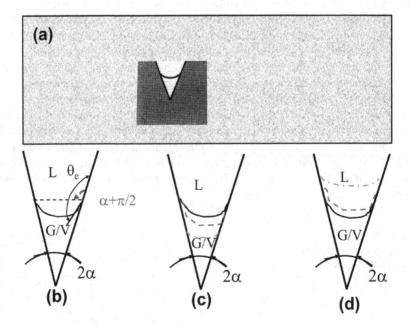

Fig. 4.4 Gas trapped inside a cavity on a hydrophobic surface

Thus, from Sect. 3.6 and the present discussion on nucleation in this chapter, we conclude this portion by summarizing the main aspects of nucleation:

- Liquids (and vapours) undergo phase change at pressure (or temperature) corresponding to the saturated values only if there exists enough nuclei in the liquid (or vapour).
- In the absence of big-sized nucleus, the phase change from liquid to vapour takes place at a pressure lower than the vapour pressure (corresponding to liquid temperature) and this pressure difference is known as liquid tension.
- Similarly, phase change from liquid to vapour takes place at temperatures above saturation temperature (at a given liquid pressure) and this excess temperature is termed as superheat.
- Superheat and liquid tension are related through Clausius–Clapeyron equation.
- If the phase change from vapour to liquid (condensation) takes place at temperatures lower than that of saturation, the temperature difference is known as subcooling.
- Homogeneous nucleation theory suggests that liquid tension should be of the order of 1000 bar while experimental observations indicate much lower tension.
- This discrepancy is attributed to the presence of weak spots or nuclei. This forms heterogeneous nucleation.
- Free gas bubble does not remain in the liquid for extended period of time because smaller bubbles dissolve in liquid due to surface tension and bigger bubbles rise due to buoyancy.
- Crevice model or organic skin model is a possible mechanism for stabilization of gas pocket inside a liquid.

So far, we have outlined the role and importance of nucleation to initiate phase change in cavitation and boiling/condensation. We have also highlighted the fact that heterogeneous nucleation more readily seen in almost all engineering applications, restrict the maximum tension and superheat in case of boiling of liquid (or subcooling in case of condensation of vapour) achievable.

Cavitation is influenced by both surface nuclei and nuclei present in the bulk volume of liquid. Boiling on the other hand, depends critically on surface nuclei. We take it up for discussion now.

We draw your attention to Eq. (3.22) and rewrite here for completeness

$$\frac{dP}{dT} = \frac{h_{fg}}{v_{fg} T_{sat}} \tag{a}$$

where h_{fg} is the latent heat and v_{fg} is the change in specific volume during phase change. For differential amount, this reduces to

$$\frac{P_G - P_L}{T_L - T_{sat}} = \frac{h_{fg}}{v_{fg} T_{sat}} \tag{b}$$

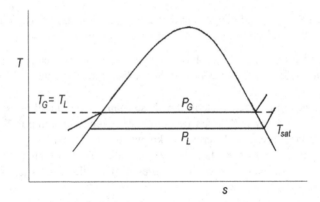

Fig. 4.5 Liquid superheat

Thus, connecting Laplace pressure difference with surface tension term (Eq. 4.1), we get an expression for the liquid superheat (see Fig. 4.5).

$$T_L - T_{sat} = \frac{2\sigma v_{fg} T_{sat}}{r h_{fg}} \tag{4.5}$$

This shows that the liquid superheat required for a bubble to exist varies inversely with bubble diameter and directly as surface tension. Thus, surface tension plays a major role in boiling. For example, the boiling behaviour of water is quite different from that of refrigerants having much less surface tension and producing much smaller bubbles.

However, this amount of superheat is not sufficient for inception of the nucleation process on a solid surface. For example, for water at atmospheric pressure, the liquid superheat predicted by Eq. (4.5) is about 3 °C, while in reality, on a solid surface, bubbles appear only at a superheat of 8 °C or more. As a consequence, the temperature profile in the vertical direction, arising out of boiling on a submerged surface, is as shown in Fig. 4.6.

The above discrepancy was explained by Hsu and Graham (1976) by a nucleate theory. They assumed that after each bubble departure some amount of vapour remains trapped within the cavity on the surface, which acts as a nucleation site. Even though cold liquid rushes towards the cavity, it cannot condense this residual vapour and the residual vapour acts as a starting process in the next bubble growth. The bubble growth and departure process is shown in Fig. 4.7.

This process of nucleation can be quantitatively assessed by considering a transient conduction in the liquid after a bubble departs, given by

$$\frac{T - T_{sat}}{T_w - T_{sar}} = 1 - \frac{n}{\delta} - \frac{2}{\pi} \sum_{n=1}^{\infty} \frac{1}{n} \sin n\pi \left(\frac{n}{\delta}\right) e^{-\frac{n^2 \pi^2 a\tau}{\delta^2}} \tag{4.6}$$

where δ is the thickness of the superheated boundary layer on the solid wall.

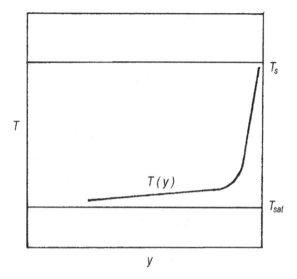

Fig. 4.6 Temperature profile for nucleate pool boiling

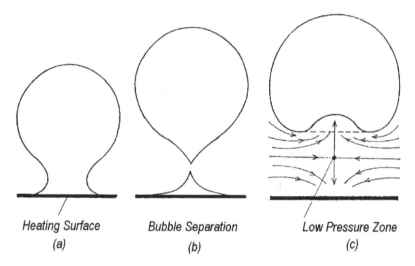

Fig. 4.7 Bubble growth and departure. (*Courtesy:* "Heat Transfer in Condensation and Boiling" by Karl Stephan, with permission from Springer)

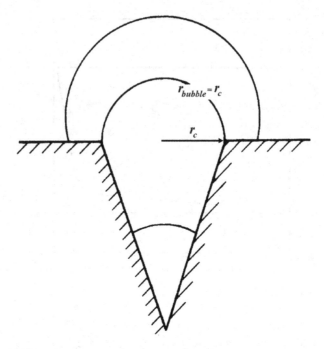

Fig. 4.8 Bubble formation on a conical cavity

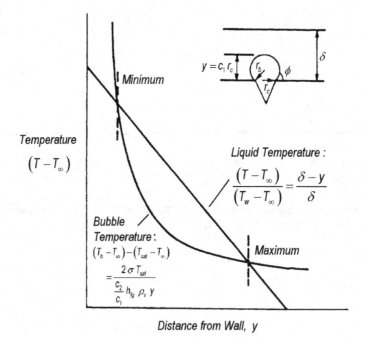

Fig. 4.9 Condition for nucleation on a surface

The liquid superheat required for nucleation on a conical cavity shown in Fig. 4.8 is given by

$$T - T_{sat} = \frac{2\sigma v_{fg} T_{sat}}{r h_{fg}} \qquad (4.7)$$

Now it is obvious that nucleation will take place only when the liquid superheat given by Eq. (4.7) is reached through transient conduction in the liquid given by Eq. (4.6). Hence, the condition of nucleation can be obtained by superposition of curves for these two equations given in Fig. 4.9.

It can be said that intersection of Eqs. (4.6) and (4.7) gives the bubble size, corresponding waiting period and required liquid superheat. The point A shows the case when waiting time is infinite giving no nucleation.

4.3 Static Stability of an Isolated Bubble

Let us revisit Eq. 4.1. But now, we consider a bubble (comprising of a mixture of vapour and gas) of initial radius R_0 in a liquid. Hence, the pressure inside this bubble is related to the outside liquid pressure by the relationship

$$p_{G0} + p_V - p_{L0} = \frac{2\sigma}{R_0} \qquad (4.8)$$

In this Eq. 4.8, the subscript 0 refers to initial value. p_V is the saturation (vapour) pressure at the liquid temperature and does not depend on the bubble size. Now, if we start reducing the liquid pressure (we may do so simply by connecting the liquid-surrounding air system to a vacuum pump), the gas bubble tends to adjust its size so that it comes to a new equilibrium. If this process is allowed to continue, will the bubble adjust itself continually? No. At one value of pressure (p_n^* in Fig. 4.10), any further reduction (even by a very small amount) of pressure will give rise to a very large increase in bubble size. The task before us is to obtain this static stability threshold of a bubble beyond which we need to consider the dynamics of bubble.

Let us start with static equilibrium relation for the bubble at a given size R when corresponding liquid pressure is P_L.

$$p_G + p_V - p_L = \frac{2\sigma}{R} \qquad (4.9)$$

We assume that a change in pressure induces a change in volume of gas inside the bubble following isothermal process. This is a reasonable assumption because the bubble is considered to be in quasi-static equilibrium, or in other words, the pressure changes slowly. Then at any radius, we can write

$$p_{G0} R_0^3 = p_G R^3 = C \text{ (say)}$$

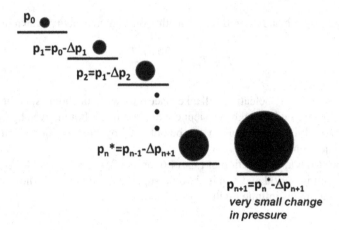

Fig. 4.10 Schematic of threshold of static stability

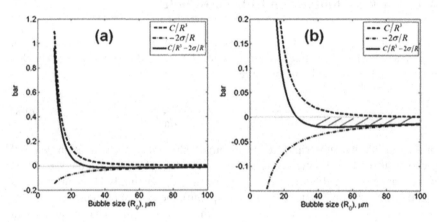

Fig. 4.11 Variation of different terms of Eq. 5.2. For these results, R_0 is taken as $10\mu m$. (**b**) is enlarged view of (**a**). The shaded region is mechanically unstable

Inserting this expression for gas pressure, we can rewrite Eq. 4.9 as

$$p_L - p_V = \frac{C}{R^3} - \frac{2\sigma}{R} \qquad (4.10)$$

In order to understand the physical significance of these terms clearly, let us plot the terms in Eq. 4.10. Figure 4.11 portrays this variation where both the terms are plotted. It is seen that in the lower asymptote ($R \longrightarrow 0$), the first term of the right-hand side dominates, while as ($R \longrightarrow \infty$) the second term dominates.

Applying the condition for explosive growth, $\frac{dR}{dP_L} \to \infty$, we get

$$R_c = \sqrt{\frac{3C}{2\sigma}} = \sqrt{\frac{3\left(p_{L0} - p_V + \frac{2\sigma}{R_0}\right) R_0^3}{2\sigma}}$$

R_c is the critical size of the bubble which will undergo an explosive growth when the ambient liquid pressure is p_{L0}. Alternately, it can be stated that for a liquid having bubble of initial size R_0, the critical pressure can be given as

$$p_{LC} = p_V - \frac{4\sigma}{3} \left[\frac{2\sigma}{3\left(p_{L0} - p_V + \frac{2\sigma}{R_0}\right) R_0^3} \right]^{1/2} \tag{4.11}$$

The hatched region in Fig. 4.11b is the mechanically unstable region which shows that a small reduction in pressure value brings about a phenomenal change in radius. It can be noted that by measuring liquid tension ($p_V - p_{LC}$) experimentally it is possible to estimate the size of bubbles present in a liquid. Of course, if a liquid has a size distribution of bubbles (and this is a usual scenario in any real-life devices), then this method will give us the size of the weakest nuclei (or the size of the largest bubble present). This method of estimating bubble size forms the basis of cavitation susceptibility meter (Oldenziel 1982; d'Agostino and Acosta 1991). Equation 4.11 gives an expression for static stability threshold, also known as Blake threshold. If the pressure drops to a lower value, assumption of quasi-static change is not valid and we need to consider dynamics of bubbles subjected to pressure changes. This is discussed in the next section.

4.4 Dynamics of an Isolated Gas Bubble: Growth and Collapse

Dynamics of a bubble due to pressure changes has been the subject of continued research ever since the development of bubble dynamics equation by Lord Rayleigh in 1917. We shall derive bubble dynamics equation based on mass and momentum conservation principles. Figure 4.12 shows the schematic of a bubble in an infinite liquid. The liquid is assumed to be inviscid and the flow is incompressible. The motion of the bubble is considered to be radial, and hence spherical symmetry condition is used for the mathematical analysis.

Let us consider a spherical shell of liquid at a distance r from the centre of the bubble. This shell has a thickness dr such that $dr \ll r$. If the bubble grows in response to the change in pressure sufficiently away from the liquid (by reducing p_∞), then

Fig. 4.12 A cavity of
instantaneous size R in an
infinite liquid. p refers to the
liquid pressure which takes
the value of p_L on the bubble
surface and p_∞ far away
from the bubble. p_i is the
pressure inside the bubble.
For empty cavity, p_i is 0

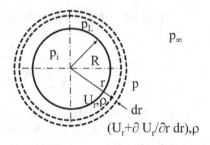

that will produce a radial velocity on the bubble wall ($\dot{R}$). This radial velocity induces
a radial velocity U_r in the liquid at a radius r. Thus, applying the conservation of
mass across the spherical shell, we can write

$$4\pi r^2 U_r \rho = 4\pi \left(U_r + \frac{\partial U_r}{\partial r} dr \right) (r + dr)^2 \tag{4.12}$$

Upon simplifying we get,

$$2\frac{dr}{r} = -\frac{dU_r}{U_r} \tag{4.13}$$

Thus,

$$U_r = \frac{C(t)}{r^2} = \frac{\partial \Phi(r,t)}{\partial r} \tag{4.14}$$

where, Φ is the velocity potential.
    At $r = R$, $U_r = \dot{R}$. Hence

$$U_r = \frac{R^2 \dot{R}}{r^2} \tag{4.15}$$

As $r \longrightarrow \infty$, U_r, $\Phi \longrightarrow 0$. This means that the influence of radial motion of the bubble does not affect the liquid at sufficiently far away. Let us consider the momentum
conservation along radial coordinates.

$$\frac{\partial U_r}{\partial t} + U_r \frac{\partial U_r}{\partial r} = -\frac{1}{\rho}\frac{\partial p}{\partial r} \tag{4.16}$$

After writing radial velocity in terms of velocity potential and manipulating, we get

$$\frac{\partial}{\partial r}\left[\frac{\partial \Phi}{\partial t} + \frac{U_r^2}{2} + \frac{p}{\rho} \right] = 0 \tag{4.17}$$

Integrating Eq. 3.9 from $r = R$ to $r \longrightarrow \infty$

$$\int\limits_{r=R}^{r\to\infty} \frac{\partial}{\partial r} \left[\frac{\partial \Phi}{\partial t} + \frac{U_r^2}{2} + \frac{p}{\rho} \right] dr = 0 \qquad (4.18)$$

Noting that $p = p_L$ at $r = R$ and $p = p_\infty$ for $r \longrightarrow \infty$, Eq. 4.18 yields

$$R\ddot{R} + \frac{3}{2}\dot{R}^2 = \frac{1}{\rho}[p_L - p_\infty] \qquad (4.19)$$

This is the most fundamental equation of dynamics of a bubble which may undergo growth or collapse (radial motion) but does not undergo any translation. Different forms of equations can be arrived at depending on the assumptions made. In the first work in this area, Rayleigh had assumed the cavity to be empty and hence $P_L = 0$. Later on Plesset, Poritsky, Noltingk and Neppiras considered the presence of vapour and permanent gas inside the bubble as well as real fluid effects like viscosity and surface tension. Equation (4.20) popularly goes by the name of Rayleigh–Plesset equation and is the most fundamental equation of bubble dynamics in cavitation.

$$R\ddot{R} + \frac{3}{2}\dot{R}^2 = \frac{1}{\rho} \left[p_V + p_{G0} \left(\frac{R_0}{R} \right)^{3k} - \frac{2\sigma}{R} - \frac{4\mu\dot{R}}{R} - p_\infty \right] \qquad (4.20)$$

Here, the gas-phase momentum equation is not considered and it is assumed that the gas expands/collapses following a polytropic law and k is the polytropic constant. Left-hand side of this equation relates with the inertia of the bubble and right-hand side depicts different forces. During growth, terms which are positive support growth and during collapse, negative terms favour collapse. We know that surface tension tends to reduce the surface area and hence tries to reduce the size of the bubble. Thus, it favours collapse of the bubble and opposes the growth. Viscosity, on the other hand, opposes both growth and collapse. These are also apparent from the knowledge of the terms with appropriate signs. When the bubble is growing, $\dot{R}$ is positive and hence viscosity term is negative while during collapse, $\dot{R}$ is negative and viscosity term is positive. Surface tension term is always negative. Vapour and gas inside the bubble opposes collapse but favours growth, as these terms are always positive.

This equation can be further extended to include thermal effects and to account for liquid compressibility. These are not covered here and the reader is urged to read specialized text books on cavitation, such as that by Brennen (1995).

Equation (4.20) is a stiff, non-linear second order differential equation which cannot be solved analytically unless simplifying assumptions are made. One important simplification that has a lot of practical importance is the assumption of low amplitude of pressure oscillation. This assumption of low pressure amplitude enables us to linearize this equation. This is now taken up for an elaborate discussion.

4.5   Linearized Bubble Dynamics

If we consider the bubble being excited by a small amplitude pressure oscillations, then it is natural to expect that the corresponding change in bubble radius will also be small and this can be expressed mathematically as a small perturbation (ξ) of bubble radius around the initial radius (R_0)

$$R(t) = R_0 + \xi(t), \quad \frac{|\xi|}{R_0} \ll 1 \tag{4.21}$$

Upon substitution of this Eq. (4.21) in Eq. (4.17) and simplifying the resultant equation by neglecting higher order terms ($\xi^2, \xi\dot{\xi}, \ldots$), we get a linear second order differential equation of the form

$$M\ddot{\xi} + C\dot{\xi} + K\xi = F(t) \tag{4.22}$$

where

$$M = \rho R_0, \quad C = \frac{4\mu}{R_0} \quad \text{and} \quad K = \left[\frac{3kp_{G0}}{R_0} - \frac{2\sigma}{R_0^2}\right] \tag{4.23}$$

Thus, linear resonant frequency of a bubble of size R_0 is given by

$$\omega_0 = \sqrt{\frac{k}{m}} \Rightarrow f_0 R_0 = \frac{1}{2\pi}\sqrt{\frac{3kp_{G0}}{\rho} - \frac{2\sigma}{\rho R_0}}$$

$$= \frac{1}{2\pi}\sqrt{\frac{3k}{\rho}\left(p_{L0} + \frac{2\sigma}{R_0} - p_V\right) - \frac{2\sigma}{\rho R_0}} \tag{4.24}$$

For big bubbles, surface tension terms can be neglected and $p_{L0}(\sim p_{atm}) > p_V$. Thus, we get

$$f_0 R_0 = \frac{1}{2\pi}\sqrt{\frac{3kp_{atm}}{\rho}} \Rightarrow f_0 R_0 \approx 3\,\text{ms}^{-1}$$

where the numerical value of $f_0 R_0$ is obtained under the assumption of isothermal volume change. On the other hand, if the bubble size is small, surface tension term dominates and we get

$$f_0 R_0 = \frac{1}{2\pi}\sqrt{\frac{2\sigma}{\rho R_0}(3k - 1)}$$

It may be interesting to compare the size of resonant radius with the acoustic wavelength in liquid at a given frequency. Using $\lambda = 2\pi c/\omega_0$, hence $2R_0/\lambda \sim 0.004$.

Thus, bubble size is quite small compared to the wavelength. For bubbles in the radius range of $100\,\mu m$ usually observed in natural water, this corresponds to a resonant frequency of 30 kHz.

In real-life situations like cavitation or boiling, the idealized modelling of bubble dynamics carried out in this chapter, like an isolated bubble or spherical bubble assumptions, may not be valid as will be discussed in subsequent chapters. However, the analysis carried out here forms the basis of theoretical understanding for those phenomena.

Examples

Problem 1: A bubble, of diameter $20\,\mu$, in water is in equilibrium. Determine the gas pressure inside the bubble if it is known that surface tension is 0.07 N/m, atmospheric pressure is 1 bar (you may neglect hydrostatic pressure variation inside water) and vapour pressure of water is 3500 Pa.

Solution: We know that the condition of static stability is given by Eq. 4.8. In this equation, $R_0 = 10\,\mu m$, $p_{L0} = 1$ bar, $\sigma = 0.07$ N/m. Thus, $p_{G0} = p_{L0} + \frac{2\sigma}{R_0} - p_V = 1.11$ bar.

Problem 2: For bubble considered in Example Problem 1 above, determine the pressure above which dynamics need to be considered.

Solution: We know that any pressure the threshold pressure is given by Blake threshold (Eq. 4.11)

$$p_{LC} = p_V - \frac{4\sigma}{3} \left[\frac{2\sigma}{3 \left(p_{L0} - p_V + \frac{2\sigma}{R_0} \right) R_0^3} \right]^{1/2} \tag{4.25}$$

Thus, the static stability threshold value, $p_{LC} = 1.58$ kPa.

Problem 3: For the bubble considered in Example Problem 1 above, determine the linear resonant frequency. You may consider isothermal oscillation and density of water as $1000\,kg/m^3$.

Solution: The relationship of linear resonant frequency with bubble size is given in Eq. 4.24. Thus, we have:

$$f_0 R_0 = \frac{1}{2\pi} \sqrt{\frac{3k}{\rho} \left(p_{L0} + \frac{2\sigma}{R_0} - P_V \right) - \frac{2\sigma}{\rho R_0}} \tag{4.26}$$

Thus, resonant frequency $(f_0) = 284$ kHz.

Exercises

1. A venturimeter is used as a cavitation susceptibility meter and can be used to determine (indirectly) the size of the largest nuclei (bubble) by measurement of pressure. The venturimeter has an inlet diameter of 18 mm and a throat diameter of 3 mm. The venturimeter is connected to an upstream tank whose pressure is 1 bar. Cavitation event is detected when the throat velocity is 14 m/s. Determine the size of the largest bubble that can act as the weakest nuclei.

2. The bubble dynamics equation for a gas bubble encapsulated with a thin protein layer is given by

$$R\ddot{R} + \frac{3}{2}\dot{R}^2 = \frac{1}{\rho}\left[p_{G0}\left(\frac{R_0}{R}\right)^{3k} - \frac{2\sigma}{R} - \frac{4\mu\dot{R}}{R} - \frac{4\kappa^s\dot{R}}{R^2} - p_0 - p_A\sin(\omega t) \right]$$

(4.27)

Here, ρ and μ are liquid density and viscosity, σ and κ^s are the properties of the interface, viz., surface tension and surface dilatational viscosity, respectively. Other symbols have their usual meaning. Derive an expression for the linear resonant frequency of this bubble. Does the protein encapsulation increase/decrease the resonant frequency of this bubble when compared with a free bubble?

References

Apfel RE (1971) A novel technique for measuring the strength of liquids. J Acoust Soc Amer 49(1B):145–155

Brennen CE (1995) Cavitation and bubble dynamics. Cambridge university press.

d'Agostino L, Acosta AJ (1991) A cavitation susceptibility meter with optical cavitation monitoring-part one: design concepts. J Fluids Eng 113(2):261–269

Epstein PS, Plesset MS (1950) On the stability of gas bubbles in liquid gas solutions. J Chem Phys 18(11):1505–1509

Fox FE, Herzfeld KF (1954) Gas bubbles with organic skin as cavitation nuclei. J Acoust Soc Amer 26(6):984–989

Harvey EN, Barnes DK, McElroy WD, Whiteley AH, Pease DC, Cooper KW (1944) Bubble formation in animals. I. Physical factors. J Cell Comp Physiol 24(1):1–22

Hsu YY, Graham RW (1976) Transport processes in boiling and two-phase systems, including near–critical fluids. Washington.

Oldenziel DM (1982) A new instrument in cavitation research: the cavitation susceptibility meter. ASME J Fluids Eng 104:136–142

Trevena DH (1984) Cavitation and the generation of tension in liquids. J Phys D: Appl Phys 17(11):2139

Chapter 5
Cavitation

In Chap. 4, we have outlined the role and importance of nucleation to initiate phase change in cavitation and boiling/condensation. We have also highlighted the fact that heterogeneous nucleation, more readily seen in almost all engineering applications, restricts the maximum tension in case of cavitation (superheat and subcooling, respectively, in case of boiling and condensation) achievable. In this chapter, we shall extend the discussion on the response of a bubble to pressure changes. This forms the core of cavitation. But what is cavitation?

Cavitation is the formation and subsequent dynamic life of a bubble in a liquid. By *dynamic life of bubble*, we mean growth and collapse. Many times this phenomenon is brought about by lowering the pressure locally below the critical pressure corresponding to the size of the nucleus present. In this context, it is worth revisiting Fig. 3.5. In most of the applications, cavitation does not bring about a change in temperature (notable exceptions being cavitation in cryogenic fluids or near the boiling point). Hence, if we follow an isotherm during pressure reduction, then it reaches point 'f' from a point 1 along $1f$. However, as mentioned in Sect. 3.6, phase change does not start at point 'f' in the absence of impurities. Instead, this phase change takes place at a point E, and cavitation initiates only after this pressure reduction below the saturation pressure. The reduction in pressure can be brought about by fluid flow (e.g., flow inside a venturi, Fig. 5.1) or by the application of piezoelectric or magnetostrictive transducers (Fig. 5.2). The former type of cavitation is called hydrodynamic cavitation and later one is termed as acoustic cavitation. Excellent reviews (for example, Flynn 1964; Plesset and Prosperetti 1977) and dedicated monographs (for example, Knapp et al. 1970; Leighton 1994; Brennen 1995) are available, and interested readers will do well to go through these for a more detailed account of cavitation.

© The Author(s) 2023
S. K. Das and D. Chatterjee, *Vapor Liquid Two Phase Flow and Phase Change*, https://doi.org/10.1007/978-3-031-20924-6_5

Fig. 5.1 Hydrodynamic
cavitation produced in a
venturi

Fig. 5.2 Acoustic cavitation
produced in a spherical flask
with the help of piezoelectric
crystals

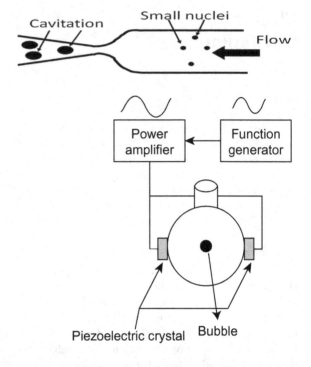

5.1 Emergence of Studies on Cavitation

Cavitation as a topic of research has attracted the attention of scientists and engineers
over the past several decades. The first type of cavitation observed was the formation of bubbles in liquids supersaturated with a gas (as in carbonated beverages)
(Tomlinson 1867).

Interest in cavitation began with the development of high-speed turbines in the
mid-1800s. With this advance came the means of moving an object (such as a propeller) rapidly through a fluid. In fact, the term *cavitation* was first used in connection
with the observed propeller inefficiency in the torpedo boat destroyers (Thornycroft
and Barnaby 1895). In 1917, Lord Rayleigh gave the first theoretical treatment of the
collapse of a vapour-filled cavity under constant external pressure (Rayleigh 1917).
It was felt that the cavities, which form on the low-pressure side of a propeller blade,
collapse with such violence that the resulting pressure shock created in water erodes
the surface of the propeller. Later high-performance hydraulic machines were also
observed to suffer similar damage due to cavitation activity.

In the middle of the nineteenth century, the first attempts to measure the tensile
strength of liquids started. The tensile strength of a liquid may be defined as the
limiting negative pressure (tension) that a liquid can withstand before it fragments
or ruptures to form a new stable phase. Around 1850, Berthelot measured the tensile
strength of water and found the maximum value on the order of 50 bar (Blake

1949a, b). The highest tensile strength (275 bars) was measured by Briggs (1950). Apfel (1971) used a drop levitation technique for determining the threshold. This novel technique helped in obtaining a very clean, almost mote-free liquid. His results agreed reasonably well with the theoretical prediction at an elevated temperature.

It is observed that all the experiments done with water samples at room temperature fell much short of the theoretically predicted homogeneous nucleation threshold limit of water (which is thousands of bars). This discrepancy has been attributed to the fact that no liquid can be as pure as postulated in the theory and inhomogeneities are always present in a bulk liquid sample. This contamination weakens the molecular bonding. This is because the adhesive force between the liquid molecules and these inhomogeneities may be much less than the cohesive force between two liquid molecules. For example, pockets of undissolved gas or a hydrophobic solid surface can serve as weak spots preventing liquids from sustaining high tension.

Such weak spots are examples of cavitation nuclei, and the process of nucleation, as explained in Chap. 4, is called heterogeneous nucleation. Understanding the role of nuclei in any cavitation process has been central to the study of both acoustic and hydrodynamic cavitation and has been discussed by many investigators, for example, Holl (1970) in the case of hydrodynamic cavitation and Flynn (1964) for acoustic cavitation.

For a known pressure field experienced by a certain liquid, vapourous cavitation will take place if the minimum pressure experienced is less than the critical pressure. In experiments, the method of detecting cavitation relies either on visual observation under proper lighting (like stroboscopic lighting) or on monitoring the acoustic signal. In acoustic cavitation, only the threshold pressure is indicated; whereas hydrodynamic cavitation is characterized by a non-dimensional number, called the inception cavitation number (σ_i). The cavitation number as applicable to test conditions in a water tunnel is defined as

$$\Sigma = \frac{p_{up} - p_v}{\frac{1}{2}\rho V_{ref}^2}$$

where p_{up} and V_{ref} are freestream pressure and characteristics velocity, respectively. However, it must be mentioned that other equivalent definitions for Σ are also possible under different test conditions.

5.2 Effects of Cavitation

The next question that may come to our mind is why should we study cavitation? This is because, cavitation not only appeals to us because of its rich physics but also because of the effects produced by cavitation. Are these effects beneficial or harmful? The answer depends on the type of cavitation—hydrodynamic or acoustic—we are talking about. Acoustic cavitation is known mainly for beneficial reasons and some of the uses are listed below. There is an excellent review on the different applications of acoustic cavitation by Suslick et al. (1999) and readers are encouraged to read that review for more details.

A brief outline of a few applications of acoustic cavitation is provided below.

Ultrasonic cleaner: Acoustic cavitation is known to produce cleaning effects that find many industrial applications. The dependence of the threshold level on the acoustical and geometrical parameters of these devices suggests a deeper understanding is required for the phenomenon.

Sonochemical reaction: Acoustic cavitation is known to not only accelerate chemical reactions (even at room temperatures) but also to promote chemical reactions which would not otherwise occur. In a sonochemical reactor (which, simply speaking, is a vessel that allows ultrasonic excitation of liquids by the piezoelectric crystal), bubbles grow and collapse. High temperatures (hot spots) may form inside the bubbles. These microbubbles, therefore, act as microreactors and bring about desired chemical reactions. Bubble dynamics dictate whether conditions needed for a chemical reaction are achieved or not and hence the a bubble at different operating conditions of frequency and pressure amplitude plays an important role.

Dispersion and emulsification of solids and liquids in liquids: Acoustic cavitation is also known to bring about dispersion and emulsification. For example, in one experiment by Olaf, oil, grease and wax were dispersed in trichloroethylene (Barger 1964).

Production of nanoparticles: Nanometer-sized solids often exhibit properties distinct from those of the bulk. Such nanoparticles are finding uses in many different applications and can be produced by different ways. One such route to produce nanoparticles is to make use of extreme conditions of pressure and temperature produced by acoustic cavitation.

Medicinal applications: Ultrasonically induced acoustic cavitation leads to the formation of gas-filled microspheres (like air-filled bubble encapsulated by protein/lipid layer) through sonochemical reactions. These microspheres play important roles in contrast agents employed for ultrasonic detection of cancerous cells, for use in targeted drug delivery, etc.

In the context of hydrodynamic cavitation, as experienced by a mechanical or a naval engineer, the effects of cavitation are harmful and it is important to know how best it is to avoid cavitation or reduce the harmful effects of cavitation. Some of these effects are as follows:

- Loss in the performance of hydraulic machineries.
- Pitting and surface damage.
- Noise.
- Vibration of components.
- Liquid instability (POGO instability).

We shall discuss each of the first three aspects in some detail in the section on hydrodynamic cavitation.

Now we will turn our attention to the interaction of nuclei and an acoustic field, a subject that has attracted renewed attention in recent times with the discovery of the fascinating phenomenon of single bubble sonoluminescence (Gaitan 1990).

5.3 Acoustic Cavitation

Acoustic cavitation is the phenomenon in which a nucleus when subjected to an acoustic pressure field shows attributes associated with cavitation. Thus, central to the understanding of any acoustic cavitation event is the knowledge of bubble–acoustic field interaction.

5.3.1 Interaction of a Bubble with an Ultrasonic Field

The first-order bubble motion under the influence of an ultrasonic pressure field is described by the well-known Rayleigh–Plesset (RP) equation (already derived and presented in Chap. 4 as Eq. 4.17). The Rayleigh–Plesset equation in relation to acoustic cavitation has been extensively dealt with in excellent reviews on the subject, for example in Flynn (1964) and Plesset and Prosperetti (1977). For the sake of completeness, the same equation is given here once again. This equation is given by

$$R\ddot{R} + \frac{3}{2}\dot{R}^2 = \frac{1}{\rho}\left[p_v + p_g - \frac{2\sigma}{R} - \frac{4\mu\dot{R}}{R} - p_\infty(t)\right] \qquad (5.1)$$

In this equation, R, $\dot{R}$ and $\ddot{R}$ refer to instantaneous bubble radius, wall velocity and wall acceleration, respectively. p_g is the gas pressure within a bubble, and $p_\infty(t) = p_0 - P_A \sin(\omega t)$ is the time-varying pressure field far away from the bubble to which a bubble responds. μ and ρ are liquid density and viscosity, respectively. This non-linear second-order differential equation can be solved for initial conditions $R(t = 0) = R_0$ and $\dot{R}(t = 0) = 0$. We can classify the bubble response to the external acoustic field on the basis of the magnitude of bubble-radius oscillations as follows (Brennen 1995):

- For *very small* pressure amplitudes, the response is linear. Hence, the RP equation can be linearized which gives the natural frequency (ω_n) of the bubble oscillation (also see Eq. 4.21 for derivation):

$$\omega_n = \left[\frac{1}{\rho R_0^2}\left\{3k(P_\infty - P_v) + 2(3k - 1)\frac{\sigma}{R_0}\right\}\right]^{\frac{1}{2}} \qquad (5.2)$$

where k is the polytropic constant and $k = 1$ and 1.4 correspond to the isothermal and adiabatic limits, respectively.

- If the acoustic pressure amplitude is *moderately high*, a bubble will undergo non-linear oscillations. If the bubble exhibits stable oscillations, this phenomenon is termed as stable acoustic cavitation. Many interesting non-linear phenomena are associated with this type of bubble motion, like rectified diffusion, bubble movement due to Bjerknes force, etc.

- With a further increase in the pressure amplitude, a bubble may grow to a size many times larger than its initial radius. In such a case, the explosive growth is followed by a violent collapse and under this condition, the bubble is said to undergo *transient* cavitation. Apfel (1981) has suggested that if the maximum bubble size attained during oscillation, R_{max}, equals or exceeds $2.3R_0$, then the resultant motion will be transient; otherwise, the bubble will go through stable oscillations.

Figure 5.3 shows the variation of bubble radius with time when a $10\,\mu$m-bubble is driven at 20 kHz frequency under different excitation amplitudes. We find that at a low-pressure amplitude of 0.1 bar, the bubble oscillates gently about its initial radius while at the highest pressure amplitude of 0.12 bar, the bubble grows to several times its initial radius, thereby confirming Apfel's criterion of transient cavitation based on bubble oscillation in an acoustic field. A very rapid collapse can be observed at this pressure amplitude in contrast to that at $P_A = 0.1$ bar. All the effects of cavitation, discussed earlier, result from this violent bubble collapse. Intermediate acoustic pressure amplitude exhibits stable oscillation.

Understanding both stable and transient bubble motions are necessary for an effective control of acoustic cavitation in different applications. Hence, we consider in further detail the physical processes of rectified diffusion and Bjerknes force which can become important in case of stable bubble oscillation.

5.3.1.1 Rectified Diffusion

In the absence of any acoustic excitation, as mentioned in Chap. 4, a gas bubble will get dissolved due to surface tension effects, and a typical bubble size variation with time is shown in Table 5.1 (Epstein and Plesset 1950). However, in the presence of an ultrasonic pressure field, this picture may get altered due to the phenomenon of rectified diffusion. A bubble, in such a case, may start growing under the influence of an external varying pressure field. This is possible due to *area* and *shell* effects. In the course of an acoustic cycle, gas diffuses into a bubble, undergoing stable oscillation, during the rarefaction and diffuses out of it during the compression phase. As the mass transfer is dependent on the surface area of the bubble, more gas diffuses in during bubble expansion than that goes out during bubble compression. This will result in a net growth of the bubble over time and is possible when the acoustic pressure amplitude is such that the bubble growth takes place overcoming the driving out of gas due to the Laplace pressure difference present (as it happens with a static bubble

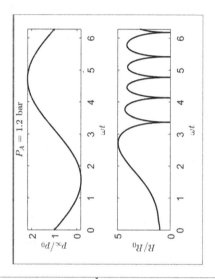

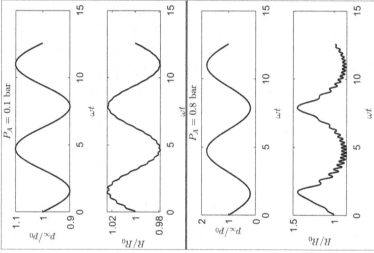

Fig. 5.3 Dynamics of a 10 μm-bubble subjected to an acoustic field whose driving frequency is 20 kHz of varying magnitude

Table 5.1 Dissolution of a bubble in the absence and the presence of surface tension (adapted from Epstein and Plesset 1950)

	$R_0 = 10\,\mu m$		$R_0 = 100\,\mu m$		$R_0 = 1000\,\mu m$	
$f = C/Cs$	T (s)	Ts (s)	T (s)	Ts (s)	T (s)	Ts (s)
0	1.25	1.17	1.25	1.24	1.25	1.25
0.25	1.67	1.46	1.67	1.64	1.67	1.66
0.5	2.5	1.96	2.5	2.41	2.5	2.49
0.75	5	2.99	5	4.6	5	4.95
1	∞	6.63	∞	58.8	∞	580

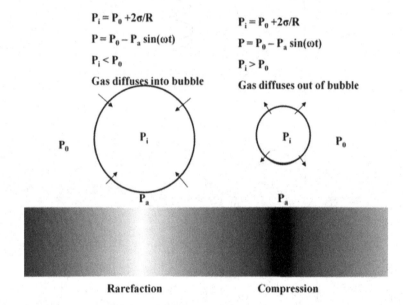

Fig. 5.4 Schematic showing gas diffusion into or out of an oscillating bubble: the *area effect*. (*Courtesy:* "The Growth of Bubbles in an Acoustic Field by Rectified Diffusion" by Thomas Leong, Muthupandian Ashok Kumar and Sandra Kentish, Handbook Ultrason. Sonochem., 2016, with permission from Springer)

discussed in Chap. 4). This is shown schematically in Fig. 5.4. Another effect, namely the *shell effect*, comes into the picture due to the oscillating bubble and it concerns the concentration gradient in the neighbourhood of the oscillating bubble. As the bubble shrinks during compression, the shell thickness increases, while it reduces as the bubble expands during the expansion phase. As a result, the concentration gradient of gas is lower when the bubble is in compression, but higher during expansion. Thus, this effect also results in a net accumulation of mass into the bubble. Figure 5.5 exemplifies this effect. It may be mentioned here this was first shown by Eller and Flynn (1965).

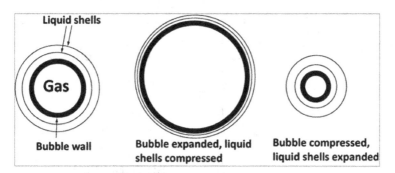

Fig. 5.5 Schematic showing gas diffusion into or out of an oscillating bubble: the *shell effect* (*Courtesy:* "The Growth of Bubbles in an Acoustic Field by Rectified Diffusion" by Thomas Leong, Muthupandian Ashok Kumar and Sandra Kentish, Handbook Ultrason. Sonochem., 2016, with permission from Springer)

We now present a brief trajectory of the research taken in this niche area. Rectified diffusion was first recognized by Harvey et al. (1944), but an early theoretical treatment of the subject is due to Blake (1949a, b). However, it was only partially correct as it took into consideration only the area effect. Hsieh and Plesset (1961) and Eller and Flynn (1965) gave theoretical expressions for the threshold for diffusive stability taking into consideration both area and shell effects. Crum (1980) has reported his experimental observations on bubble growth due to rectified diffusion. In an excellent review, he has pointed out that the numerical results based on solving the bubble dynamics equations match with the experimental data better than the approximate analytical results and also that these analytical expressions fail to account for harmonic resonances and the corresponding rapid bubble growth (Crum 1984). Fyrillas and Szeri (1994) computed the net mass flux of dissolved gas towards or away from a bubble by splitting the problem into two parts—an oscillatory part, that changes on fast time scales, and a smooth part changing on the slow diffusive time scale. However, it was found that even with this sophistication in the analytical approach, the experimental results for growth rates do not compare well with their theoretical results. This could possibly be due to the presence of surfactants on the bubble surface as mentioned by Crum (1984). Following Fyrillas and Szeri (1994), the expression for a threshold can be represented by

$$\frac{C_\infty}{C_0} - \frac{< P(t) >_{t,4}}{P_\infty} = 0 \qquad (5.3)$$

Here, C is the concentration of dissolved gas in a liquid. Subscript ∞ denotes a location inside the liquid far away from the bubble while 0 refers to the saturation condition at liquid temperature. $< q(t) >_{t,4}$, weighted average, is given by

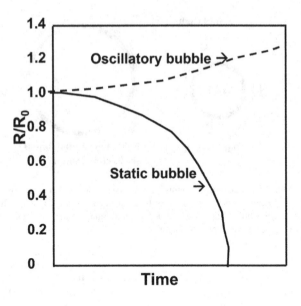

Fig. 5.6 Schematic showing gas diffusion into or out of bubbles

$$\frac{\int_0^T q(t)R^4 dt}{\int_0^T R^4 dt}$$

Depending on the sign of the left-hand side terms of Eq. 5.3, a bubble will grow or shrink: positive value leads to growth and the reverse happens when it is negative. Figure 5.6 shows a schematic comparing the temporal change of radii of two bubbles, one static and another oscillatory, and thereby highlights the effect of surface tension-driven gas diffusion in the case of a static bubble and a rectified diffusion-enabled growth in the other case. In his work, Chatterjee (2003) has used the piezoelectric crystal to produce a non-uniform acoustic pressure field superimposed on the flow inside a pipe. Based on his observation, he has classified the pipe cross section into three regions (Fig. 5.7):

- Zone I: near the pipe axis where acoustic pressure is high and bubbles undergo transient oscillation.
- Zone II: in the middle portion between the centreline and the wall of the pipe where the bubbles oscillate stably and undergo growth due to rectified diffusion.
- Zone III: near the wall where the acoustic pressure amplitude is low, and though the bubble is oscillating, the surface tension-driven outward gas mass transfer dominates.

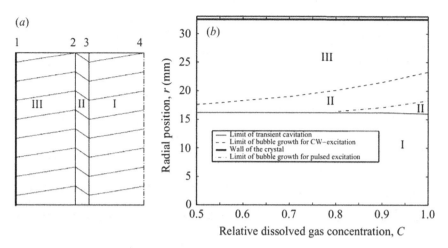

Fig. 5.7 Response of a bubble subjected to a non-uniform acoustic pressure field superimposed on the hydrodynamic flow field. The left picture shows the zones and the right picture brings out the variation quantitatively for different dissolved gas concentration levels. (*Courtesy:* Some investigations on the use of ultrasonics in travelling bubble cavitation control by Chatterjee, D. and Arakeri, V. H., J. Fluid Mechanics, 2004, with permission from Cambridge University Press)

5.3.1.2 Bjerknes Forces

Another bubble activity which is important when the driving pressure amplitude is not very high is the movement of a bubble due to the Bjerknes forces. Bjerknes forces are of two types. The first type, called the primary Bjerknes force, arises due to the interaction of a bubble with an imposed spatially non-uniform acoustic field. Due to this force, a bubble tends to move towards the pressure antinode if its size is smaller than the resonant bubble radius (in other words, $\omega < \omega_n$). On the other hand, it will move towards the pressure node, if $\omega > \omega_n$. However, as shown by Parlitz et al. (1999), this is only valid for cases where linearized approximation holds good. The primary Bjerknes force (F_{B1}) is given by the following expression:

$$F_{B1} = - < \nabla P_A(t) V(t) >_T \tag{5.4}$$

In the above equation, $V(t)$ denotes instantaneous bubble volume, $\nabla P_A(t)$ refers to the gradient of the acoustic pressure amplitude at the bubble location and $< \cdot >_T$ denotes an average over a time period T. Another type of Bjerknes force, termed as the secondary Bjerknes force, arises due to the presence of more than one bubble in a liquid subjected to acoustic excitation. The details can be found in the work of Parlitz et al. (1999). The bubble motion due to Bjerknes forces together with rectified diffusion can play an important role in influencing the bubble dynamics when a bubble is undergoing a stable oscillatory motion. Chatterjee (2003) has shown the effect of a non-uniform pressure field on bubble motion under the combined acoustic and hydrodynamic pressure fields. In his analysis, he has considered the primary

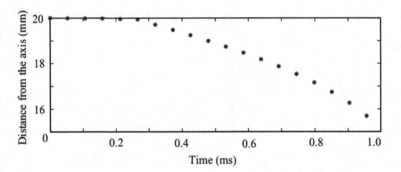

Fig. 5.8 Trajectory of a bubble subjected to a non-uniform acoustic pressure field superimposed on the hydrodynamic flow field. (*Courtesy:* Some investigations on the use of ultrasonics in travelling bubble cavitation control by Chatterjee, D. and Arakeri, V. H., J. Fluid Mechanics, 2004, with permission from Cambridge University Press)

Bjerknes force, drag force and added mass force. Figure 5.8 shows the trajectory of a bubble subjected to these forces, and it can be seen that the bubble moves towards the higher acoustic pressure region.

5.3.1.3 Transient Cavitation

Transient cavity motion is responsible for both destructive and constructive features associated with cavitation. Flynn and Church (1988) have referred to the condition in which a bubble goes into transience in the first acoustic cycle as prompt transient cavitation; on the other hand, if the bubble exhibits transient motion after a few cycles, it is termed as an oscillatory transience (Fig. 5.9).

It may so happen that a nucleus in a bulk volume of a liquid moves towards a pressure antinode under Bjerknes force and then grows there due to rectified diffusion till its size exceeds that corresponding to the transient cavitation threshold. During the collapse phase of transient cavitation, a nucleus may exhibit shape instability, thereby, leading to the formation of several smaller nuclei due to fragmentation. In this way, the nuclei number can proliferate. Hence to sustain acoustic cavitation in a stagnant liquid, a continuous supply of nuclei is not required. This is one of the major differences between hydrodynamic and acoustic cavitation; for in the former, in most situations, a continuous supply of nuclei is essential for its sustenance.

Here are two classic experiments which bring out the fragmentation of bubbles. In the work of Nyborg and Hughes (1967), bubbles were destroyed in an acoustic field produced by a magnetostrictive transducer. Nyborg and Hughes (1967) observed the phenomenon of bubble annihilation in an acoustic field produced by a magne-tostrictive transducer in a vessel with transparent walls. They found that although large bubbles were formed continually at a rapid rate, their total number did not increase with time. As it was not possible for the bubbles to go out of the photographic frame, they concluded that some of these bubbles were disappearing. In

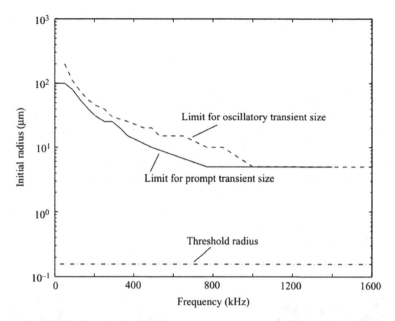

Fig. 5.9 Typical frequency dependence of the bounds for prompt and oscillatory transient cavities. The threshold radius denotes the Blake threshold. (*Courtesy:* Dhiman Chatterjee, PhD Thesis, 2003, IISc Bangalore)

another classic paper, Harrison (1952) had reported the disappearance of bubbles in a cavitating flow through a venturi. Barger (1964) had reported the fragmentation of bubbles through multiple cavitation interactions. In his experiments, he measured an increase in the threshold value for a water sample by subjecting it to repeated acoustic cavitation.

Flynn and Church (1984) indicated that the maxima they observed in iodine release under pulsed ultrasonic excitation could be linked to the dynamics of nuclei size distribution present in a liquid. They suggested that the optimum pulse width was related to bubble fragmentation and nuclei survival issues. This, therefore, gives us an idea that the manipulation of nuclei size and number in a liquid sample is possible if one can select pulse characteristics, pressure amplitude or frequency of excitation judiciously. So far we have described the role of nuclei in cavitation and their interaction with an acoustic field. Now we turn our attention to hydrodynamic cavitation.

5.4   Hydrodynamic Cavitation

Hydrodynamic cavitation, or cavitation induced by fluid flow, can be broadly classified into the following types:

a. **Travelling bubble cavitation**: *In this type of cavitation, individual and approximately spherical- or hemispherical-shaped bubbles form in the liquid. These bubbles grow, shrink and collapse as they travel in the liquid along with the flow. These bubbles will be approximately spherical in size (Fig. 5.10a).* The dynamics of these bubbles can be captured from the knowledge of the pressure variation in the flow. These are the simplest to analyse as the interaction of the bubble with the flow can be represented as one-way coupled with fluid flow dictating the bubble dynamics. Here, real fluid effects like flow separation or vortices do not play any significant role.

b. **Fixed cavitation**: *This type of cavitation may form under the developed cavitation stage when the liquid detaches from the surface of a rigid boundary of*

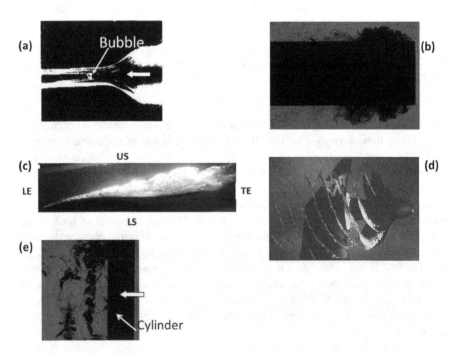

Fig. 5.10 Different types of hydrodynamic cavitation. **a** Travelling bubble cavitation in a venturi throat; **b** Fixed cavitation on a blunt cylinder. (*Courtesy:* Mr. R. Rahul, IIT Madras); **c** Supercavitation on S-shaped hydrofoil. (*Courtesy:* T. Micha Premkumar, PhD thesis, 2012, IIT Madras); **d** Vortex cavitation in a propeller (Courtesy for propeller cavitation: "Surge instability on a cavitating propeller" by M. E. Duttweiler and C. E. Brennen, J. Fluid Mechanics, 2002, with permission from Cambridge University Press); **e** Wake cavitation behind a cylinder. (*Courtesy:* Pankaj Kumar, PhD thesis, 2017, IIT Madras)

Fig. 5.11 Photograph showing cavitation inception, developed condition and schematically depicts cavitation hysteresis (*Courtesy:* T. Micha Prem Kumar, PhD Thesis, 2012, IIT Madras)

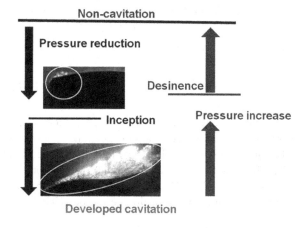

a submerged object and gives rise to the formation of a partial cavity attached to the boundary of the object (Fig. 5.10b). This type of cavitation often has the appearance of a glassy, transparent front portion, and at the end where it terminates, the appearance is more like a boiling surface. It is not necessary that the cavity thus formed will terminate on the body similar to that shown in Fig. 5.10b. It may extend well beyond the length of the object such that the object may appear covered by myriads of the bubble. This latter structure (Fig. 5.10c) is known as supercavitation. Supercavitation is very important for naval applications.

c. **Vortex cavitation**: Cavities form in the vortex cores formed in shear-flow regions. Perhaps, the most famous example is the tip-vortex cavitation (Fig. 5.10d). Vortex cavitation can also occur in submerged jets or in the wake of objects (Fig. 5.10e).

It has already been mentioned that hydrodynamic cavitation is brought about by a reduction in pressure. Thus, a flow is divided into three conditions from the point of view of cavitation: when pressure is reduced gradually from non-cavitating flow, a stage comes when the first signs of cavitation are observed. This sign could be in the form of an acoustic signal above the background flow noise or could be the first traces of bubbles observed visually. This condition is known as incipience or cavitation inception. As pressure is reduced further, developed cavitation occurs when a large pocket of cavitating bubbles is seen and a significant increase in noise level is heard. If the pressure is gradually increased from developed cavitation, it is seen that the extent of cavitation reduces and eventually cavitation disappears. The condition at which cavitation just disappears is termed as desinence or desinent cavitation. Though it may appear that cavitation inception and desinence occur at the same pressure value, if the liquid does not contain enough nuclei, then desinence occurs at a pressure higher than the inception condition, and this difference is known as cavitation hysteresis. This is depicted in Fig. 5.11. It may be noted that the inception of cavitation is closely linked with water quality and hence may vary from facility to facility. Hence, the state of cavitation is often defined in terms of cavitation number at inception (Σ_i) in that facility and one of the simplest ways to express the flow condition with relation to inception is Σ / Σ_i.

5.5 Factors Influencing Cavitation Inception

Apart from cavitation number (Σ), other factors which influence cavitation process can be broadly classified into two categories: parameters concerning liquid quality (like number and size distribution of nuclei and their likely sources) and factors that affect the flow field (velocity, pressure distribution, body size and shape, viscous effects, etc.) Rood (1991) has given a coherent summary of the inception dynamics of cavitation and a much of the present description is adapted from this excellent review.

5.5.1 Importance of Nuclei

In an earlier section, the importance of nuclei to initiate acoustic cavitation has been mentioned. It has also been noted that hydrodynamic cavitation, unlike acoustic cavitation, in general, requires a continuous supply of nuclei. Another difference is that in acoustic cavitation the primary source of nuclei is normally from the bulk volume of the liquid; whereas, in hydrodynamic cavitation surface nuclei, besides the freestream ones, can also be of importance. Surface properties or the way these surfaces are treated, before an experiment on cavitation is conducted, seem to influence the inception cavitation number as well as the appearance of cavitation in the developed stage. Holl (1970) has given a detailed description of the various sources of cavitation nuclei. He conducted cavitation experiments on hemispherical-nosed bodies with different materials. It was seen that teflon cavitated most easily whereas a delay in cavitation onset in the case of glass was observed.

The important role played by the nuclei population in hydrodynamic cavitation makes it imperative to measure and characterize water quality. This information is also important in deducing scaling laws to extrapolate model tests to full-scale operation. These topics are the subject of an excellent review by Billet (1986). Optical methods such as photography or holography which image a sample volume in the liquid allow one to detect nuclei whose size range is larger than $10\,\mu$m. These methods can discriminate between dirt particles and bubbles but have the disadvantage that results are not immediately and continuously available. Other methods, like optical or acoustical radiation scattered by a nucleus, can operate and provide results continuously but cannot differentiate between bubbles and particles. In spite of the significant advances in the quantitative characterization of nuclei in different environments, there remains a question as to how to relate these to the actual cavitation inception conditions.

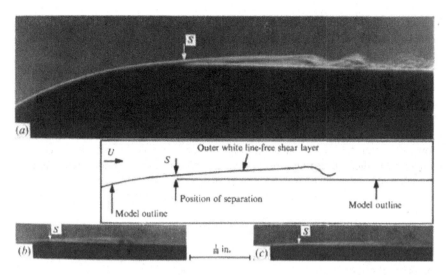

Fig. 5.12 Schlieren image of non-cavitating flow over a hemispherical nose showing point of separation. Flow is from left to right. (*Courtesy:* "Viscous effects on the position of cavitation separation from smooth bodies" by V. H. Arakeri, J. Fluid Mechanics, 1975, with permission from Cambridge University Press)

The above limitation is overcome by the use of a venturi as a cavitation susceptibility meter (CSM), an instrument first introduced by Oldenziel (1982). Cavitation susceptibility is defined as the smallest tensile stress in a liquid at which some bubbles explode. d' Agostino and Acosta (1991) discuss at length the various aspects associated with the design of a CSM. These CSMs can reliably measure the critical tension of the weakest nuclei and their concentration upto a certain value, after which the venturi throat gets saturated with nuclei and the flow chokes. There are additional problems associated with a CSM, like the possibility of the formation of sheet cavitation in the diffuser region.

The free and total air content of a liquid sample can be taken as a gross measure of nuclei content. Therefore, monitoring of the total air content is important (Peterson 1972). This can be done using a Van Slyke apparatus or a dissolved oxygen meter.

5.5.2 Viscous Effects

Viscous effects on cavitation inception were demonstrated by Arakeri and Acosta (1973) using Schlieren photography. A brief summary of this *real* fluid effect is presented here. For ease of description, cavitation inception studies using different types of bodies are considered.

Axisymmetric Bodies: Huang (1979) found that for bodies having natural transition without laminar separation, cavitation originated immediately downstream of the

location of the minimum pressure. Explosive growth was observed further downstream in the region of transition to turbulence where very large pressure fluctuations were measured. For bodies with laminar separation, Arakeri and Acosta (1973)
observed that inception occurs in the turbulent reattachment zone on hemispherical-
nosed bodies. Arakeri (1975) observed that cavitation originates in the separated flow
region, downstream of the minimum pressure point. Figure 5.12 shows the Schlieren
photograph of non-cavitating flow past a hemispherical nose profile where the point of
separation as well as some features of the reattachment point are visible. Figure 5.13
shows the extent of cavitation within the separated region for different cavitation
numbers. Another interesting aspect of developed cavitation on bodies with laminar
separation like the hemispherical geometry discussed here is the shifting of the point

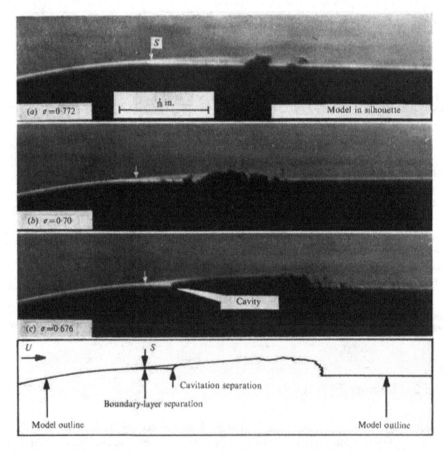

Fig. 5.13 Image of cavitating flow over a hemispherical nose under different cavitating conditions.
Flow is from left to right. (*Courtesy:* "Viscous effects on the position of cavitation separation from
smooth bodies" by V. H. Arakeri, J. Fluid Mechanics, 1975, with permission from Cambridge
University Press)

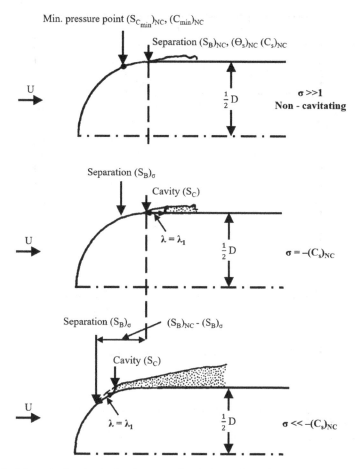

Fig. 5.14 Schematic showing the development of cavitation with a reduction in test section pressure at a constant water tunnel speed. Advancement of the point of separation in developed cavitation conditions is visible. Flow is from left to right. (*Courtesy:* "Viscous effects on the position of cavitation separation from smooth bodies" by V. H. Arakeri, J. Fluid Mechanics, 1975, with permission from Cambridge University Press)

of flow separation to an upstream location under developed cavitation in comparison to that observed in non-cavitating flow conditions. This is exemplified in Fig. 5.14. To establish the flow physics related to the laminar separation region, experiments were conducted on axisymmetric headforms with backward-facing steps by Arakeri (1979). He observed that cavitation inception on a headform took place in the shear layer as bubble ring cavitation (close to the surface) or travelling bubble cavitation (away from the surface).

Vortex Cavitation: Arndt (1981) showed that for nuclei trapped in the vortices inside the boundary layer, explosive growth occurred at mean static pressures higher than

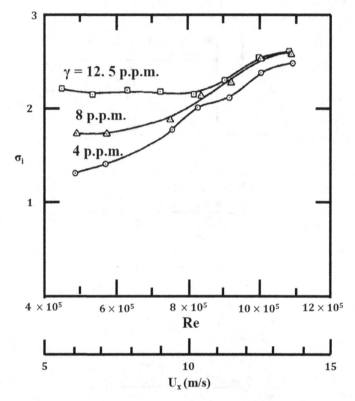

Fig. 5.15 Dependence of tip-vortex cavitation inception on Reynolds number and dissolved gas content. (*Courtesy:* "Some observations of tip-vortex cavitation" by R.E. A. Arndt et al., J. Fluid Mechanics, 1991, with permission from Cambridge University Press)

the critical pressure. The decade, 1980–1990, saw a lot of research where the finer aspects of the effect of vortices and turbulent shear on cavitation were investigated systematically. Katz (1981) focussed his investigation on the microscale cavitation in flows characterized by laminar separation with free shear layer transition and subsequent flow reattachment. He found that cavitation inception took place in the turbulent separated flow close to the reattachment region. Katz and O'Hern (1986) and O' Hern (1990) explained cavitation inception in the shear layer behind a sharped-edged plate.

Another type of vortex cavitation, tip-vortex cavitation, is a feature common to lifting surfaces. Hub-vortex and blade-shroud gap vortex cavitation are other examples. Arakeri et al. (1986) demonstrated the effect of Reynolds number on vortex core growth and cavitation inception. Arndt et al. (1991) brought out the effect of Reynolds number and dissolved gas concentration on the tip-vortex cavitation inception as shown in Fig. 5.15.

The above discussion indicates that the nuclei size distribution as well as their interaction with the different viscous flow characteristics are responsible for the onset of cavitation. Once cavitation begins, effects as indicated below are a natural consequence.

5.6 Effects of Hydrodynamic Cavitation

The effects of hydrodynamic cavitation have already been listed. In this section, we shall discuss in some detail three of these effects that trouble hydraulic engineers significantly. These are noise, damage and the reduction in the performance of turbomachines.

5.6.1 Cavitation-Induced Noise

In 1894 Reynolds had reported about the emission of sound from cavitating flows. In fact, right from the beginning of research in cavitation, noise emission has received as much attention as the drop in the performance of turbomachines. This is because during the World Wars, it was found that submarines get easily detected by the enemy because of noise emitted by bubbles. Stealth is a major concern for naval applications and research has been carried out to control cavitation or at least its noise. Not just in the realm of technology, but even in nature, there are many examples where the presence of bubble-induced noise can be traced. It is found that raindrops falling in a liquid generate that sound because of a contracting bubble entrained in the liquid (Prosperetti and Oguz 1993) or as a snapping shrimp catches its prey the cacophony produced is also due to imploding bubble!

In a real scenario of cavitation, myriads of bubbles may be present as shown in some of the photographs in Fig. 5.10. But in this section, we shall discuss about noise due to a single cavitating bubble. This is because this study gives us an idea of basic sound generation and also provides a scaling of sound measurements. We shall carry out the analysis based on Fig. 4.7.

Based on mass conservation we have shown, in Eq. 4.12, that $U_r = R^2 \dot{R}/r^2$. Integrating the momentum equation (Eq. 4.13) of the liquid from the bubble wall ($r = R$) to some distance inside the liquid away from the bubble centre ($r = r$), we get

$$\frac{p_L - p_r}{\rho} = -\frac{R^2 \ddot{R} + 2R\dot{R}^2}{r} + \frac{R^4 \dot{R}^2}{2r^4} + R\ddot{R} + \frac{3}{2}\dot{R}^2 \tag{5.5}$$

It may be noted that $p_R = p_L$ in Eq. (5.5) similar to that made in Eq. (4.16). Combining Eq. (4.16) with Eq. 5.5, we get

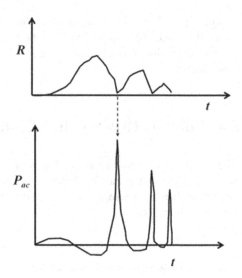

$$\frac{p_r - p_\infty}{\rho} = \frac{R^2 \ddot{R} + 2R\dot{R}^2}{r} - \frac{R^4 \dot{R}^2}{2r^4} \tag{5.6}$$

For large values of $r \gg R$, neglecting the second term, we get

$$p_{ac}(r, t) = p_r - p_\infty = \rho \frac{R^2 \ddot{R} + 2R\dot{R}^2}{r} = \frac{\rho}{4\pi r} \ddot{V}\left(t - \frac{r}{c}\right) \tag{5.7}$$

In Eq. (5.7), p_{ac} is the excess pressure term and is the sound pressure emitted by an oscillating bubble and $\ddot{V}\left(t - \frac{r}{c}\right)$ is the bubble volume acceleration as noted by an instrument at a delayed time in a location r away from the centre of the bubble and c is the speed of sound in the liquid. This volume oscillation signifies a monopole source of the sound.

Spectral energy received by a spherical shell at a radius r from the centre of the bubble is given by

$$E_{ac} = 4\pi r^2 \int_0^\infty \frac{p_{ac}^2(t)}{\rho c} = \frac{4\pi \rho}{c} \int_0^\infty \frac{(R^2 \ddot{R} + 2R\dot{R}^2)}{\dot{R}} dR \tag{5.8}$$

It can be shown that most of the noise is produced during the collapse of a bubble and not during its growth (Fig. 5.16). The features of cavitation noise that distinguish it from other flow noise are sudden onset and a very fast increase at cavitation condition just below inception but a slower increase with a further increase in cavitation extent. There is a shift in the peak frequency of the noise spectrum towards lower frequencies with an increase in the extent of cavitation. These features can be obtained using single bubble cavitation noise theories.

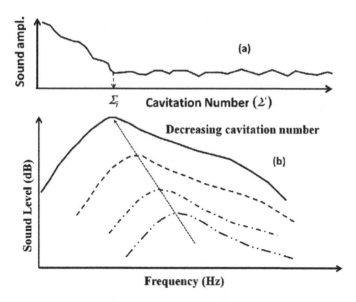

Fig. 5.17 a Variation of sound amplitude as cavitation number (Σ) is reduced. **b** Frequency spectrum of cavitation-induced sound at different cavitation numbers

Figure 5.17a shows the schematic of sound amplitude increase rapidly as inception (Σ_i) takes place and then keeps on increasing as the cavitation number is reduced further. If one measures the frequency at which peak amplitude is observed for different cavitation numbers, then it can be shown that the frequency decreases as shown in Fig. 5.17. This is because as the extent of cavitation increases, the sizes of the bubbles increase.

5.6.2 Cavitation-Induced Erosion and Surface Damage

We have discussed so far single bubble dynamics and have shown how experimentally observed features of cavitation-induced noise be explained with the help of rather simplistic single spherical bubble dynamics. However, the same spherical dynamics cannot be used to explain cavitation damage. This is because, unlike the assumption of a bubble in an infinite (or very large body of liquid in comparison to the bubble size), cavitation damage necessarily means the presence of a solid surface in close proximity to the bubble. The presence of the solid surface therefore breaks the spherical symmetry assumption made earlier, and as a result, the bubble dynamics changes. Figure 5.18 shows the change in the bubble size with time in the presence of a solid wall. It can clearly be seen that the formation of microjet approaches the solid wall.

Initial sphere

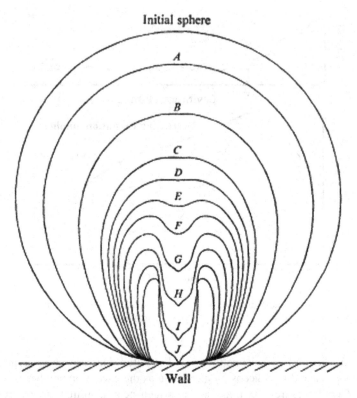

Fig. 5.18 Variation of bubble size and shape with time in the presence of the wall. (*Courtesy:* "Collapse of an initially spherical vapour cavity in the neighbourhood of a solid boundary" by M. S. Plesset and R. B. Chapman, J. Fluid Mechanics, 1971, with permission from Cambridge University Press)

In the initial days (1940–50) of research on cavitation damage, it was thought that high-speed bubble wall collapse produces shockwaves in the liquid and this high pressure is responsible for the damage produced. However, later on, primarily through the works of Benjamin and Ellis (1966), the formation of microjet was proved beyond doubt and this led to the shift in the focus to microjet-induced cavitation damage. Finally, it was left to Tomita and Shima to show detailed photographs of bubble collapse near a solid wall, and the initial bubble location with respect to the wall was varied systematically in their experiments. This is shown in Fig. 5.19 for different separating distances, given in terms of normalized distance, $\frac{L}{R_{max}}$, where R_{max} is the maximum bubble size attained. It can be seen that if the separating distance is small (Fig. 5.19), microjet forms and hits the surface. Once the liquid jet strikes the wall, it moves as a radial one. However, if the microjet does not reach the surface, then while the microjet comes out of the bubble, the bubble assumes a toroidal shape. On the other hand, when the distance is too less (bubble virtually resting on the surface), the microjet formation is absent and the bubble crushes like a cake. Thus, it is seen

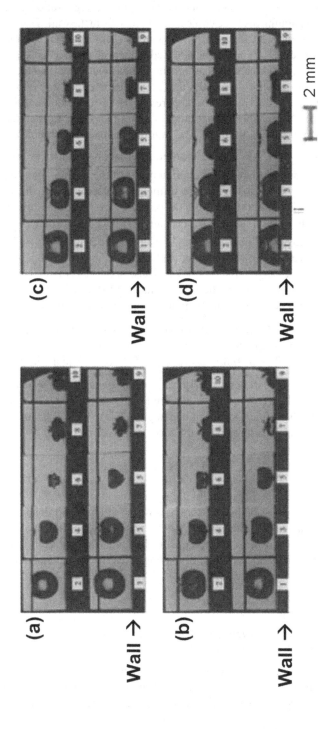

Fig. 5.19 Series of photographs showing the development of the microbubble collapsing at different distances away from a solid boundary. **a** $\frac{L}{R_{max}} = 1.23$, **b** $\frac{L}{R_{max}} = 1.10$, **c** $\frac{L}{R_{max}} = 0.94$, **d** $\frac{L}{R_{max}} = 0.77$. (*Courtesy:* "Mechanisms of impulsive pressure generation and damage pit formation by bubble collapse" by Y. Tomita and A. Shima, J. Fluid Mechanics, 1986, with permission from Cambridge University Press)

that the formation of the microjet and its possibility of striking a solid wall depends crucially on the separating distance of the initial bubble from the wall. This is brought out more clearly in Fig. 5.20. Figure 5.20a shows the variation of the non-dimensional bubble base diameter ($B^*_{Vmin} = \frac{B_{Vmin}}{R_{max}}$), attached to the wall, with a non-dimensional separating distance. It can be seen that the non-dimensional base diameter shows a higher value between $0.4 < \frac{L}{R_{max}} < 0.8$ and drops off on either side of this range. This is because the outward radial flow suppresses the inward motion of the bubble surface. Figure 5.20b describes the variation of non-dimensional impact time of the bubble for varying stand-off distances. This non-dimensional impact time is given as $\Delta \tau_j = \frac{\Delta T_j}{R_{max}} \left(\frac{p_\infty - p_v}{\rho} \right)^0 .5$. ΔT_j denotes the time difference between the time of impact of the liquid jet and the instant of the minimum size of the bubble. Thus, a positive value of this quantity refers to the microjet hitting the surface earlier to the minimum size formation and is of importance from the point of view of surface erosion due to microjet formation. It can be seen that this quantity is negative beyond $\frac{L}{R_{max}}$ value of 1.2 and, further, the peak is observed for $\frac{L}{R_{max}}$ value in the range of 0.6–0.7. Finally, in Fig. 5.20c, the velocity of the microjet is plotted, and as expected, it also shows a higher value for the range $0.6 < \frac{L}{R_{max}} < 1.2$. Thus, in the case of cavitating liquid, it is clear that microjet cannot be the sole mechanism by which cavitation damage takes place. This has been categorically pointed out by Shima (1997) as

(a) $L/R_{max} < 0.3$ and $L/R_{max} \geq 1.5$, shock-wave is the dominant mechanism;
(b) $0.6 \leq L/R_{max} \leq 0.8$, microjet is major concern;
(c) $0.3 \leq L/R_{max} \leq 0.6$ and $0.8 \leq L/R_{max} \leq 1.5$, both are equally important.

Now, why is microjet so dangerous from material damage perspective? This is because microjet velocity is very high and the liquid can no longer be assumed to be incompressible. This microjet phenomenon gives rise to a water hammer and the pressure loading in the small region of the surface is given by

$$P_{WH} = \rho C_j c$$

where C_j is the microjet velocity. Taking some typical numbers may help us to understand the seriousness of this pressure rise. Work of Plesset and Chapman (1971) or Tomita and Shima (1986) suggest that C_j can be quite high ~ 100 m/s, which puts $P_{WH} \sim 100$ MPa and this acts for a very short duration. The surface is thus loaded and unloaded repeatedly and this type of loading gives rise to material deformation and pitting of the surface (Fig. 5.21).

5.6.3 Cavitation in Turbomachines

Cavitation leads to three effects in turbomachines: performance deterioration, material damage and increased noise and vibration of components. In this section, we shall focus on the performance deterioration. One parameter important for understanding

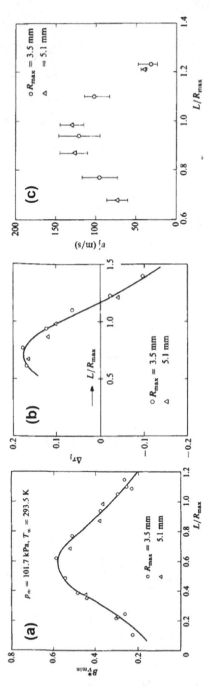

Fig. 5.20 Effect of separating distance on **a** non-dimensional bubble base diameter, **b** impact time of liquid jet on the solid boundary and **c** microjet velocity. (*Courtesy:* "Mechanisms of impulsive pressure generation and damage pit formation by bubble collapse" by Y. Tomita and A. Shima, J. Fluid Mechanics, 1986, with permission from Cambridge University Press)

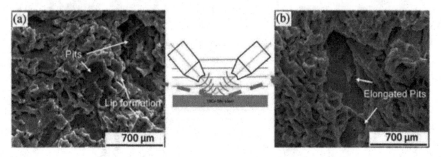

Fig. 5.21 Cavitation-induced damage. (*Courtesy:* C. Syamsundar, PhD Thesis, 2016, IIT Madras)

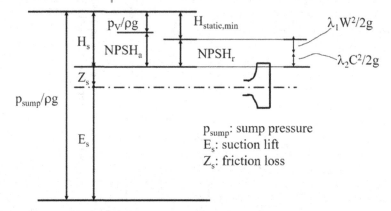

Fig. 5.22 Definition of Net Positive Suction Head (NPSH) used in centrifugal pumps

cavitation in turbomachines is Net Positive Suction Head (NPSH). We shall begin with a discussion of NPSH. In order to appreciate this term better, we need to keep in mind two things: (a) cavitation, for most engineering liquids, can be assumed to start when local pressure reaches vapour pressure, and (b) user of a pump has no information about the flow inside the pump and the designer or manufacturer of a pump does not know the flow inside the pipe connecting the pump with the sump. Because of this difference in information available, we need to define two kinds of NPSH: NPSHa and NPSHr. NPSHa stands for NPSH available and this is estimated by a user of the pump. NPSHr means NPSH required and is provided by the designer. In order to ensure that no cavitation occurs, the user should guarantee that in the pumping system, NPSHa $\geqslant$ NPSHr. We shall now establish this.

Let us consider a pump connected to a sump and located at an elevation of E_s above the sump level (Fig. 5.22). Then the suction pressure ($p_{suction}$) in the pipe just outside the inlet flange can be expressed as

$$\frac{p_{suction}}{\rho g} = \frac{p_{sump}}{\rho g} - \frac{V_{suction}^2}{2g} - E_s - Z_s \qquad (5.9)$$

where Z_s is the pipe friction losses between the sump and pump inlet and $V_{suction}$ is the velocity of the liquid at the location of measurement of $p_{suction}$. Now the minimum pressure occurs inside the pump at the suction (inlet) side of the pump impeller and we can assume that this minimum pressure (p_{min}) is related to suction pressure by

$$\frac{p_{min}}{\rho g} = \frac{p_{suction}}{\rho g} - \frac{\Delta p}{\rho g} \tag{5.10}$$

where Δp is the pressure drop taking place between the suction flange and inlet edge of the vanes and is a function of W_1 and C_1, where W_1 is the relative velocity of the flow at the pump impeller inlet and C_1 is the absolute velocity at the impeller inlet. It may be assumed that $V_{suction} \approx C_1$. This pressure loss term Δp is known to the manufacturers and quite often this is given as

$$\frac{\Delta p}{\rho g} = \lambda_1 \frac{W_1^2}{2g} + \lambda_2 \frac{C_1^2}{2g} \tag{5.11}$$

where $\lambda_{1,2}$ are constants that depend on the design of the pump. Thus, combining Eqs. (5.9) to (5.11), we get

$$\frac{p_{min}}{\rho g} = \frac{p_{sump}}{\rho g} - \frac{V_{suction}^2}{2g} - E_s - Z_s - \left(\lambda_1 \frac{W_1^2}{2g} + \lambda_2 \frac{C_1^2}{2g} \right) \tag{5.12}$$

Some of these terms in Eq. 5.12 are known only to the user and some others are known only to the designer of pumps. Rearranging Eq. 5.12 by taking terms known to respective groups, and together with the knowledge that $p_{min} \geq p_{vap}$ in order to avoid cavitation, we get

$$\frac{p_{sump}}{\rho g} - E_s - Z_s - \frac{p_v}{\rho g} \geq \lambda_1 \frac{W_1^2}{2g} + \lambda_2 \frac{C_1^2}{2g} \tag{5.13}$$

The terms which are known to the users of the pumps, that is the terms in the left-hand side, are termed as NPSHa ($\frac{p_{sump}}{\rho g} - E_s - Z_s - \frac{p_v}{\rho g}$) while those in the right side are called NPSHr ($\lambda_1 \frac{W_1^2}{2g} + \lambda_2 \frac{C_1^2}{2g}$) and we could show that NPSHa $\geq$ NPSHr and the condition of equality indicates that the critical condition for the onset of the effects of cavitation. Figure 5.22 shows schematically the relationships derived so far. It may be noted that this critical condition is not the onset of cavitation but the condition under which the performance gets affected. This aspect is discussed later.

Examination of the expressions for NPSHa and NPSHr indicate that, with an increase in flow rate, NPSHa decreases as flow losses increase (when other conditions are held constant) while NPSHr increases as both velocity terms are dependent directly on the flow rate. This is depicted in Fig. 5.23 which also shows the regions of cavitating and non-cavitating operating conditions.

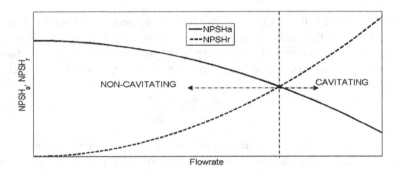

Fig. 5.23 Variation of NPSH; available and required; used in centrifugal pumps

In order to test cavitation in pumps in a laboratory, we need to reduce NPSHa below that of NPSHr. Looking at the terms given in NPSHa indicates that NPSHa can be reduced by reducing sump pressure or by increasing static lift (E_s), frictional loss (Z_s) or by increasing temperature of the liquid and thereby increasing the vapour pressure. However, maintaining elevated fluid temperature or reduced sump pressure is not easy to achieve in the laboratory. Hence, more common methods of bringing in cavitation in pumps in laboratories are to change E_s or Z_s. Of the two, we feel that changing losses may not be well-suited for cavitation research. The usual method of increasing fluid losses in a pipeline (without changing the pipe itself) is carried out by closing a valve further. Now closing a suction valve upstream of the pump may lead to cavitation in the valve, and bubbles thus produced will change the quality of the liquid.

So, in this book, we recommend the method of increasing the suction height and the results that could be obtained from those experiments. Figure 5.24 shows a schematic of a test facility where cavitation in the pump could be tested in a laboratory. In this experiment, first a non-cavitating condition is established and pressure at the suction and delivery sides of a centrifugal pump as well as flow rate are measured. Then gradually the level of water in the tank is decreased by shifting water from the main tank to a secondary tank with the help of an auxiliary pump while maintaining flow rate and rotational speed constant. At a given flow rate, for a fixed NPSHa, the head is estimated from the measurement of pressure, velocity, etc., according to the relationship

$$H = \frac{p_{down} - p_{up}}{\rho g} + \frac{V_{down}^2 - V_{up}^2}{2g} + \left(z_{down} - z_{up}\right) \qquad (5.14)$$

where subscripts 'up' and 'down' refer to the pressure measurement locations in upstream and downstream pipes as shown in Fig. 5.24.

A typical non-cavitating pump characteristics curve obtained at a given rotational speed of a radial-flow pump is shown as curve AB in Fig. 5.25. If the NPSHa is reduced such that it falls below NPSHr and cavitation sets in, then for a given NPSHa = NPSHcav < NPSHr, curve AC is obtained and it shows that beyond a particular

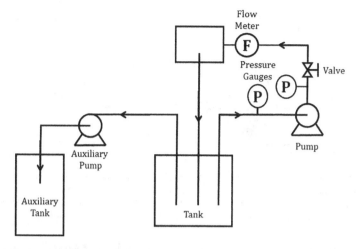

Fig. 5.24 Laboratory experiment for characterizing cavitation in centrifugal pumps: reduction in NPSHa by a change in the tank (sump) water level

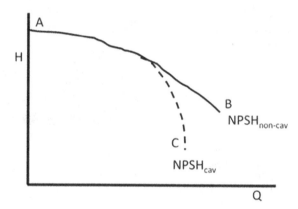

Fig. 5.25 Effect of cavitation on the performance of centrifugal pump

flow rate, the head drops very sharply and one can find out percentage head drop at a given flow rate. A more convenient way of conducting experiments on cavitation in pumps is to keep flow rate constant as NPSHa is gradually reduced and a typical curve is shown in Fig. 5.26. Here also, as in Fig. 5.25, this head difference is noticeable.

This significant head drop takes place not at the onset of cavitation when the first traces of bubbles are formed inside the impeller. This happens when the cavity (comprising of several bubbles) occupies a large portion of the impeller flow passage (Fig. 5.26). A typical picture of the cavitating region inside a radial-flow pump impeller is shown in Fig. 5.27.

It should be kept in mind that critical cavitation condition in the pump depends on the parameter we use for investigation. This is brought out in Fig. 5.28 which shows that noise detection is the most sensitive method of detecting cavitation. The pump manufacturers and users are primarily interested in the performance deterioration in

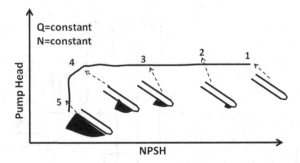

Fig. 5.26 Effect of NPSH on the performance of the centrifugal pump. Points 1 to 5 denote five conditions of operation. (1) non-cavitation, (2) cavitation inception, (3) developed cavitation, (4) three percent head drop (corresponding to critical NPSH) and (5) breakdown condition. Also shown are schematics of cavity lengths over impeller blades for each of these five conditions of operation

Fig. 5.27 Photograph of cavitation inside a radial-flow pump impeller. (*Courtesy:* Experimental facility of Prof. S. Kumaraswamy and Dr. S. Christopher of IIT Madras)

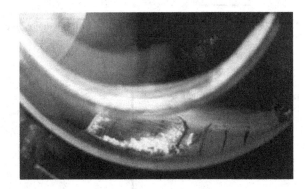

Fig. 5.28 Variation of different cavitation indicators as NPSH is reduced. **A**: NPSH value at which noise level increases over the background noise level, **B**: NPSH value at which cavity makes its visual appearance and **C**: NPSH value at which 3 percent drop in head noted

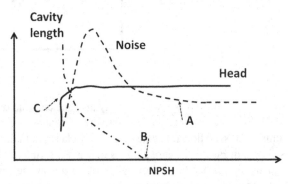

pumps due to cavitation and hence they consider a 3 percent head drop (equivalently, 4 mm length of the cavity inside the impeller) as the critical cavitation condition (Fig. 5.28).

In the case of hydraulic turbines, derivation of critical cavitation condition (Σ_c) can be carried out in the same way as done in the case of a pump. It should be borne in mind that, unlike a pump, cavitation starts inside the turbine rotor (impeller) at the

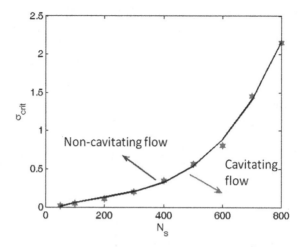

Fig. 5.29 Variation of critical cavitation for the hydraulic turbine as a function of specific speed (N_s in rpm)

exit edge. Critical cavitation number, given in terms of atmospheric pressure at the site of the hydropower plant (p_{atm}), vapour pressure (p_V) and draught tube height above the tailrace level (E_S), is given in Eq. (5.15). This term is termed as critical Thoma's cavitation factor. This critical Thoma's cavitation factor is dependent on turbine geometry, and a typical variation is shown in Fig. 5.29.

$$\Sigma_c = \frac{\frac{p_{atm}}{\rho g} - \frac{p_V}{\rho g} - E_S}{H_{net}} \tag{5.15}$$

Examples

Problem 1: A centrifugal pump rotating at 1000 rpm delivers 160 litres/s of water against a head of 30 m. The atmospheric pressure is 0.1 MPa (absolute) and vapour of water is 3 kPa (absolute). The length (L_s) and diameter (D_s) of the suction pipe are 10 m and 300 mm, respectively. The suction pipe friction factor (f) is 0.02.

The critical cavitation factor $NPSH_c/H_{noncav}$ and the pump specific speed ($N_s = N\sqrt{(\dot{V})}/H^{3/4}$) are related by

$$\frac{NPSH_c}{H_{noncav}} = 1.03 \times 10^{-3}(N_s)^{4/3}$$

If the height of the impeller from the free surface in the sump is 7.5 m, determine whether cavitation will occur in the pump or not.

Solution: Based on the given data, $N_s = 31.2$ m. Hence, $NPSH_c/H_{noncav} = 0.1$. Hence, $NPSH_c = 3.04$ m. For this critical condition, $NPSHa = NPSH_c = 3.04$. Hence,

$$\frac{p_{sump}}{\rho g} - E_s - Z_s - \frac{p_v}{\rho g} = 3.04$$

where

$$Z_s = f\frac{L_s}{D_s}\frac{V_s^2}{2g}$$

The average velocity of flow in a suction pipe (V_s) can be obtained from

$$V_s = \frac{4\dot{V}}{\pi D_s^2}$$

Thus, we get for the critical condition, $E_s = 6.67$ m. Any height of the pump above this level will result in the lowering of the NPSHa below NPSHc. Hence, for a height of 7.5 m, the pump will suffer from cavitation.

Exercises

1. When a laboratory test was carried out on a pump, it was found that, for a pump total head of 36 m at a discharge of 0.05 m³/s, cavitation began when the sum of the static pressure plus the velocity head at the inlet was reduced to 3.5 m. The atmospheric pressure was 760 mm of mercury and the vapour pressure of water was 2 kPa. If the pump is to operate at a location where atmospheric pressure is reduced to 600 mm of mercury and the vapour pressure of water is 850 Pa, and develop the same total head and discharge, find out the maximum height of the suction line. Assume the head losses in the suction line to be the same in both the cases.

2. Find a suitable speed (N) at which a centrifugal pump with the following specifications should be run so that cavitation is avoided. Flowrate, $\dot{V}$, is 0.1 m³/s, specific work $W(= gH) = 200$ m²/s² (or, pump head developed, $H = 20.4$ m), suction line height from the supply surface to impeller is 4.5 m, Atmospheric pressure is 10 m water, vapour pressure is 2.0 m water and Suction line head loss is 0.5 m water. $NPSH_c/H$ and the specific speed of the pump

$$\left(N_s = \frac{N\sqrt{(\dot{V})}}{H^{3/4}}\right)$$

are related through

$$\frac{NPSH_c}{H_{noncav}} = 12.2 \times 10^{-4}(N_s)^{4/3}$$

3. Calculate the maximum height of a straight draught tube for a Francis Turbine running at 125 rpm and producing 12 MW under a head of 24.5 m. Atmospheric pressure is 10 m of water. Take vapour pressure as 0.20 m water. Thoma cavitation factor and the specific speed for turbines

$$\left(N_s = \frac{N\sqrt{(\dot{P_c})}}{H^{5/4}} \right)$$

are related through $\Sigma_c = 0.319 \times 10^{-5}(N_s)^2$.

References

Apfel RE (1971) A novel technique for measuring the strengths of liquids. J Acoust Soc Amer 49:145–155

Apfel RE (1981) Acoustic cavitation prediction. J Acoust Soc Amer 69:1624–1633

Arakeri VH (1979) Inception of cavitation from a backward facing step. In: Morgan WB, Parkin BR (eds) International symposium on cavitation

Arakeri VH, Higuchi H, Arndt REA (1986) Analysis of recent tip vortex cavitation inception data. In: Proceeding of the 21st American two tank conference, Washington DC

Arakeri VH (1975) Viscous effects on the position of cavitation separation from smooth bodies. J Fluid Mech 68(4):779–799

Arakeri VH, Acosta AJ (1973) Viscous effects in the inception of cavitation on axisymmetric bodies. Trans ASME J Fluids Engng 95:519–527

Arndt REA (1981) Recent advances in cavitation research. Adv Hydrosc 12:1–77

Arndt REA, Arakeri VH, Higuchi H (1991) Some observations of tip-vortex cavitation. J Fluid Mech 229:269–289

Barger JE (1964) Thresholds for acoustic cavitation. Technical report, 59, Acoustics research Laboratory, Harvard University

Benjamin TB, Ellis AT (1966) The collapse of cavitation bubbles and the pressures thereby produced against solid boundaries. Phil Trans R Soc Lond A 260:221–240

Billet M (1986) Importance and measurement of cavitation nuclei. In: Arndt REA, Stefan HG et al (eds) Advances in aerodynamics, fluid mechanics and hydraulics

Blake FG (1949a) The tensile strength of liquids — a review of the literature. Technical report 9, Acoustics Research Laboratory, Harvard University

Blake FG (1949b) The onset of cavitation in liquids. Technical report, 12, Acoustics Research Laboratory, Harvard University

Brennen CE (1995) Cavitation and bubble dynamics. Oxford University Press

Briggs LJ (1950) Limiting negative pressure of water. J Appl Phys 21:921

Chatterjee D (2003) Some investigations on the use of ultrasonics in hydrodynamic cavitation control. PhD Thesis, IISc Bangalore

Crum LA (1980) Measurements of the growth of air bubbles by rectified diffusion. J Acoust Soc Amer 68:203–211

Crum LA (1984) Rectified diffusion. Ultrasonics 22:215–223

d' Agostino L, Acosta AJ (1991) A cavitation susceptibility meter with optical cavitation monitoring – part one: design concepts. Trans ASME J Fluids Engng 113:261–269

Eller A Flynn HG (1965) Rectified diffusion during nonlinear pulsations of cavitation bubbles. J Acoust Soc Amer 37(3):493–503

Epstein PS, Plesset MS (1950) On the stability of gas bubbles in liquid-gas solutions. J Chem Phys 18:1505–1509

Flynn HG (1964) Physics of acoustic cavitation in liquids. Phy Acoust (Ed by Mason WF) 1B:58–172

Flynn HG, Church CC (1984) A mechanism for the generation of cavitation maxima by pulsed ultrasound. J Acoust Soc Amer 76:505–512

Flynn HG, Church CC (1988) Transient pulsations of small gas bubbles in water. J Acoust Soc Amer 84:985–998

Fyrillas MM, Szeri A (1994) Dissolution or growth of soluble spherical oscillating bubbles. J Fluid Mech 277:381–407

Gaitan DF (1990) On pressure developed in a liquid during collapse of spherical cavity. PhD thesis, University of Mississippi

Harrison M (1952) An experimental study of single bubble cavitation noise. J Acoust Soc Amer 24(6):776–82

Harvey EN, Barnes DK, McElroy WD, Whiteley AH, Pease DC, Cooper KW (1944) Bubble formation in animals. I. Physical factors. J Cell Comp Physiol 24(1):1–22

Holl JW (1970) Nuclei and cavitation. Trans ASME J Basic Engng 92:681–688

Hsieh DY, Plesset MS (1961) Theory of rectified diffusion of mass into gas bubbles. J Acous Soc Amer 33:206–215

Huang TT (1979) Cavitation inception observations on axisymmetric headforms. In: Morgan WB, Parkin BR (eds) International symposium on cavitation

Katz J (1981) Cavitation inception in separated flows. PhD Thesis, California Institute of Technology

Katz J, O'Hern TJ (1986) Cavitation in large scale shear flows. Trans ASME J Fluids Engng 108:373–376

Knapp RT, Daily JW, Hammitt FG (1970) Cavitation. McGraw Hill, New York

Leighton TJ (1994) The acoustic bubble. Academic, Cambridge

Nyborg WL, Hughes DE (1967) Bubble annihilation in cavitation streamers. J Acoust Soc Amer 42:891–894

O' Hern TJ (1990) An experimental investigation of turbulent shear flow cavitation. J Fluid Mech 215:365–391

Oldenziel DM (1982) A new instrument in cavitation research: the cavitation susceptibility meters. Trans ASME J Fluids Engng 104:136–142

Parlitz U, Mettin R et al (1999) Spatio-temporal dynamics of acoustic cavitation bubble clouds. Phil Trans R Soc Lond A 357:313–334

Peterson FB (1972) Hydrodynamic cavitation and some considerations of the influence of free gas content. In: Ninth symposium of naval hydrodynamics

Plesset MS, Chapman RB (1971) Collapse of an initially spherical vapour cavity in the neighbourhood of a solid boundary. J Fluid Mech 47(2):283–290

Plesset MS, Prosperetti A (1977) Bubble dynamics and cavitation. Ann Rev Fluid Mech 9:145–185

Prosperetti A, Oguz HN (1993) The impact of drops on liquid surfaces and the underwater noise of rain. Ann Rev Fluid Mech 25:577–602

Rayleigh L (1917) On pressure developed in a liquid during collapse of spherical cavity. Phil Mag 34:94–98

Rood EP (1991) Review – mechanisms of cavitation inception. Trans ASME J Fluids Engg 113:163–175

Shima A (1997) Studies on bubble dynamics. Shock Waves 7:33–42

Suslick KS, Didenko Y et al (1999) Acoustic cavitation and its chemical consequences. Phil Trans R Soc A 357:335–353

Thornycroft J, Barnaby SW (1895) Torpedo boat destroyers. Inst Civil Enggs 122:51

Tomita Y, Shima A (1986) Mechanisms of impulsive pressure generation and damage pit formation by bubble collapse. J Fluid Mech 169:535–564

Tomlinson C (1867) On the so-called 'inactive' conditions of solids. Phil Mag 34:136–143 & 229–230

Chapter 6
Types of Boiling—The Pool Boiling Curve

Boiling is the process of changing the liquid to vapour at a constant temperature known as the 'saturation temperature' or 'boiling point' as discussed in Chap. 3. There are many ways in which boiling can be classified. It can be classified based on the mechanism of phase change or the geometry of fluid container. It can also be classified as simply external or internal boiling in a way similar to that done for single-phase forced convection heat transfer. Before discussing various classifications, let us also focus our attention to another phenomenon known as evaporation or 'silent boiling'.

6.1 Evaporation

Nature has got its own mechanism of turning liquid to vapour. Water from water bodies like lakes, oceans or rivers continuously turns into vapour with and without the aid of sun rays. This phase change does not require the liquid to reach saturation temperature. At almost all ambient conditions, liquid turn into vapour making the air in contact more humid. This mechanism is also used in industrial processes like that in cooling towers, comfort air conditioning and HDH (humidification-dehumidification) based desalination systems. In these cases, the vapourization takes place once the ambient temperature is higher than the saturation temperature corresponding to the partial pressure of water vapour in air. The amount of water vapour in air being small, its partial pressure is also small and hence evaporation takes place almost at all ambient temperatures except when it is too cold or when the air is already saturated with water vapour.

In practical applications, evaporation takes place under another condition. When the liquid is heated but the heat flux is such that the wall is just above saturation temperature at the given pressure, then under this condition, the wall temperature is not sufficient for vapour bubbles to form. However the temperature of a pool heated

S. K. Das and D. Chatterjee, *Vapor Liquid Two Phase Flow and Phase Change*, https://doi.org/10.1007/978-3-031-20924-6_6

Fig. 6.1 Evaporation from a
pool heated from bottom

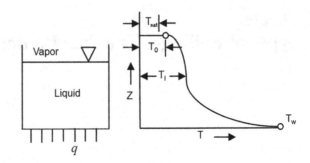

form the bottom takes a typical temperature profile as shown in Fig. 6.1. Here T_w
is the temperature of the wall and T_{sat} is the saturation temperature at the given
pressure. The temperature of the interface of air and water is T_0 which is very close
to T_{sat}. For water–air, this temperature difference ($T_0 - T_{sat}$) is very small ≈ 0.03 K.
The boundary layer where the temperature from the wall (T_w) drastically drops to
the liquid bulk temperature T_l typically has got a thickness of 1 mm. This layer is
formed primarily by liquid conduction. Due to natural convection bulk of the liquid
gets mixed and attains a nearly uniform temperature of T_l. Since the liquid is free
to move at the gas interface, boundary layer refreshes due to convection. In case of
evaporation, heat transfer takes place only at the liquid gas interface without much
agitation of the liquid and hence sometimes it is termed as 'quiet boiilng'. Since the
liquid bulk temperature T_1 changes, it is difficult to predict and hence usually the
heat transfer coefficient is defined on the basis of wall superheat $T_w - T_{sat}$, is

$$q = h(T_w - T_{sat}) \tag{6.1}$$

The heat transfer coefficient here usually follows that of natural convection giving

$$h \propto \Delta T_{sat}^{1/4} \quad \text{for laminar flow}$$
$$\propto \Delta T_{sat}^{1/3} \quad \text{for turbulent flow} \tag{6.2}$$

where $\Delta T_{sat} = T_w - T_{sat}$. Combining this with Eq. (6.1), we get

$$h \propto q^{1/5} \quad \text{for laminar flow}$$
$$\propto q^{1/4} \quad \text{for turbulent flow} \tag{6.3}$$

6.2 Classification of Boiling

Boiling can be classified in a number of ways. Most popular way of classifying
boiling is based on the geometry in which boiling takes place. This geometry-based
classification is shown in Fig. 6.2.

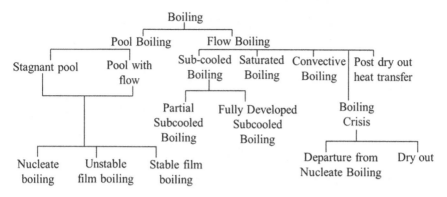

Fig. 6.2 Classification of boiling based on boiling geometry

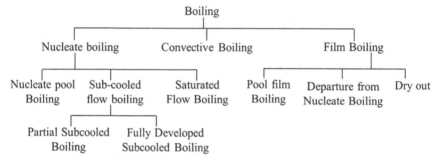

Fig. 6.3 Classification of boiling based on mechanism of boiling

We shall take up a discussion of the two major modes of boiling—pool boiling and flow boiling. However, before that let us also identify the other ways of classifying boiling. Boiling can also be classified on the basis of the mechanism of boiling as shown in Fig. 6.3.

Here nucleate boiling is a mechanism where vapour bubbles are formed on a heated surface in contact with the liquid. This can happen both with stagnant or flowing liquids on the surface. Convective boiling is the case in which the liquid forms a thin film (as inside a tube) on the solid wall and the heat is conducted through this film and finally evaporates the liquid at the vapour interface without any bubble formation. Film boiling is the case in which due to high flux a vapour film blankets the heater surface completely and as a result the heat transfer becomes very poor. This raises the temperature of the heating surface and may finally cause the melting of the solid surface itself which is known as 'physical burn out'. Since the mechanisms are not always very unique and sometimes several mechanisms coexist (such as during convective boiling, small amount of nucleation may also take place), the geometry-based classification is more common in literature.

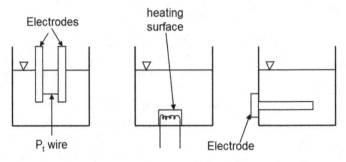

Fig. 6.4 Various configurations for studying pool boiling

The term 'pool boiling' originated from the physical situation when the liquid is stagnant in a pool and heater is inserted to the pool to heat the liquid (even the bottom surface of the pool can supply the heat, as in case of boiling water in a pot on a kitchen burner). However, the most important point is the fact that the heating surface is submerged in the liquid. Although we talk of a stagnant liquid here, due to dynamics of the bubble and the liquid circulating due to bubble motion as well as natural convection the fluid in pool boiling is not really stagnant. Moreover even if the liquid is not stagnant but flowing over the heating surface, the behaviour is, at least qualitatively, similar to pool boiling. There can be various configurations in which pool boiling can take place. Some very popular configurations are shown in Fig. 6.4. Figure 6.4a shows the configuration of a platinum wire submerged in the liquid. Electric current is passed through the wire. The wire acts both as a heater and a thermometer because the temperature coefficient of platinum is known very accurately and this eliminates the need for any separate temperature sensor. A large number of classical studies have been carried out on such configurations. Figure 6.4b shows the configuration in which a horizontal plate heater is inserted from the bottom and submerged in the pool. As we will see later in this chapter, the heating surface orientation (i.e., horizontal, vertical or inclined) makes significant difference in the results of pool boiling. The orientation is important because natural convection and buoyancy plays a major role in pool boiling. For this reason often pool boiling is also referred to as 'free convection boiling' (Stephan 1992). In this configuration effects such as surface roughness of the heater can be probed conveniently. In Fig. 6.4c boiling with cylindrical heater inserted from the side wall of the vessel is shown. For a horizontal tubular heater the surface orientation changes from bottom to top of the circular cross section and that can have important influence which is observed here. We will discuss about this in the next chapter. Most important part of the pool boiling is in the nucleate boiling regime and at high heat flux it changes to film boiling. In the nucleate boiling regime the heat transfer coefficient varies as

$$h \propto \Delta T_{sat}{}^3 \qquad\qquad (6.4)$$

Fig. 6.5 Temperature profile
in nucleate pool boiling

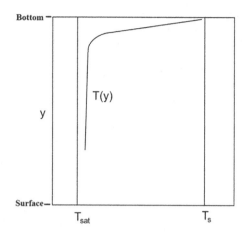

Please note that while in evaporation, heat transfer coefficient was proportional to at best $\Delta T_{sat}{}^{1/3}$, in case of nucleate boiling it is proportional to $\Delta T_{sat}{}^{3}$. This sudden rise in heat transfer coefficient is the real attraction of boiling as a heat transfer mechanism. It gives the magnitude of heat flux removal capability which a single-phase fluid fails to give.

It must be mentioned here that some authors look at pool boiling as 'external boiling' and flow boiling as 'internal boiling' drawing an analogy with external and internal convection of single-phase liquid. However, although quite logical, this terminology has not gained popularity. The temperature profile in nucleate pool boiling is similar to what is shown in Fig. 6.5 where the depth is measured from the surface of the pool.

In industrial equipments such as boiler, refrigerator, feed water heater, *etc.* boiling takes place under conditions of forced flow of fluid inside pipes or ducts. The vapourization of liquid can be partial or complete depending on application. However, the entire vapour generation in flow boiling process goes through change not only with respect to vapour or liquid content but also in the pattern in which liquid–vapour coexists. Flow boiling, thus, has several regimes such as forced convection, subcooled boiling, saturated boiling, convective boiling, dry out and spray cooling. Physically, we see various fluid structures such as bubbly flow, slug flow, churn flow, annular flow, mist flow etc. Various flow patterns and boiling regimes are shown in Fig. 6.6. We have already discussed these patterns in more details in Chap. 1. While the patterns such as bubbly or plug flow follows the nucleate boiling pattern of pool boiling closely, the convective boiling part which occurs during annular flow is of special attention because it is typical to flow boiling. For this reason, quite often, by flow boiling we essentially imply convective boiling.

Now in this figure (Fig. 6.6), we see two major changes in flow pattern. First we see transition from single-phase liquid heat transfer or with little evaporation to nucleate boiling and then from nucleate boiling to convective (annular) boiling. Comparison of these two transitions can clearly indicate the difference between

Fig. 6.6 Various flow
patterns in flow boiling.
(*Courtesy:* "Heat Transfer in
Condensation and Boiling"
by Karl Stephan, with
permission from Springer)

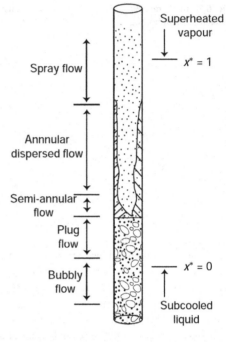

Fig. 6.7 Heat transfer
coefficient during nucleate
and convective boiling
(qualitative). (*Courtesy:*
"Heat Transfer in
Condensation and Boiling"
by Karl Stephan, with
permission from Springer)

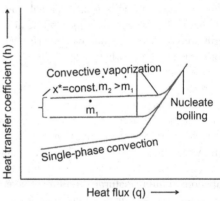

nucleate and convective boiling. Figure 6.7 shows the first transition. If we assume that the subcooled liquid (liquid below saturation temperature) enters the tube in turbulent conditions their heat transfer coefficient will depend on the liquid mass flux.

Although it will also depend mildly on heat flux at higher mass flux, we will get a higher heat transfer coefficient. However as soon as nucleate boiling starts in the tube, all the curves merge and there is a sudden rise of heat transfer coefficient.

Fig. 6.8 Boiling heat transfer in a horizontal tube. (*Courtesy:* "Heat Transfer in Condensation and Boiling" by Karl Stephan, with permission from Springer)

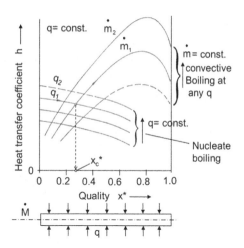

This shows that nucleate boiling has higher rate of heat transfer and is independent of mass flux. On the other hand, when we look at transition from nucleate to convective boiling we clearly see (Fig. 6.8) that nucleate boiling depends on heat flux (q_1, q_2, etc.) but the moment the constant heat flux line (say q_1) reaches the convective boiling point (with the increase of vapour quality x^*) they follow the constant mass flux.

Thus we can say that the nucleate boiling is predominantly heat flux dependent while convective boiling is mass flux dependent. In both the cases, the heat transfer coefficient may be expressed as

$$h = cq^n \dot{m}^p f(x^*) \tag{6.5}$$

where n is a function of fluid property. For nucleate boiling $0.1 < p < 0.3$ and $n \approx 0.75$, while for convective boiling $0.6 < p < 0.8$ and $n \approx 0$.

6.3 Pool Boiling Curve

Pool boiling heat transfer can be understood by plotting the submerged surface heat flux against the temperature difference between the wall of the heated surface and the liquid bulk temperature. This temperature difference, known as the 'wall superheat', is given by

$$\Delta T = T_w - T_l \tag{6.6}$$

Even though the wall temperature is taken as constant (at least for the case of constant wall temperature), recent studies (Kenning 1999) show that considerable variation of the temperature remains on the heating surface. On the other hand, under normal

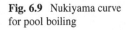

Fig. 6.9 Nukiyama curve
for pool boiling

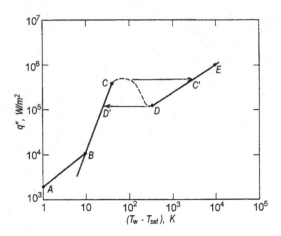

condition, the bulk liquid temperature remains the saturation temperature under the
given pressure,

$$T_l = T_{sat} \quad \text{giving,} \quad \Delta T_w = T_W - T_{sat} \tag{6.7}$$

However, boiling can also take place when the bulk liquid temperature is lower than
the saturation temperature but the liquid near the heater is sufficiently superheated.
This is called 'subcooled boiling'.

In 1934 Nukiyama (1934) carried out an experiment with a platinum wire apparatus similar to what was shown in Fig. 6.4 (leftmost arrangement). He presented the
heat flux (q) versus wall superheat (ΔT_w) curve as shown in Fig. 6.9 which is famous
as 'Nukiyama Curve'. This was probably the beginning of quantitative research in
boiling. It is important to note that in the curve both heat flux (q) and wall superheat
(ΔT_w) are plotted in logarithmic scale to cover wide range of heat flux as well as
surface temperature. This curve can be divided into various parts such as

From A to B: This is the part where heat flux is very low and hence boiling does
not take place. Natural convection is the mode of heat transfer here.

At point B: This is the point at which the boiling starts because the liquid near the
wall is enough superheated to nucleate (we will discuss this condition in the next
chapter). This is called 'the inception of nucleate boiling'.

From B to C: With the increase of wall superheat, number of bubbles nucleating
from the wall increase and heat flux also increases. This part is called the 'nucleate
boiling' regime. This is the most desired form of boiling because we get large amount
of heat removal and vapour formation here. Because of its utility largest number of
studies have been carried out in this region only. One can observe the sudden change
in the slope of the curve at the point B. We define boiling heat transfer coefficient on
the basis of wall superheat, and hence

$$q = h_{boiling}(T_w - T_{sat}) \quad \text{or} \quad h_{boiling} = \frac{q}{\Delta T_w} \qquad (6.8)$$

Thus, the slope of the Nukiyama curve (q vs ΔT_w) actually represents the boiling heat transfer coefficient. Hence, the sudden change in the slope is an indicator of the sudden change in heat transfer coefficient. From natural convection with very poor heat transfer coefficient, it changes to nucleate boiling which gives one of the highest heat transfer coefficients. Often nucleate boiling is divided into two parts. The first part (B to D') is called unstable nucleate boiling. Here, the bubbles emerge from the surface but still the bulk liquid remains little subcooled (below saturation temperature). As a result, as these bubbles rise due to buoyancy, the vapour inside the bubbles condense and the bubbles collapse inside the liquid. However as the wall temperature further rises, we get 'fully' developed nucleate boiling, (D' to C) where the bubbles continue to grow as they rise in bulk fluid which is at saturation temperature and finally burst at the liquid surface. It can be seen that transition from partial or unstable nucleate boiling to fully developed nucleate boiling coincides with formation of vapour jets on the surface. From Fig. 6.9, it can be seen that since the log-log plot of q versus ΔT_w is not a straight line, during nucleate boiling the relation between q and ΔT_w may change, i.e.,

$$q \propto \Delta T_w{}^m \qquad (6.9)$$

and the value of m is not constant it changes between B and C.

Point C: The point C which is of immense importance in boiling is known as Critical Heat Flux (CHF). It is the maximum heat flux that can be achieved by a fluid at a given pressure. For water at atmospheric pressure the value of CHF is about 1.4 MW/m^2. This phenomenon is also known by various other names such as boiling crisis, DNB (departure from nucleate boiling) or burn out. Although the heater usually does not physically burn out at this point, the term 'burn out' actually indicates that sudden rise of heater temperature at this point which eventually leads to the physical burn out of the heater at a later stage. The consequence of this phenomenon is the fact that at this point, with no further increase of heat flux, suddenly the heater wall temperature rises substantially. The reason for this can be understood very well by referring to Fig. 6.10. This is the point at which large amount of vapour formed gets stuck to the heater wall because the rate of vapour formation is more than the rate of its removal by buoyancy. This makes a large part of the heater surface exposed to the vapour only and not to the liquid. Vapour being a poor heat transfer medium offers very low heat transfer coefficient. As a result to maintain the constant wall heat flux the ΔT_w must increase, indicating a sudden rise in wall temperature. Point C is the one in which the vapour film first appears due to merger of the mushrooming vapour jets.

From C to D: This part is called 'transition boiling' or 'unstable film boiling'. Here the vapour film formed on the surface is unstable but with more and more increase in wall temperature, greater and greater fraction of the heating surface is covered with vapour resulting in decrease of heat flux progressively.

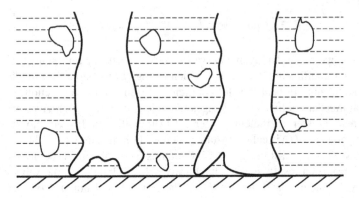

Fig. 6.10 Schematic of liquid and vapour near the CHF

Point D: Point D is the one named as 'Leidenfrost point'. This is the minimum heat flux achieved during film boiling. At this point almost the entire surface is covered by a vapour blanket and heat transfer takes place only by the mechanism known as 'film boiling'.

From D to E and beyond: After the point D, stable film boiling takes place in which the heat is removed only by the film covering the surface and the evaporation takes place at the vapour liquid interface. Due to poor heat transfer through the vapour the wall temperature remains high here. However, the heat flux again starts rising here. This is because of radiation. The wall superheat here is of the order of several hundreds of degrees (for water at atmospheric pressure it is of the order of 100 K). As a result, a large temperature difference exists across the vapour film because one side of this film is saturated liquid and the other side is heated wall. Vapour acts as a participating medium, and heat is transferred from the wall to the liquid by radiation. Due to increased radiation contribution heat flux again increases finally leading to a situation where the material of the heater fails due to excessive high temperature.

It is important to note in Nukiyama curve that there is a hysteresis. When we supply constant heat flux in the heater (typically by electric heating), the curve follows the path $ABD'CC'E$ and when we decrease heat flux it returns following the path $EC'DD'BA$ (Fig. 6.9). This means that the dotted line CD representing the transition boiling regime is never achieved. The total curve in Fig. 6.9 without hysteresis was guessed by Nukiyama can only be achieved if wall temperature rather than wall heat flux is kept uniform. The way it can be achieved is by sending condensing vapour through a tube at various pressures and the outer surface of the tube acts as a heater as shown schematically in Fig. 6.11.

Fig. 6.11 Boiling with
constant wall temperature

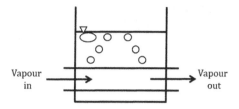

Vapour
in

Vapour
out

6.3.1 Effect of Various Parameters on Pool Boiling

Pool boiling is affected by various parameters other than the nature of fluid. As a
result the pool boiling curve of Nukiyama gets changed. We will discuss very briefly
how these parameters affect the boiling curve. Detailed discussion on these will
follow in the next chapter when we will uncover the physics of nucleate boiling.

(a) **Pressure:** It goes without saying that the system pressure plays a major role in
boiling. From the discussion on thermodynamics in Chap. 3, it is obvious that
critical properties like saturation temperature, enthalpy of evaporation and saturation enthalpies depend on pressure at which phase change takes place. Quite
often the parameter which affects the nature of boiling is the reduced pressure
P^*, which is the ratio of the absolute pressure to the critical pressure of the fluid.
We will discuss this in the next chapter. With the increase of system pressure,
the boiling curve shifts to the left as shown in Fig. 6.12. This means that at the
same wall superheat (ΔT_w), the heat flux increases. This results in higher heat
transfer coefficient meaning an efficient heat transfer. The effect of pressure has
been incorporated in many boiling correlations which we will discuss in the next
chapter.

(b) **Surface roughness:** Surface roughness is an important parameter for pool boiling. It is our experience in the kitchen that boiling water is easier in an old vessel

Fig. 6.12 Effect of pressure
on pool boiling

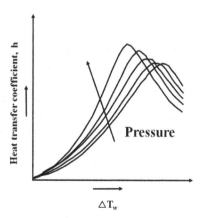

Heat transfer coefficient, h

Pressure

ΔT_w

Fig. 6.13 Effect of surface
roughness on boiling curve

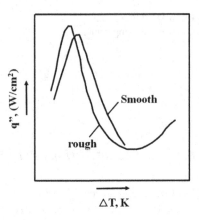

than in a new one. This is because the bottom surface of a new vessel being
smoother (less rough). Hence boiling and nucleation becomes less likely while
in an old vessel the bottom becomes rough and nucleation accelerates. The reason for this will become obvious when we discuss the process of nucleation in
the next chapter.

(c) **Surface wettability:** Surface wetting behaviour affects the pool boiling curve
significantly. Surface wettability is characterized by contact angle. It is the angle
made by a static bubble (or a droplet) on a solid surface. Higher, the contact
angle more hydrophobic is the surface and as a result better is boiling. When
the contact angle is low, the liquid wets the surface very well and a very few
nucleation sites are available on the surface. This will be discussed during the
explanation of nucleation in the next chapter. Liaw and Dhir (1989) presented a
plot for the nucleate boiling part of Nukiyama curve for different contact angles
incorporation data from Nishikawa (1974) as shown in Fig. 6.13.

(d) **Surface orientation:**   The orientation of the surface also plays a role in pool
boiling. The data of Nishikawa et al. (1984) clearly shows that. It can be seen
that as the plate orientation is changed (angle with horizontal increased), boiling
enhances. However, it not only brings out the effect of inclination but also the
distinction between upward facing and downward facing surface. It is interesting
to note from Fig. 6.14 that the effect of inclination is limited only to lower heat
flux. At higher heat flux all the curves collapse indicating no effect of inclination.
As will be discussed in the next chapter, the inclination effect comes from sliding
of bubbles during boiling and increase of nucleation density at higher heat flux
prevents bubble sliding thus reducing the effect.

Apart from the above major effects, there are many secondary effects which
also influence pool boiling such as the effects of surface contamination, thermal
conductivity of the heater material, heater geometry, liquid subcooling and even
the way tests are conducted (steady or transient). Dissolved noncondensable
gases can also influence boiling significantly. Since we normally talk of boiling
of pure liquid on earth surface, the effect of gravity and multiple fluids are not

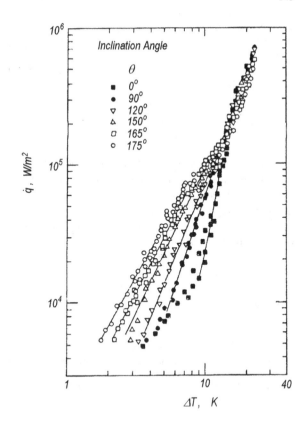

Fig. 6.14 Effect of inclination on nucleate pool boiling. (*Courtesy:* "Effect of surface configuration on nucleate boiling heat transfer" by Nishikawa et al., Int. J Heat Mass Trans., 1984, with permission from Elsevier)

considered. In space or under microgravity (reduced gravity), boiling changes completely which is a major field of research today. We will discuss about boiling of binary mixtures in a separate chapter. Boiling of suspensions is also different which we will not discuss here.

6.3.2 Effect of Liquid Velocity on External Nucleate Boiling

As indicated earlier, the characteristics of boiling when the liquid is not initially stagnant in the pool is essentially similar to that of pool boiling. Bergles and Rohsenow (1964) showed that in this case the combined forced flow and nucleate boiling can be viewed as the superposition of one on the other. The mechanism here gradually transforms from liquid forced convection to nucleate boiling as the heat flux is raised.

This particular type of nucleate boiling has got a very important application in nuclear industry. The nuclear fuel rods are arranged in bundles where the flow takes place over the fuel rods in the axial direction. This is the so-called 'rod bundle' geometry. Here the case is somewhat complex although the entire bundle is enclosed

Fig. 6.15 Two types of
nucleate fuel assembly with
7 rod

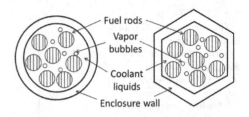

inside a tube or enclosure through which the coolant fluid flows, the boiling takes
place on the tube and not on the inside wall of the enclosure tube. This is shown
in Fig. 6.15. This is a case of external nucleate boiling with forced flow of liquid.
The film boiling in such cases are more complicated since the film on one rod may
interfere with that of other neighbouring rods. A large number of studies on boiling
in rod bundles is available in literature mainly to cater the need of nuclear industry.

Examples

Problem 1: A two phase mixture of water and air at 25 °C flows in a horizontal tube
with $x^* = 0.25$. What is the volumetric vapour content if slip is 1.1?

Solution: At 25 °C, density of water is 996.3 kg/m^3 and density of air is 1.169
kg/m^3

$$s = \frac{\omega_g}{\omega_l}$$

$$s = \frac{x^*}{1 - x^*} \frac{1 - \varepsilon}{\varepsilon} \frac{\rho_l}{\rho_g}$$

therefore

$$1.1 = \left[\frac{.25}{1 - .25} \right] \frac{1 - \varepsilon}{\varepsilon} \frac{996.3}{1.169}$$

therefore

$$\frac{1 - \varepsilon}{\varepsilon} = 0.003872$$

therefore

$$\varepsilon = \mathbf{0.996}$$

Problem 2: A two phase flow of air and water flows at atmospheric conditions through an adiabatic pipe with inner diameter of 50 mm. Quality of mixture is 0.25. Find volume flow rate of air if total volume flow rate is 1.5 lpm.

Solution:

$$x^* = \left[\frac{\varepsilon^*}{\varepsilon^* + (1 - \varepsilon^*)\frac{\rho_l}{\rho_g}}\right]$$

and $\varepsilon^* = v_g/v$, therefore

$$0.25 = \left[\frac{\varepsilon^*}{\varepsilon^* + (1 - \varepsilon^*)845.7}\right]$$

therefore

$$\varepsilon^* = 0.996 = \frac{v_g}{v_l}$$

therefore

$$v_g = 1.494 \text{ lpm}$$

Problem 3: Water is boiled at the rate of 30 kg/hr in a polished copper pan 250 mm diameter at atmospheric pressure. Assume nucleate boiling conditions; calculate the temperature of the bottom surface of the pan?

Solution: Given data:
Evaporation rate $(m) = 30$ kg/hr.
Diameter of pan $(d) = 250$ mm
Surface fluid contact $(C_{fc}) = 0.013$.
For thermo-physical properties of water at 100 °C and 1 atmospheric pressure, $T_{sat} = 100°C$, $\rho_l = 958.4$ kg/m^3, $\rho_v = 0.5955$ kg/m^3, $C_{pl} = 4220$ J/kgK, $\mu_l = 279 \times 10^{-6}$Ns/m^2, $Pr = 1.75$, $h_{fg} = 2257$kJ/kg, $\sigma = 58.9 \times 10^{-3}$N/m, $n = 1$ for water. Surface heat flux

$$(q_s) = \frac{Q}{A} = m \times \frac{h_{fg}}{A}$$

$$q_s = \frac{30}{3600} \times \frac{22571000 \times 4}{3.14 \times 0.25^2}$$

By using the Roshenow correlation, surface heat flux

$$(q_s) = \mu_l h_{fg}[g(\rho_l - \rho_v)/\sigma]^{0.5}[C_{pl}(\Delta te)/(C_{fc})h_{fg}(Pr)^n]^3$$

Rearranging the equation and find out the surface temperature of pan,

$$(\Delta te) = (qs)/\mu_l h_{fg} [\sigma/g(\rho_l - \rho_v)/]^{.5^{0.33}} [(C_{fc}) h_{fg} (Pr)^n / C_{pl}]$$

$$(\Delta te) = (383354.564/629.703)[0.0589/9.81(957.8045)]^{0.5^{0.33}} \times 12.1674$$

$$(\Delta te) = 14 \,°\mathrm{C}$$

Surface temperature of pan $(T_{sur}) = T_{sat} + \Delta te = 100 + 14 = 114 \,°\mathrm{C}$.

Exercises

1. A vertical evaporator tube with $L = 3.5\,\mathrm{m}$, inner diameter of $12\,\mathrm{mm}$ has water flowing from bottom to top. The inlet pressure is $5.5\,\mathrm{MPa}$, subcooled with temperature of $210\,°\mathrm{C}$. It is initially heated to saturation temperature and then partly vaporized. The wall constant heat flux is $7.58 \times 10^5\,\mathrm{W/m^2}$. Calculate the length at which water reaches saturation temperature at its central core.
2. Water at a temperature of $250\,°\mathrm{C}$ and pressure of $6\,\mathrm{MPa}$ flows into a vertical tube of inner diameter of $20\,\mathrm{mm}$. The mass flow rate is $1000\,\mathrm{kgs/m^2}$. The heat flux due to condensing steam outside is constant and is $8 \times 10^5\,\mathrm{W/m^2}$. How long is the tube for x^* at the outlet to be 0.25? Neglect the pressure drop.
3. A nickel-coated heater element with a thickness of $20\,\mathrm{mm}$ and a thermal conductivity of $50\,\mathrm{W/m.k}$ is exposed to saturated water at atmospheric pressure. A thermocouple is attached to the back surface, which is well insulated. Measurement at a particular operating condition yields electrical power dissipation in the heater element of $7 \times 10^7\,\mathrm{W/m^3}$ and a temperature measured in the thermocouple of $270\,°\mathrm{C}$. Calculate the surface temperature (T_s) and the heat flux at the exposed surface.

References

Bergles AE, Rohsenow WM (1964) The determination of forced-convection surface-boiling heat transfer. J Heat Transf 86(3):365–372
Dhir VK, Liaw SP (1989) Framework for a unified model for nucleate and transition pool boiling. J Heat Transf 111(3):739–746
Nishikawa (1974) Handbook of ph. Ch. pp 116–119
Nishikawa K, Fujita Y, Uchida S, Ohta H (1984) Effect of surface configuration on nucleate boiling heat transfer. Int J Heat Mass Transf 27:1559–1571
Nukiyama S (1934) Film boiling water on thin wires. Society of Mechanical Engineering, 37
Stephan K (1992) Heat transfer in condensation and boiling. Springer, Berlin, p 84

Chapter 7
Heat Transfer Mechanisms and Correlations in Nucleate Pool Boiling

In the last chapter, we have seen that as nucleation starts in pool boiling, there is a sudden increase in the heat removal rate indicated by a large increase in the heat transfer coefficient. It is important that we investigate the reason for such a huge rise in heat transfer characteristics in boiling compared to single phase heat transfer. It goes without saying that the attraction of boiling as a mode of heat transfer is this large heat transfer coefficient and hence the underlying reasons and mechanisms for this need to be understood to use and control their potential.

The heat transfer during boiling undergoes a transformation as heat flux is increased which has been described through the Nukiyama curve given in Chap. 6. The evaporation changes into nucleate boiling leading to critical heat flux and finally film boiling. From the viewpoint of high heat transfer rate as well as the safety of the equipment, nucleate boiling appears to be the ideal one for application.

Nucleate boiling is the most important part of process equipment. The order of phenomena in the bubble life cycle in nucleate boiling are nucleation, bubble growth, departure, bubble collapse, waiting period for next renucleation and bubble growth, and this is shown diagrammatically in Fig. 7.1. Hence, we need to understand these processes in some detail, and that will be taken up now. However, it must be borne in mind that these processes by themselves alone cannot explain the large heat flux observed in nucleate pool boiling. There have been efforts to bring out a complete model of heat transfer for nucleate boiling for the last half a century. In the following section, we will try to trace the history of these investigations which will reveal the real nature of various classes of heat transfer correlations available for them.

© The Author(s) 2023
S. K. Das and D. Chatterjee, *Vapor Liquid Two Phase Flow and Phase Change*, https://doi.org/10.1007/978-3-031-20924-6_7

Fig. 7.1 Bubble cycle from
nucleation through growth,
departure and re-nucleation
responsible for controlling
nucleate boiling heat flux

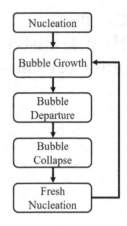

7.1 Nucleate Boiling—Role of Bubble Life Cycle

As shown in Fig. 7.1, the nucleation process marks the first step in the formation of
bubbles during nucleate boiling. The nucleation process due to heating and subsequent bubble growth has already been explained in Chap. 4 and is not repeated here.
Therefore, in this section, we start with bubble departure.

Bubble Departure
The departure of the bubble influences boiling heat transfer. A balance of surface
tension and buoyancy can give this diameter. Departure diameter signifies the size
of a bubble on a horizontal surface just when the buoyancy force trying to lift the
bubble off from the surface just exceeds the surface tension force holding it on the
surface (Fig. 7.2). This gives the force balance at departure as

$$F_b - W - F_s = 0 \tag{7.1}$$

where

buoyant force, $F_b = \frac{4}{3}\pi R_b^3 \rho_l g$,
weight, $W = \frac{4}{3}\pi R_b^3 \rho_g g$ and
surface tension force, $F_s = 2\pi (R_c)\sigma \sin\theta = 2\pi (R_b \sin\theta)(\sigma \sin\theta)$.

Here, the bubble volume is taken as completely spherical.

Substituting the above forces in Eq. (7.1), we get the expression for departure
diameter,

$$\frac{4}{3}\pi R_b^3 \, g(\rho_l - \rho_g) = 2\pi (R_b)\sigma \sin^2\theta$$

$$R_b = \sqrt{\frac{3}{2}\sigma \frac{\sin^2\theta}{g(\rho_l - \rho_g)}}$$

Fig. 7.2 Bubble at the point
of departure

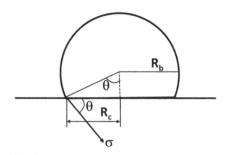

For small θ, $\sin \theta \simeq \theta$,

$$R_b = 1.22\theta \sqrt{\frac{\sigma}{g(\rho_l - \rho_g)}}$$

Changing the angle from radians to degrees and radius to diameter, we get the expression known as the Fritz formula (1935),

$$D_b = 2.449\theta \sqrt{\frac{\sigma}{g(\rho_l - \rho_g)}} \tag{7.2}$$

Here, θ is the contact angle of the bubble on the surface. The following points may be remembered in this regard:

- When surface tension forces are dominant, the departure shape is spherical.
- When inertial forces are dominant, it is hemispherical.
- When both forces are comparable, it is of oblate shape.

At higher wall superheat effects of inertia and convection become significant and the departure diameter is given by

$$D_b = 2.5\theta \sqrt{\frac{\sigma}{g(\rho_l - \rho_g)}} \left[1 + \left(\frac{Ja}{Pr_l}\right)^2 \frac{1}{Ar_l} \right]^{\frac{1}{2}} \tag{7.3}$$

where the Archimedes number and the Jacob number are

$$Ar = \frac{g}{v_l^2} \left[\frac{\sigma}{\rho_l g} \right]^{0.5}$$

and $Ja = \left(\frac{\rho_f C_{pf} \Delta T}{\rho_v h_{fg}}\right)$

$$5 \times 10^{-7} \leq \left(\frac{Ja}{Pr_l}\right) \frac{1}{Ar_l} \leq 10^1$$

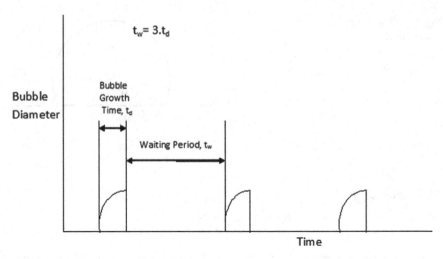

Fig. 7.3 Bubble departure frequency

In the above correlations, a length scale called the Laplace constant is seen,

$$L' = \sqrt{\frac{\sigma}{g(\rho_l - \rho_g)}} \qquad (7.4)$$

This parameter appears in many boiling correlations where other length scales are absent. The next important parameter is the departure frequency. This is the inverse of the time period between nucleations as given by

$$f = \frac{1}{t_d + t_w} \qquad (7.5)$$

where t_d is the departure time and t_w is the waiting period. Figure 7.3 shows these periods. Jakob (1949) was the first to measure departure frequency photographically. He suggested a correlation

$$f D_b = C \qquad (7.6)$$

Later on, Zuber reasoned that bubble grows rapidly; one bubble is just one diameter behind the other as shown in Fig. 7.4 (t_w is approximately zero) giving

$$x = D_b$$
$$D_b = \frac{v_b}{f}$$
$$f D_b = v_b$$

Fig. 7.4 Liquid column
supported by bubble

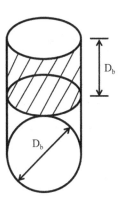

where v_b is the bubble rise velocity and $t_w = 0$. For a finite waiting period, v_b can be estimated as follows:

$$x = v_b(t_w + t_d),$$

$$v_b = fD_b\left(\frac{t_w + t_d}{t_d}\right) \tag{7.7}$$

The net force experienced by the bubble in the buoyancy force is given by

$$F_B = \frac{4}{3}\pi R_b^3(\rho_l - \rho_g)g \tag{7.8}$$

We consider that the bubble has to support a liquid element of length equal to bubble rise velocity. Then the force given on the liquid element or the rate of change of momentum on it is

$$\frac{d(mv_b)}{dt} = \frac{4}{3}\pi R_b^3(\rho_l - \rho_g)g$$

where m is the mass of the liquid column and v_b is the bubble rise velocity. For constant bubble velocity

$$\dot{m}v_b = \frac{4}{3}\pi r_b^3(\rho_l - \rho_g)g$$

Replacement of the mass rate of the column is given by

$$\dot{m} = \left(\frac{\pi}{4}D_b^2 v_b\right)\rho_l$$

From Eq. 7.2,

$$D_b = C_1\sqrt{\frac{\sigma}{g(\rho_l - \rho_g)}}$$

We get

$$v_b = 1.18 \left(\frac{\sigma g \Delta \rho}{\rho_l^2} \right)^{\frac{1}{4}}$$

Usually, the frequencies for the hydrodynamic (initial) and thermodynamic (later) periods are different. Separate correlations are arrived at for two regions:

1. Hydrodynamic region: This is the theory with buoyancy and drag forces. Here

$$f D_b \left(\frac{t_w + t_d}{t_d} \right) = v_b = 1.18 \left[\frac{\sigma g (\rho_l - \rho_v)}{\rho_l^2} \right]^{1/4}$$

2. Thermodynamic region

$$f D_b^2 = \text{constant}$$

The formula, given by Malenkov (1971), covers the entire region as

$$f = \frac{1}{D_b \pi} \left(1 + \frac{q}{\rho_G h_{fg} \omega} \right) \omega$$

where

$$\omega = \left[\frac{d_b g (\rho_L - \rho_G)}{2(\rho_L + \rho_G)} + \frac{2\sigma}{D_b (\rho_L + \rho_G)} \right]^{1/2}$$

Thus, departure frequency is a strong regime and property-dependent parameter.

7.2 Heat Transfer Models in Nucleate Pool Boiling

It is obvious that boiling is a statistical phenomenon. A large number of bubbles of various sizes nucleate, grow and burst converting liquid to vapour. As a result, a complete deterministic theory where a smooth continuous change of flow and thermal parameters takes place (example: single phase heat transfer associated with boundary layer theory) is not available for it. Hence, the effort has always been to analyse a single 'model bubble' and its surrounding such that the average behaviour can be predicted and understood by extending the single bubble behaviour through statistical averaging. The following sections describe the models proposed with this idea in chronological order. It can be seen that analysis of nucleating, bubble growth and bubble dynamics act as important tools to construct these heat transfer models.

7.2.1 Vapour–Liquid Exchange Model

Probably, Forster and Greif (1959) was the first one to try a heat transfer model for nucleate boiling. They assumed that the hot liquid boundary layer formed by heat transfer from the wall to the liquid is pushed up by the growing bubble underneath as shown in Fig. 7.5.

It postulates that the vapour bubble exchanges heat with this hot layer and grows. It essentially looks at the vapour bubble acting as a pumping piston to break the hot liquid layer and bring the cooler liquid in contact with the heated surface making this cooler liquid become hot. The same hot layer has been named as the relaxation layer by Van Stralen et al. (1975). In this model, the Jacob number (to be discussed later) attains a high value (≈ 100) which prompted Forster and Grief to conclude that the liquid vapour heat exchange mechanism is the main contributor to the high rate of heat transfer. Although this was the first effort to explain the heat transfer process, there is enough controversy about this model. Firstly, Bankoff and Mason (1962) pointed out a discrepancy in the heat flux calculation procedure. More importantly, according to this model, the tip of the bubble is supposed to have the highest temperature, while most of the measurements confirm that it is in fact the lowest temperature of the bubble which often causes condensation at the tip.

7.2.2 Bubble Agitation Model

There was an effort to explain the high heat transfer coefficient by liquid mixing. The near-wall hot liquid was postulated to continuously mix with the cold liquid far away. Hsu and Graham (1961, 1976) demonstrated such mixing with the help of shadow graphs and Schlieren photography. However, Kast (1964) indicated that although such mixing may contribute to heat transfer, this alone cannot be attributed to the high heat transfer coefficient of the nucleate pool boiling.

Fig. 7.5 Liquid–vapour exchange model

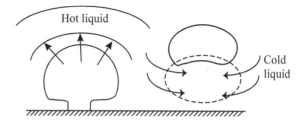

7.2.3 Microlayer Evaporation Theory

Moore and Mesler (1961) brought out a revolutionary theory which postulated that underneath the bubble, there exists a thin microlayer of liquid which evaporates very rapidly (Fig. 7.6). Microlayer measurement shows that on a boiling surface, at times, the temperature may fall by 10 °C–17 °C within a time of the order of 2 ms. Such a rapid decrease in temperature can only be explained by quick evaporation of the microlayer. Subsequently, Sharp (1964) proved the existence of a microlayer using optical measurement techniques.

Microlayer, although is certainly an important mechanism for heat transfer, cannot alone explain or even estimate the amount of heat transfer during nucleate boiling. Based on microlayer evaporation, a heat transfer model can be constructed to give heat flux as

$$q = (ah_{fg}) \left(\frac{2\pi R_g T_{sat}}{M} \right) (p_l - p_g) \qquad (7.9)$$

where

a = coefficient of evaporation
h_{fg} = latent heat of evaporation
R_g = gas constant
T_{sat} = saturation temperature
M = molecular weight
p_l, p_g = liquid, gas pressure.

Hsu and Graham (1976) demonstrated that in case we assume evaporation in vacuum ($p_g = 0$) for both water and mercury, the heat flux predicted by the above model still falls short of the actual heat flux.

7.2.4 Transient Conduction-Based Models

This model proposed by Mikic and Rohsenow (1969) identified transient conduction as the dominating heat transfer mechanism along with natural convection. The model

Fig. 7.6 Microlayer evaporation model

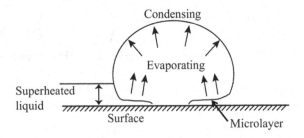

considers that subsequent to every bubble departure from the surface, the cooler liquid rushes to replenish it. Subsequently, a hot liquid layer is formed by transient conduction. It is assumed that once the bubble departs, it carries a part of the hot liquid layer and the layer gets ruptured. Cold liquid replenishes this hot layer, and the hot layer is reformed by transient conduction. The series of events postulated are schematically shown in Fig. 7.7. Han and Griffith (1965) clearly demonstrated the series of events postulated above. The increase of liquid temperature by transient conduction, leading to the reformation of the hot layer, provides enough liquid superheat for the nucleation site to be reactivated and the next bubble comes up. Han and Griffith (1965) proposed that the area of the hot layer, which gets ruptured, has a diameter twice the bubble diameter. Taking this into consideration, the model for heat flux can be built as

$$q = \frac{K^2}{2}\sqrt{\pi(k\rho c_p)_l f D_b{}^2 N_a \Delta T} + \left(1 - \frac{K^2}{4}N_a\pi D_b{}^2\right)h_{nc}\Delta T \qquad (7.10)$$

where

$K = 2$ (the diameter of influence relative to bubble diameter)
k = liquid thermal conductivity
D_b = bubble departure diameter
ρ_l = liquid density
C_p = liquid specific heat
N_c = nucleation site density

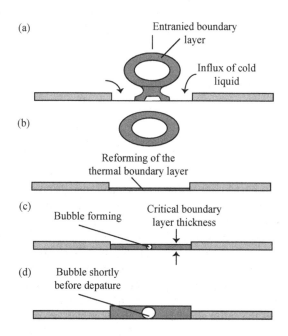

Fig. 7.7 Model for heat transfer during nucleate boiling, according to Han and Griffith (1965) (*Courtesy:* "The mechanism of heat transfer in nucleate pool boiling—Part II: The heat flux-temperature difference relation" by Han Chi-Yeh and Peter Griffith, Int. J. Heat and Mass Trans., with permission from Elsevier)

(a)   Entranied boundary layer

Influx of cold liquid

(b)

Reforming of the thermal boundary layer

(c)   Bubble forming   Critical boundary layer thickness

(d)   Bubble shortly before depature

Fig. 7.8 Drift current model
proposed by Beer (1969)
(*Courtesy:* "Heat Transfer in
Condensation and Boiling"
by Karl Stephan, with
permission from Springer)

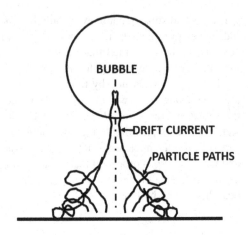

f = bubble departure frequency
ΔT = wall superheat
h_{nc} = natural convection heat transfer coefficient.

In this model, the first term on the right-hand side gives the heat transfer due to bubble departure and subsequent transient conduction, while the second term stands for natural convection in the area of the surface where nucleation does not take place.

The major drawback of this model is that it does not consider evaporation at all. While for a certain range of parameters transient conduction may be the dominant contributor to heat transfer mechanism, it is hard to believe that heat transfer due to evaporation does not contribute at all. This model was extended by Beer (1969) to include drift current as shown in Fig. 7.8.

7.2.5 Composite Models for Heat Transfer

The models discussed above all contain truth but only partially. The mechanisms of transient conduction, microlayer evaporation, natural convection, hot layer reformation, etc., all play roles indeed in the process of boiling heat transfer. However, they cannot be individually adequate to dissipate the boiling heat flux. This is the reason why the above theories and models could only explain data of a particular type and range. None of the above models could either predict or reproduce a large amount of experimental data available in the literature. The situation is similar to what the present authors describe as blind men's description of an elephant. Once few blind men were taken to feel how an elephant is. One of them touched the trunk and exclaimed "Elephant is like a pipe", the other touched the legs and disagreed, "No, it is like a pillar". The other touched the tusk and said "It is for sure like a spear" and so on. None of them was incorrect, but they felt only a part of the elephant and not the entire elephant. The above models are also true in a similar way.

To get an entire picture of the heat transfer process in boiling, we need to put all these mechanisms together and build up a composite model. The most important challenge of building such a composite model is the fact that the various mechanisms are not involved in the same space and during the same time duration. Take, for example, microlayer evaporation. It only takes place as long as the bubble remains attached to the surface. After bubble departure, the transient conduction and hot layer reformation dominate. Similarly, in the spaces where nucleation does not take place, natural convection is the mechanism of heat transfer. Thus, evaluating the various contributions of heat transfer over various times and spaces during the boiling process is a challenging task. The estimation of bubble growth, departure frequency, waiting period, departure diameter and nucleation site density is extremely useful in deciding the spatial and temporal spread of the various mechanisms of heat transfer in nucleate boiling.

Judd and Hwang (1976) were the first ones to try to bring together the transient conduction-based Mikic and Rohsenow (1969) theory and the microlayer evaporation theory. They proposed a composite model in which transient conduction, natural convection as well as microlayer evaporation were incorporated to suggest the average heat flux as

$$q = \frac{K^2}{2}\sqrt{\pi(k\rho c_p)_l f D_d{}^2 N_a \Delta T} + \left(1 - \frac{K^2}{4} N_a \pi D_d{}^2\right) h_{nc}\Delta T$$
$$+ h_{ev}\Delta T N_a \left(\frac{\pi}{4}D_d^2\right) \tag{7.11}$$

Here, h_{ev} = evaporative heat transfer coefficient. The other quantities are as described in Eq. (7.10). It should be noted that two terms are taken at some area of influence which are given by K, where

$$K = \frac{\text{diameter of influence of the area around a bubble}}{\text{bubble departure diameter}} \tag{7.12}$$

Han and Griffith (1965) found from their experiments that $K \approx 2$. It can also be seen that the transient conduction term is active only during a part of the cycle (hence the frequency, f is important), while the evaporative heat transfer coefficient h_{ev} is averaged over the entire cycle although it remains active only as long as the microlayer exists. The success of this model was better than other models. However, this also could not conclusively prove the mechanisms because there were large uncertainties in quantities like h_{ev} which made the model somewhat empirical. Yet, as a first effort this composite model showed the way ahead and was more logical.

Moving forward, Benjamin and Balakrishnan (1996) presented a model which was not only physically consistent and logical but also could predict heat flux for a range of liquids in nucleate boiling without the necessity of any arbitrary or adjustable quantity. They also considered the total heat flux during nucleate boiling to be composed of three components:

(i) Microlayer evaporation (q_{ME});
(ii) Transient conduction during hot liquid layer reformation (q_c);
(iii) Natural convection in the area not influenced by bubbles (q_{nc}).

The novelty of the model lies in the fact that it uses accurate time scales over which the above three phenomena take place. The above three fluxes, unlike Judd and Hwang (1976), have not been simply added here but have been weighted by appropriate time scales keeping the realistic physical series of events in boiling in mind. The heat flux is combined as

$$q_{tot} = \frac{q_{ME}t_b + q_c t_w}{t_b + t_w} + q_{nc} \qquad (7.13)$$

where

t_b = bubble growth period (time for bubble departure)
t_w = waiting period
q_{tot} = heat flux observed in nucleate boiling.

It has to be understood that the heat fluxes (q_{ME}, q_c, q_{nc}) are calculated over unit boiling area, and hence the amount of area involved in them is taken into consideration. These contributions are evaluated as follows:

1. Benjamin and Balakrishnan (1996) considered the heat flux due to microlayer evaporation per unit time as the mass of total vapour generated per unit area per unit time multiplied by the latent heat of evaporation. Now total vapour generated per unit area can be obtained by the product of vapour mass per bubble, the number of bubble nucleation sites per unit area (or nucleation site density) and the frequency of bubble generation. Thus

$$q_{ME} = \left(\frac{1}{6}\pi D_b^3 \rho_v\right) h_{fg} N_a f \qquad (7.14)$$

where

ρ_v = density of vapour
h_{fg} = latent heat of vaporization
N_a = nucleation site density
f = bubble departure frequency.

They used bubble departure diameter equation from a well-known bubble growth equation,

$$D_b(t) = B Ar^{0.135} \left(\frac{kJa}{\rho_f C_{pf}}t\right)^{0.5} \qquad (7.15)$$

where B is constant ≈ 1.55 for water, Ar is Archimedes number

$$Ar = \left(\frac{g}{\nu_f^2}\right)\left(\frac{\sigma}{\rho_f g}\right)^{0.5} \qquad (7.16)$$

σ = surface tension
ν_f = kinematic viscosity
ρ_f = liquid density
Ja = Jacob number.

$$Ja = \left(\frac{\rho_f C_{pf} \Delta T}{\rho_v h_{fg}} \right) \tag{7.17}$$

where C_{pf} is liquid specific heat, ΔT is temperature difference and ρ_v is vapour density. We know that bubble departure diameter can be obtained by equating surface tension force to the buoyancy force on the bubble (Eq. (7.12)). Substituting that bubble departure diameter in Eq. (7.15) can give the time required for bubble departure t_d, which can be used in Eq. (7.13). Now, the waiting period is related to departure time. Van Stralen et al. (1975) proposed that the time (t_w) between the departure of one bubble and the incipience of the next bubble at the same nucleation site is given by

$$t_w = 3t_d$$

The bubble departure frequency is the inverse of the total cycle time,

$$f = \frac{1}{t_d + t_w} \tag{7.18}$$

Thus, all the quantities of Eq. (7.13) can be evaluated and the heat flux due to microlayer evaporation q_{ME} calculated.

2. Coming to the heat flux due to the transient conduction following the disruption of the hot liquid layer (Fig. 7.7), (q_c), it must be recognized that it will take place only during the waiting period because during the bubble growth, the nucleation site is occupied by the bubble. Now, the heat transfer during the transient conduction can be given by the well-known transient conduction equation

$$Q = \int_0^{t_w} \frac{(k_t \Delta T)}{\sqrt{\pi \alpha t}} dt = 2 \frac{(k_t \delta T)}{\sqrt{\pi \alpha t_w}} t_w \tag{7.19}$$

where k_f is liquid thermal conductivity and α is liquid thermal diffusivity.

Han and Griffith (1965) postulated that bubble on its departure takes with it liquid from an area called the area of influence. They claimed that this area is four times the projected area of the bubble at the time of departure (this means the diameter of area of influence is 2 D_d which gives $K = 2$, as explained for Eq. 7.23). Judd and Hwang (1976) could match their prediction with the experimental data only for $K = 1.8$ and hence with this value the heat flux can be calculated as

$$q_c = 2 \sqrt{\frac{k_f \rho_f c_{pf}}{\pi t_w}} \Delta T \left(N_a K \frac{\pi D_d^2}{4} \right) f \tag{7.20}$$

3. Finally, the heat transfer due to natural convection can take place for only the area outside the area of influence. The area of natural convection can be given by

$$A_{nc} = 1 - \left[\frac{3}{4} N_a \left(K \frac{D_d^2}{4} \right) + \frac{1}{4} N_a \left(\frac{\pi D_d^2}{4} \right) \right] \qquad (7.21)$$

The three terms on the right-hand side can be explained as follows. Unity stands for unit heat transfer area from which the two areas (second term, area of influence and third term area where bubbles are sitting) are subtracted to give the area over which natural convection takes place. Because $t_w = 3t_d$, at any instant of time three fourth of the nucleation sites will be in the waiting period, while one-fourth will be nucleating. This justifies the 3/4 and 1/4 as coefficients of the area of influence and nucleating area. For finding out the heat transfer coefficient for natural convection, widely used Churchill and Chu (1975) correlation can be used

$$Nu = \left[0.6 + 0.387 (Gr_l Pr C)^{1/6} \right]^2 \qquad (7.22)$$

where

$$C = \left[1 + \left(\frac{0.559}{Pr} \right)^{9/16} \right]^{-16/9}$$

$$h_{nc} = \frac{Nu k_f}{L}$$

Nu = Nusselt number
Pr = Prandtl number of the liquid
Gr_l = Grashof number.

Finally, natural convection heat flux can be given by

$$q_{nc} = h_{nc} A_{nc} \Delta T \qquad (7.23)$$

It can be noted that in the equation while at the nucleation site-weighted average with respect to time t_w and t_d are taken, for natural convection time weighting is not done. This is because unlike the nucleation site which gets disrupted by bubble departure, the area under natural convection remains undisturbed unless heat flux is changed. This gives the form of Eq. (7.13) where the various heat flux terms can be calculated from Eqs. (7.14), (7.20) and (7.23). The expression used for nucleation site density by Benjamin and Balakrishnan was

$$\frac{N}{A} = 218.8 (Pr)^1 .63 \left(\frac{1}{\gamma} \right) \theta^{-0.4} (\Delta T)^3 \qquad (7.24)$$

where γ is called surface liquid interaction parameter and defined by

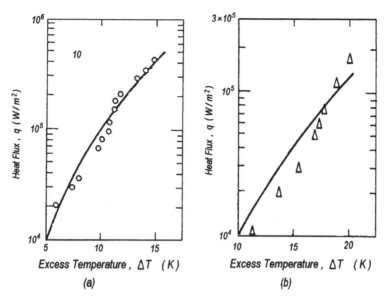

Fig. 7.9 Agreement of Benjamin and Balakrishnan (1996) model with experiments for **a** Water and **b** Acetone (*Courtesy:* R. J. Benjamin, Ph.D. thesis, under the supervision of Prof. A. R. Balakrishnan, 1996, IIT Madras)

$$\gamma = \left(\frac{K_w \rho_w C_{pw}}{k_f \rho_f C_{pf}} \right)^{1/2}$$

where k_w= Thermal conductivity of wall, ρ_w = density of wall and C_{pW} = specific heat of wall. The agreement of the above model of Benjamin and Balakrishnan for various fluids was found to be excellent as shown in Fig. 7.9.

7.2.6 Heat Transfer Mechanism Inclined Surfaces

While discussing the effects of various parameters on nucleate boiling in Chap. 6, we have shown that inclination of the heating surface has a remarkable influence on nucleate boiling heat transfer, particularly at lower heat flux. The reason for this is the fact that unlike horizontal surfaces where the bubbles grow and directly depart from the surface, in an inclined surface, the bubbles grow and then start sliding on the surface. Finally, it leaves the surface after a certain distance.

This is shown schematically in Fig. 6.15. It can be noted in the figure that for a downward-facing surface, the bubbles cannot leave the surface while for a vertical surface they can slide some way up and can then finally leave the surface. The other case is that of a horizontal cylindrical heater. In this case, although the heater is a

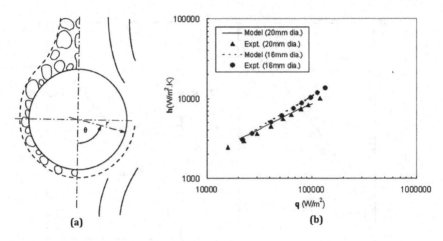

Fig. 7.10 a Sliding bubbles on a horizontal cylindrical heater. **b** Das and Roetzel model (2002) for horizontal heater

horizontal one, the boiling surfaces are inclined and the inclination changes with the circumferential position. This gives rise to bubble sliding as shown in Fig. 7.10a where the sliding of bubbles has been schematically shown. Das and Roetzel (2002) presented a simple model to take care of the sliding bubbles on the curved surface of the horizontal heater. Figure 7.10b shows the comparison of this model with experimental data of Barthau and Hahne (2000). This model uses the Benjamin and Balakrishnan model (1996) but adds the contribution of sliding bubbles to Eq. (7.13).

A more complete model for bubble sliding was presented by Sateesh et al. (2005). This model takes the history of the sliding bubbles as shown in Fig. 7.11. It shows that the bubble after nucleating grows up, then departs from the site but continues to slide and grow further. Finally, it lifts off from the surface. It is important to account for each of these processes.

Various forces act on the sliding bubble. These are the forces of buoyancy, liquid drag and surface tension. Using the expression for surface tension and buoyancy, we get

$$F_s = \pi \sigma R \sin \theta_m (1 - \cos \theta_m) \qquad (7.25)$$

and

$$F_B = \left(\frac{\pi R^3}{3}\right)(1 + \cos \theta_m)^2 (2 - \cos \theta_m)(\rho_f - \rho_g)g \qquad (7.26)$$

where

F_s = surface tension force
F_B = buoyancy force
R = instantaneous radius of the bubble.

Fig. 7.11 History of the
sliding bubbles (*Courtesy:*
G. Sateesh, Ph.D. thesis,
2007, IIT Madras)

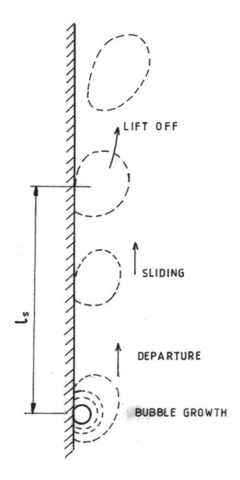

It can be seen that for a sliding bubble, there are two contact angles, advancing
contact angle (θ_a) and receding contact (θ_r). The mean contact angle (θ_m) is defined
as

$$\cos \theta_m = \frac{1 + \cos \theta_2}{2}$$

At bubble departure, the buoyancy force balances surface tension

$$F_B = F_s \tag{7.27}$$

Combing Eqs. (7.14) and (7.16) with it gives a departure diameter of

$$D_b = \sqrt{\frac{12N\sigma}{M_g(\rho_f + \rho_g)}} \tag{7.28}$$

Cornwell and Schüller (1982) showed that a pseudo-static force between surface tension, drag and buoyancy during sliding (assuming net force to be zero or in other words a constant velocity of sliding) gives a sliding velocity of

$$U_b = \sqrt{\frac{2\pi}{C_d}\left\{\frac{M(\rho_f - \rho_g)gr}{3\rho_f} - \frac{N\sigma}{r\rho_f}\right\}} \qquad (7.29)$$

Combining bubble departure diameter (Eq. 7.2) with the bubble growth Eq. (7.15) gives the time for bubble departure as

$$t_d = \frac{12N\sigma C_{pf}}{M_g J_a B^2 Ar^{0.27}k_f} \qquad (7.30)$$

After evaluating the above quantities, the heat flux due to bubble sliding can be calculated. There are two extra contributions due to bubble sliding—first the microlayer evaporation continues during bubble sliding, second due to bubble sliding the hot liquid layer will be disrupted and transient conduction will take place to reform this layer. Thorkcraft et al. (1998) reported a lift-off diameter of about 1.6 times the departure diameter, as

$$D_l = 1.6D_b \qquad (7.31)$$

where D_l = lift-off diameter.

To find out the sliding distance, the average value can be assumed to be half the distance between the two nearest nucleation sites. Thus

$$L_s = 0.5\sqrt{\frac{A}{N_a}} \qquad (7.32)$$

This can be taken as the velocity times the time integrated between the departure and lift-off time

$$\int_{t_d}^{t_l} U_b dt = 0.5\sqrt{\frac{A}{N_a}} \qquad (7.33)$$

where t_l = lift-off time which can be evaluated from Eq. (7.22). Number of sliding bubbles per unit area can be given as

$$n_s = N_a f t_s \qquad (7.34)$$

where t_s = sliding time = $t_l - t_d$. With the above expressions, the heat flux due to microlayer evaporation during sliding can be given as

$$q_{mes} = \frac{1}{6}\pi(D_l^3 - D_d^3)\rho_v h_{fg} N_a f \qquad (7.35)$$

During bubble sliding, the area swept by the bubble during a time d_t is $D(t)U_b dt$. Hence, the area for natural convection is $K D(t)U_b dt$ which on integration gives

$$q_{tcs} = 2\sqrt{\frac{k_f \rho_f c_{pf}}{t_w}} \Delta T \Delta N_a t_w f \int_{t_u}^{t_l} K D(t)U_b dt \qquad (7.36)$$

Sliding bubbles are assumed to be present as long as there is no bubble interaction. Hence, the total heat flux during nucleate boiling with sliding bubbles can be given as

$$q_{tot} = \frac{q_{ME} t_d + q_c t_w}{t_d + t_w} + q_{nc} + (q_{mes} + q_{tcs}) \qquad (7.37)$$

where q_{mes} is evaporation heat flux during sliding and q_{tsc} is transient conduction heat flux during sliding. The computed result using this model with the reported nucleation site density given by Wang and Dhir (1993) shows a good match with the experimental results for water boiling at atmospheric pressure as shown in Fig. 7.12.

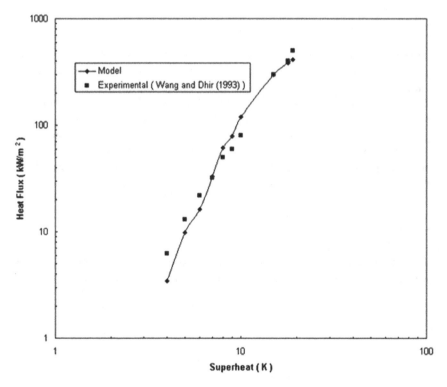

Fig. 7.12 Heat flux prediction by Sateesh et al. (2005) and comparison with experiments of Wang and Dhir (1993) (*Courtesy:* G. Sateesh, Ph.D. thesis, 2007, IIT Madras)

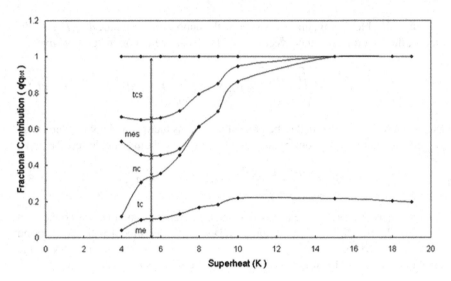

Fig. 7.13 Fractional contributions of various mechanisms (*Courtesy:* G. Sateesh, Ph.D. thesis, 2007, IIT Madras)

However, more insight can be received by plotting the various contributions to the total heat flux as shown in Fig. 7.13. It can be seen that at low wall superheat for the total heat flux for water boiling at 1 atmosphere, microlayer evaporation and bubble sliding contribute significantly and hence in this region orientation of the surface has a significant influence on the boiling characteristics. However, as the superheat goes up, the sliding components as well as natural convection reduce due to the unavailability of area for them, and the major contributions come from transient conduction. Obviously, this is the reason why Mikic and Rohsenow (1969) could build and validate their model based on transient conduction alone.

It is clear from their figure that in various zones of superheat, various contributions to heat transfer become important and only a composite model like the above can bring out such a picture. This can be even more obvious if we take another case of R134a boiling at 0.05 P_c (P_c = critical pressure) as given by Barthau and Hahne (2000). If the above model is applied to these data, we see that the microlayer evaporation and not the transient conduction dominates as given in Fig. 7.14.

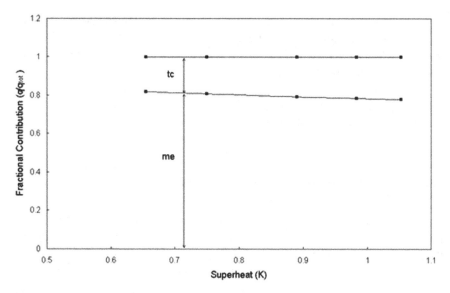

Fig. 7.14   Contribution of heat flux for R134 boiling at 0.05 P_c

7.3   Heat Transfer Correlations in Nucleate Boiling

The previous sections will help to understand the mechanisms of heat transfer during nucleate pool boiling. However, it is more important to have correlations which accurately predict nucleate boiling heat transfer. It is very clear from the preceding discussion that the irregular and statistical nature of the surface condition and (as a consequence) nucleation prohibits the development of consistent analytical theory for boiling heat transfer similar to that of single phase theory (e.g., the boundary layer theory). The major problems here are

1. The proper characterization of the surface condition. It has been shown that surface roughness alone cannot uniquely characterize a surface with respect to boiling.
2. The wetting behaviour of the liquid is dependent not only on the surface condition but also on the nature of the fluid, dissolved gases, etc.

Many investigations have suggested correlations of the form

$$q = c \Delta T_w^{1.2} N_a^{0.33} \tag{7.38}$$

where

ΔT_w is wall superheat.
N_a is nucleation site density.

However, it is known that nucleation site density, N_a, depends on the wall superheat, but the correlation is not known explicitly and hence such correlations are not of any practical relevance. Thus from a practical approach, three types of correlations are available in texts. They are

  (i) Intuitive physics-based correlations.
 (ii) Observed parameter-based correlations.
(iii) Purely empirical correlations.

7.3.1 Intuitive Physics-Based Correlations

The basis of these types of correlations is the limited understanding of the boiling process, but the introduction of the parameters is either intuitive or by analogical treatment with single phase heat transfer. The major dilemma of such an approach is whether to take the surface condition in direct or indirect manner or to ignore it altogether.

The first attempt to such a model was made by Jakob and Linke (1933). They tried to put a dimensionless equation of the form

$$\frac{hb}{k_l} = 42.4 \left(\frac{q}{\rho_v h_{fg} w} \right)^{0.8} \tag{7.39}$$

where

$$
\begin{aligned}
h &= \text{nucleate boiling heat transfer coefficient} \\
b &= \text{Laplace constant} = \sqrt{2\sigma / g(\rho_l - \rho_v)} \\
\sigma &= \text{Surface tension} \\
q &= \text{heat flux} \\
\rho_l, \rho_v &= \text{liquid, vapour density} \\
h_{fg} &= \text{latent heat} \\
w &= \text{a constant having the dimension of velocity suggested to be} \\
&\qquad \text{equal to } f D_d.
\end{aligned}
$$

Jacob and Linke took the value of w to be 0.0778 m m/s for water and CCl_4 from their measurements. Although not explicitly stated, this equation implicitly recognizes two facts:

(a) The boiling correlation is of convection type, i.e., $Nu = f(Re, Pr)$.
(b) The primary length scale of boiling is related to bubble departure diameter (through Laplace constant).

Probably the most popular correlation for nucleate pool boiling is that of Rohsenow (1952). This correlation derives its inspiration from the above correlation of Jakob and Linke (1933). Rohsenow identified in a manner similar to Jacob that boiling is primarily a convective process with liquid and vapour movement. Naturally, this suggests a convective type heat transfer correlation given by

$$Nu = C Re^m Pr^n \qquad (7.40)$$

where

    Nu is Nusselt number
    Re is Reynolds number
    Pr is Prandtl number.

However, the question arises how can we define the Reynolds number and what are the length scales associated with the Nusselt number as well as the Reynolds number? Obviously, unlike single phase heat transfer where the size and shape of the heating surface influence heat transfer (because the nature of boundary layer growth depends on them), in boiling heat transfer, the flux does not depend on the dimension of the heater. Hence, the only obvious length scale that we find is the bubble departure diameter. Hence, boiling Nusselt number can be given by

$$Nu_b = \frac{h D_b}{k_l} \qquad (7.41)$$

where

$$D_d = C_1 \beta \sqrt{\frac{2\sigma}{g(\rho_l - \rho_v)}}$$

This gives

$$Nu_b = C_1 \frac{h}{k_l} \beta \sqrt{\frac{2\sigma}{g(\rho_l - \rho_v)}} \qquad (7.42)$$

Similarly, for deriving expressions for the Reynolds number, we use the bubble departure diameter and vapour velocity. This essentially means that the boiling Reynolds number is the ratio of the bubble inertia to the liquid viscous force. This is intuitive because since the bubbles force through the liquids creating drag, the liquid viscous effect and the inertia of the vapour are logically important. This gives boiling Reynolds number as

$$Re_b = \frac{\rho_v v_b D_d}{\mu_l} \qquad (7.43)$$

    The bubble velocity can be taken as the average vapour velocity. This is the rate of vapour volume generated per unit area,

$$V_b = \frac{\text{vapour volume per sec}}{\text{Area of surface}}$$

$$= N_a f \pi \frac{D_d^3}{6} = \left(\frac{q}{h_{fg}}\right)\frac{1}{\rho_v} \tag{7.44}$$

where

q = heat flux
N_a = nucleation site density
f = frequency of bubble departure.

Hence, using the Fritz formula for bubble departure diameter, we get

$$Re_b = C_1 \frac{q}{\mu_l h_{fg}} \beta \sqrt{\frac{2\sigma}{g(\rho_l - \rho_v)}} \tag{7.45}$$

From his experimental results, Rohsenow proposed

$$Nu_b = C_2 (Re_b)^{2/3} Pr_l^{0.7} \tag{7.46}$$

Please note that since the Prandtl number incorporates viscous effects, he used a liquid Prandtl number. Later, he also found that it is convenient to put it in the form

$$\frac{Re_b Pr_l}{Nu_b} = C_3 Re_b^m Pr_l^n \tag{7.47}$$

Using the definition of Re_b (Eq. 7.45) and Nu (Eq. 7.42) and collecting all the constants including β into one C_{sf}, we get

$$\frac{c_{pl}(T_w - T_{sat})}{h_{fg}} = C_{sf} \left[\frac{q}{\mu_l h_{fg}}\sqrt{\frac{\sigma}{(g(\rho_l - \rho_v))}}\right]^{0.33} \left(\frac{C_{pl}\mu_l}{k_l}\right)^{1.7} \tag{7.48}$$

For water, the exponent of the last term is taken as 1.0 instead of 1.7. It has to be noted for this equation the heat transfer coefficient is taken as

$$h = \frac{q}{(T_w - T_{sat})}$$

Equation (7.48) is the famous Rohsenow correlation. The performance of this correlation against experimental data is shown in Fig. 7.15.

Rohsenow called C_{sf} the surface interaction parameter. Since β is absorbed in C_{sf}, it depends on the contact angle of the bubble and has a different value for each combination of liquid and surface. The typical value of C_{sf} is given in Table 7.1. It is important to note from the Rohsenow correlation that for a given value of ΔT_w, heat flux varies as C_{sf}^{-3}. The value of C_{sf} can be found (Table 7.1) to vary between 0.0025 and 0.015 which is approximately one order of magnitude. This means that depending

Fig. 7.15 Rohsenow's
Correlation compared
against experimental data

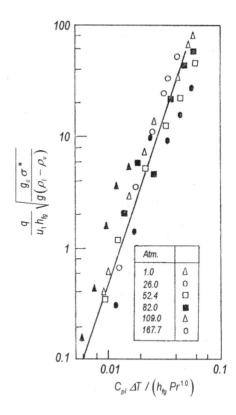

on the liquid surface combination, the heat flux can vary three orders of magnitude.
Obviously, Rohsenow correlation takes the surface effect through the factor C_{sf}. The
cases in which surface parameters are neglected are essentially empirical correlations
which will be discussed later.

7.3.2 Observed Parameter-Based Correlation

From the large number of experimental studies, we know that the boiling heat transfer
coefficient is primarily in function of heat flux. We also know the other parameters
which influence pool boiling heat transfer. Thus, pool boiling heat transfer depends
on (1) heat flux; (2) liquid properties; (3) pressure; and (4) surface characteristics of
heater.

The German Heat Atlas (VDI Warmlatlas 1996) suggests correlations as the ratio of
heat transfer coefficients taking the above factors into consideration. Primarily the
correlation used is

$$h = Cq^n \tag{7.49}$$

Table 7.1 Values of C_{sf} for Rohsenow's correlation

Fluid-heating-surface combination	C_{sf}
Water–Copper	0.031
Water–Platinum	0.013
Water–Brass	0.0060
Water–Emery-polished Copper	0.0128
Water–Ground and Polished Stainless Steel	0.0080
Water–Chemically etched Stainless Steel	0.0133
Water–Mechanically polished Stainless Steel	0.0132
Water–Emery-polished and Paraffin-treated Copper	0.0147
Water–Scored Copper	0.0068
Water–Teflon pitted Stainless Steel	0.0058
Carbon tetrachloride–Copper	0.013
Carbon tetrachloride–Emery-polished Copper	0.0070
Benzene–Chromium	0.010
n-Butyl alcohol–Copper	0.00305
Ethyl alcohol–Chromium	0.027
Isopropyl alcohol–Copper	0.00225
n-Pentane–Chromium	0.015
n-Pentane–Emery-polished Copper	0.0154
n-Pentane–Emery-polished Nickel	0.0127
n-Pentane–Lapped Copper	0.0049
n-Pentane–Emery-rubbed Copper	0.0074
35% K_2CO_3–Copper	0.0054
50% K_2CO_3–Copper	0.0027

Now the pressure dependence is brought by assuming C to be a pure function of pressure, or in dimensionless form of reduced pressure (P^*)

$$C = C_0 F(P^*) \qquad (7.50)$$

where

$$P* = \frac{P}{P_{cr}}$$

where P is boiling pressure, and P_{cr} is critical pressure. The pure pressure effect $F(P^*)$ is taken relative to a base value of reduced pressure P_o^* (usually equal to 0.1). Thus

$$F\left(P_o^*\right) = 1 \qquad (7.51)$$

Combination of Eqs. (7.50) and (7.51) gives

$$h = C_o F(P^*)q^n \tag{7.52}$$

At a chosen value of temperature and pressure (P_o^*), the heat transfer coefficient is h_o given by (using Eq. (7.52))

$$h_o = C_o F(P_o^*)q_o^n = C_o q_o^n \tag{7.53}$$

Dividing Eq. (7.52) by Eq. (7.53) gives

$$\frac{h}{h_o} = F(P^*) \left(\frac{q}{q_o}\right)^n \tag{7.54}$$

The pressure function was developed from experimental data by Danilowa, Haffner and Gorenflow (VDI 1996) as

$$F(P^*) = 1.2P^{*0.27} + \left(2.5 + \frac{1}{1 - P^*}\right) P^* \quad \text{for fluids other than water} \tag{7.55}$$

$$F(P^*) = 1.73P^{*0.27} + \left(6.1 + \frac{0.68}{1 - P^*}\right) P^{*2} \quad \text{for water} \tag{7.56}$$

However, in reality pressure effect cannot be completely separated from the heat flux effect and the heat flux exponent depends on pressure, as

$$n = 0.9 - 0.3P^{*0.3} \quad \text{for fluids other than water} \tag{7.57}$$

$$n = 0.9 - 0.3P^{*0.15} \quad \text{for water} \tag{7.58}$$

The next modification is made to Eq. (7.54) by incorporating the effect of surface roughness as

$$\frac{h}{h_o} = F(P^*) \left(\frac{Ra}{Ra_o}\right)^{0.133} \left(\frac{q}{q_o}\right)^{n(P^*)} \tag{7.59}$$

Usually, the reference value of h_o is taken from

$$q_o = 20000 \text{ W/m}^2$$
$$P_o^* = 0.1$$
$$Ra_o = 0.4 \text{ μm}$$

The performance of the VDI correlation is shown in Fig. 7.16. The extension of this correlation to finned tubes can be made by changing $n(P^*)$ to $n_f(P^*)$

$$n_f(P^*) = n(P^*) - 0.1\frac{h}{t_l} \tag{7.60}$$

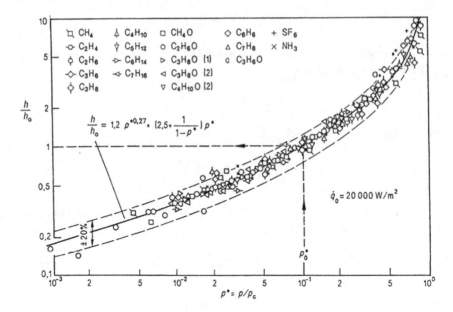

Fig. 7.16 Performance of the VDI correlation

where

t_l = fin spacing
h = fin height
$n(P^*)$ = is for the unfinned tube.

However, the pressure function also gets changed to $F_f (P^*)$

$$F_f(P^*) = F\left(\frac{P^*}{\sqrt{\psi}}\right) \tag{7.61}$$

where

$$\psi = \frac{\text{Finned surface area}}{\text{Plain tube surface area}}$$

Cornwell type correlation for tubes: Cornwell and Houston (1994) suggested a simpler correlation for pool boiling on horizontal tubes in which the tube diameter and not bubble diameter is chosen as characteristic length. He has also used VDI type reduced pressure function as

$$Nu_b = A F(P^*) Re_b^{0.67} Pr^{0.4} \tag{7.62}$$

$$A = 9.7 P_{cr}$$

Fig. 7.17 The Cornwell
correlation

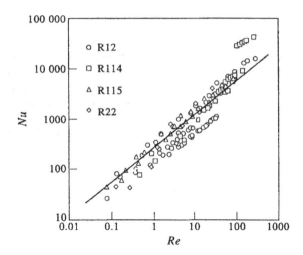

where P_{cr} is in bar

$$F(P^*) = 1.8P^{*0.17} + 4P^{*1.2} + 10P^{*1.0}$$

This correlation is valid for

Pressure: 0.001–$0.8P_{cr}$
Heat flux: 0.1–$0.8q_c$

Tube diameter: 8–50mm
Surface: machined

In this correlation, Re_b and Nu_b are defined as

$$Re_b = \frac{qD}{\mu_L h_{fg}}$$

$$Nu_b = \frac{hD}{k_L}$$

where D = tube diameter. This is depicted in Fig. 7.17.

7.3.3 Empirical Correlations

A large number of empirical correlations can be suggested from a variety of experiments. Some of these correlations are dimensionally consistent (i.e., independent of the units used) but in some others only prescribed units can be used. A well-known empirical pool boiling correlation is the Stephan and Preusser (1979) correlation,

$$Nu = 0.1 \left[\frac{q_o d_b}{k_L T_s}\right]^{0.67} \left[\frac{\rho_G}{\rho_L}\right]^{0.156} \left[\frac{h_{fg} d_b^2}{\alpha_L^2}\right]^{0.371} \left[\frac{\alpha_L^2 \rho_L}{\sigma d_b}\right]^{0.35} \left[\frac{\mu_L C_{PL}}{k_L}\right]^{-0.16}$$

(7.63)

where d_b is the bubble departure diameter. The correlation suggested by Mostinski (1963) reads as

$$h = 0.106 \, P_{cr}^{0.69} q^{0.7} \, f(P^*)$$

(7.64)

where

$$f(P^*) = 1.8 P^{*0.17} + 4 P^{*1.2} + 10 P^{*1.0}$$

(7.65)

Here, h is in $W/m^2 K$ and P_c in bar. The correlation suggested by Cooper (1984) is

$$h = A P^{*(0.12 - 0.2 \log_{10} \varepsilon)} \, (-\log_{10} P^*)^{-0.55} \, M^{-0.5} \, q^{0.67}$$

(7.66)

where

$M = $ molecular weight
$\epsilon = $ surface roughness in microns
$A = 55$ for copper plate or stainless steel cylinders
    $= 93.5$ for copper cylinders
h is in $W/m^2 K$
q is in W/m^2.

Examples

Problem 1: Give quantitative details suggesting why boiling in Microwave Ovens is dangerous. Also, suggest safety measures.

Solution: In a microwave, while boiling in smooth containers, due to the absence of nucleation sites, the amount of superheat is very high. The superheated state is unstable and can very rapidly turn into liquid at the boiling point, producing a substantial quantity of vapour. The latent heat of the vaporization of water is $L = 2.23$ MJ/kg. The specific heat capacity of water is $C = 4.2$ kJ/kg. Suppose we heat 1 kg of water at 100 °C to 101 °C, i.e., superheat by 1 °C. When it begins to boil, it will very quickly cool to 100°C and the heat liberated turns water into steam. Cooling this water by 1 °C gives 4.2 KJ which is enough to evaporate $C/L = 0.0018$ kg of water. This is only 1.9 ml of water, but it turns into 3 l of steam. These 3 litres of steam evolved quite suddenly, so the water is ejected violently from the container, and hence is dangerous. To avoid this, before putting water into the oven, insert a non-metal object with a surface that is rough and use a container which is a little scratched and not perfectly smooth. Reheating increases the superheat. Hence avoid reheating as well.

Problem 2: Water on a polished stainless steel surface ($C_{sf} = 0.0132$) is boiling at atmospheric pressure. Determine the heat flux at 106 °C. Also find the critical heat flux for nucleate boiling. (Given: σ at 100 °C is $58.8 \times 10^{-3} \approx$ N/m.)

Solution: From steam tables,

$$h_{fg} = 2257 \text{ kJ/kg}, \ \rho_l = 958 \text{ kg/m}^3, \ \rho_v = 0.6 \approx \text{kg/m}^3$$

From property tables,

$$C_{pl} = 4211 \text{ J/kg°}C, \ P_{rl} = 1.75, \ \mu_l = 2.77 \times 10^{-4} \text{ kg/ms}$$

We know that

$$\frac{C_{pl} \, \Delta T}{h_{fg} \ \text{Pr}_l^n} = C_{sf} \left[\frac{q}{\mu_l h_{fg}} \sqrt{\frac{\sigma'}{g(\rho_l - \rho_v)}} \right] \quad (n = 1 \quad \text{for water})$$

and we get q from the above equation as

$$q = \left[\frac{(4211) \, (106 - 100)}{(0.0132) \, (2.25 \times 10^6) \, (1.75)} \right]^3 (2.83)(2.257 \times 10^6)$$

$$\times \left[\sqrt{\frac{(958 - 0.6) \, 9.81}{58.8 \times 10^{-3}}} \right]$$

$$q = 29.28 \text{ kW/m}^2 \text{ [assuming nucleate boiling]}$$

$$q_{critical} = \frac{\pi}{24} \rho_v \, h_{fg} \left[\frac{\sigma'g \, (\rho_l - \rho_v)}{\rho_v^2} \right]^{1/4} \left(1 + \frac{\rho_v}{\rho_l} \right)^{1/2}$$

$$q_{critical} = \frac{\pi}{24} \times 0.6 \times 2257000 \times \left[\frac{(0.00588)(9.81)(958 - 0.6)}{0.6^2} \right]^{1/4}$$

$$\times \left(1 + \frac{0.6}{958} \right)^{1/2}$$

$$q_{critical} = 621.74 \text{ W/m}^2 \ (> \text{ heat flux})$$

so, our assumptions are correct.

Exercises

1. A thin wire of unknown material has a diameter of 1 mm and is 200 mm in length. It is submerged horizontally in water at atmospheric pressure and saturation temperature. The wire has a voltage drop of 10 V across it and a 25A current flowing through it to maintain its temperature at 115 °C. Determine $C_{s,f}$ of the wire material, assuming Rohsenow correlation is valid and $n = 1$.
2. A metallic copper plate is heated to a temperature of 100 °C. It is then submerged in water vertically at atmospheric pressure and maintained at the same temperature. What is the heat transferred per unit area?
3. A long, 1mm diameter wire passes electrical current dissipating 3150 W W/m and reaches a surface temperature of 126 °C. When submerged in water at 1 atm. What is the boiling heat transfer coefficient? Estimate the value of the correlation coefficient $C_{s,f}$.

References

Bankoff SG, Mason JP (1962) Heat Transfer from the surface of a steam bubble in a turbulent subcooled liquid stream. AIChE J 8(1):30–33 (2)

Barthau G, Hahne E (2000) Nucleation site density and heat transfer in nucleate pool boiling of refrigerant R134a in a wide pressure range. In: Proceedings of 3rd European Thermal Science Conference Heidelberg, Germany, 2000, vol 2, pp 731–736

Beer H (1969) Contribution to heat transfer in boiling. Progr Heat Mass Transf 2:311–370

Benjamin RJ, Balakrishnan AR (1996) Nucleate pool boiling heat transfer of pure liquids at low to moderate heat fluxes. Int J Heat Mass Transf 39(12):2495–2504

Churchill SW, Chu HHS (1975) Correlating equations for laminar and turbulent free convection from a vertical plate. Int J Heat Mass Transf 18:1323–1329

Cornwell K, Schüller RB (1982) A study of boiling outside a tube bundle using high speed photography. Int J Heat and Mass Transf 25(5):683–690

Cornwell K, Houston SD (1994) Nucleate pool boiling on horizontal tubes: a convection-based correlation. Int J Heat Mass Transf 37:303–9

Cooper MG (1984) Heat flow rates in saturated nucleate pool boiling-a wide-ranging examination using reduced properties. In Advances in heat transfer, 1984, vol 16, pp 157–239. Elsevier

Das SK, Roetzel W (2002) Heat Transfer model for pool boiling on a horizontal tube. In: Presented at the international heat transfer conference, Grenoble

Forster HK, Greif R (1959) Heat transfer to boiling liquid, mechanism and correlations. Trans ASME J Heat Transfer 81:43–53

Fritz W (1935) Berechnung des Maximavolumens von Dampfblasen. Phys Z 36:379–384

Han CY, Griffith P (1965) The mechanism of heat transfer in nucleate pool boiling-I and II. Int J Heat Mass Transf 8(6):887–914

Hsu YY, Graham RW (1961) An analytical and experimental study of the thermal boundary layer and ebullition cycle in nucleate boiling. N. p, Web, United States

Hsu YY, Graham RW (1976) Transport processes in boiling and two-phase systems, including near-critical fluids. Chaps 5 and 6, Hemisphere, New York

Jakob M (1949) Heat transfer, vol 1, Chap 29. Wiley, New York

Jakob M, Linke W (1933) The heat transfer from a horizontal plate to boiling water. Res Field Eng A 4(2):75–81

Judd RL, Hwang KS (1976) A comprehensive model for nucleate pool boiling heat transfer including microlayer evaporation. Trans ASME J Heat Transf 98:623–629

Kast W (1964) Bedeutung der Keimbildung und der instationären Wärmeübertragung für den Wärmeübergang bei Blasenverdampfung und Tropfenkondensation. Chemie Ingenieur Technik 36: 933–940

Malenkov IG (1971) Detachment frequency as a function of size for vapor bubbles. J Appl Phys 20:704–708

Mikic BB, Rohsenow WM (1969) A new correlation of pool boiling data including the effect of heating surface characteristics. J Heat Transfer 91:245–250

Moore FD, Mesler RB (1961) The measurement of rapid surface temperature fluctuations during nucleate boiling of water. AIChE J 7:620–624 (2)

Mostinski IL (1963) Calculation of heat transfer and critical heat flux in boiling liquids based on the law of corresponding state. Teploenergetika 10:66–71

Rohsenow WM (1952) A method of correlating heat transfer data for surface boiling of liquids. Trans ASME 74:969–976

Sateesh G, Das Sarit K, Balakrishnan AR (2005) Analysis of pool boiling heat transfer: effect of bubbles sliding on the heating surface. Int J Heat Mass Transf 48:1543–1553

Sharp RR (1964) The nature of liquid film evaporation during nucleate boiling. National Aeronautics and Space Administration. N. p, Web

Stephan K, Preusser P (1979) Heat transfer and critical heat flux in pool boiling of binary and ternary mixtures. Ger Chem Eng (Engl. Transl.); (Germany, Federal Republic of) 2(3). Springer Berlin, Heidelberg

Van Stralan SJD, Sohal MS, Cole R, Sluyter WM (1975) Bubble growth rates in pure and binary systems: combined effect of evaporation and relaxation micro-layer. Int J Heat Mass Transf 18:453–467

Thorncroft GE, Klausnera JF, Mei R (1998) An experimental investigation of bubble growth and detachment in vertical up flow and down flow boiling. Int J Heat Mass Transf 41(23):3857–71

VDI-Wärmeatlas (1996) Editors Verein Deutscher Ingenieure VDI-Gesellschaft Verfahrenstechnik und Chemieingenieurwesen (GVC), https://doi.org/10.1007/978-3-540-32218-4

Wang CH, Dhir VK (1993) Effect of surface wettability on active nucleation site density during pool boiling of water on a vertical surface. J Heat Transf 115(3):659–669

Zuber N (1959) Hydrodynamic aspects of boiling heat transfer. AECU-4439

Chapter 8
Pool Boiling Crisis, Critical Heat Flux and Film Boiling

In Chap. 6, we have seen that during nucleate boiling the heat flux increases with wall superheat. However, the Nukiyama curve (Fig. 6.10) clearly shows that this increase in heat flux is not unbounded. It increases significantly, but reaches a peak (point 'C' of Fig. 6.10) at a particular wall superheat. This peak heat flux is known in boiling literature as 'Critical Heat Flux' or CHF. Often this phenomenon is called 'Pool Boiling Crisis'. Another term quite often used to describe this phenomenon is 'Burnout', indicating that the sudden rise in wall superheat at this point (meaning a large increase in heater temperature) may cause burnout of the heater. However, this is a term used to describe the phenomenon, and a physical burnout may not happen at this point. We have indicated in Chap. 6 that the covering of the heating surface by a vapour film resulting from the merging of the vapour bubbles is the main cause of the CHF to occur. This film is inherently unstable, and the instability causes the heat flux to decrease first reaching the minimum heat flux point known as the Leidenfrost point (Fig. 6.10), then leading to stable film boiling. Thus, it is important to know and calculate the critical heat flux for any boiling equipment so that it does not fail due to burnout. In this chapter, we try to understand the physical phenomena leading to the occurrence of critical heat flux and the method to calculate it under various conditions.

Beyond the Leidenfrost point, stable film boiling takes place. Although in boiling equipment this is avoided, in certain conditions such as accidents in nuclear reactors or overloading of refrigeration equipment this may occur. Hence, there is a need to develop a comprehensive theory and an understanding of film boiling as well. What makes the situation more complex in film boiling is the role of radiation due to the large temperature difference across the vapour film. In this chapter, we will also look at these aspects of film boiling and the associated phenomena.

© The Author(s) 2023
S. K. Das and D. Chatterjee, *Vapor Liquid Two Phase Flow and Phase Change*, https://doi.org/10.1007/978-3-031-20924-6_8

8.1 Vapour Instability-Based CHF Model

8.1.1 Model for Horizontal Heater

In 1981, John Lienhard IV proposed a simple theory which describes CHF on horizontal surfaces excellently. The theory is based on the instability of vapour layers. Two types of instabilities are said to be responsible for CHF:

1. Instability of the vapour film covering the heating surface.
2. Instability of the vapour jets rising through the liquid on the vapour layer.

Figure 8.1 shows a schematic of these two types of instabilities.

It must be noted that the vapour layer on the heating surface is drawn to be wavy in Fig. 8.1. This is indeed true because in this case the liquid, which is orders of magnitude denser than the vapour (at ambient pressure), is supported by the vapour layer. A heavy fluid on a light fluid is always an unstable structure (think of a heavy person on the shoulder of a thin fellow and vice versa, which one do you think is more stable?). This will form what is classically known as the Taylor instability, and the wave is known as the Taylor wave. The wavelength of the Taylor wave is given by (Bellman and Pennington 1954)

$$\lambda_T = C \left[\frac{\sigma}{(\rho_l - \rho_g)g} \right]^{\frac{1}{2}} \tag{8.1}$$

The value of C depends on the dimensionality of the wave as well as the surface geometry, for example, $C = 2\pi\sqrt{3}$ for 1D wave on a horizontal rod, and $C = 2\pi\sqrt{6}$ for 2D wave on a flat plate.

It is also important to note that the Taylor wavelength is related to the Laplace constant used for assessing the bubble departure and heat transfer correlation used in Eq. 7.17 in Chap. 7:

$$L' = \left[\frac{\sigma}{(\rho_l - \rho_g)g} \right]^{\frac{1}{2}}$$

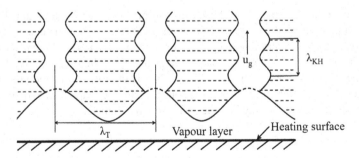

Fig. 8.1 Instabilities during film boiling

Fig. 8.2 Repeated cell
structure of vapour jets on
flat surface. (*Courtesy:*
"Boiling Condensation and
Gas-Liquid flow" by P. B.
Whalley, with permission
from Oxford University
Press)

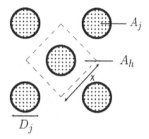

If we idealize the situation and consider that the vapour jets are equal in diameter
and equispaced on a 2D flat plate, the plan view of the jets will look as shown in
Fig. 8.2.

In this repeated design, the average distance between the jets must be equal to the
Taylor wavelength, λ_T. For 2D flat plate with $C = 2\pi\sqrt{6}$ in Eq. 8.1, this length can
be given by

$$\lambda_T = 2\pi\sqrt{6}\left[\frac{\sigma}{(\rho_l - \rho_g)g}\right]^{\frac{1}{2}} \qquad (8.2)$$

It can be readily seen that the diagonal pitch between the vapour columns x (shown
in Fig. 8.2) can be given by

$$x = \frac{\lambda_T}{\sqrt{2}} = 2\pi\sqrt{3}\left[\frac{\sigma}{(\rho_l - \rho_g)g}\right]^{\frac{1}{2}} \qquad (8.3)$$

It may be noted that the value of x is λ_T in the case of 1D wave (say on a horizontal
rod).

The second type of instability during the occurrence of CHF is the instability of
the vapour jets. This is demonstrated in Fig. 8.1 which shows how the liquid with high
density tries to squeeze a jet from the sides causing instability. This type of instability
is called the Kelvin–Helmholtz instability. This squeezing does not remain in one
place but moves upward with the jet.

Now let us examine under what condition the vapour jet will be able to sustain.
In other words, when can the squeezing of the vapour jet by the liquid completely
snap the vapour column? We have seen that (Chap. 4) the pressure difference across
a vapour–liquid curvature can be given by

$$\Delta p \propto \frac{\sigma}{r}$$

Considering that the radius of curvature of the Kelvin–Helmholtz wave is proportional to its wavelength, we can say

$$\Delta p \propto \frac{\sigma}{\lambda_{KH}}$$

or

$$\Delta p = K_1 \frac{\sigma}{\lambda_{KH}} \qquad (8.4)$$

where λ_{KH} is the Kelvin–Helmholtz wavelength as shown in Fig. 8.1 and K_1 is a constant.

For sustaining a jet, the dynamic pressure of the jet must be higher than this pressure difference, giving

$$\Delta p \leqslant \frac{1}{2} \rho_g u_g^2$$

or

$$K_1 \frac{\sigma}{\lambda_{KH}} \leqslant \frac{1}{2} \rho_g u_g^2$$

where ρ_g and u_g are the density and velocity of the vapour jets. Hence, we can get

$$u_g \geqslant \left(\frac{2K_1\sigma}{\rho_g \lambda_{KH}} \right)^{\frac{1}{2}} \qquad (8.5)$$

Turner (1973) indeed showed that $K_1 = \pi$ and hence

$$u_g \geqslant \left(\frac{2\pi\sigma}{\rho_g \lambda_{KH}} \right)^{\frac{1}{2}} \qquad (8.6)$$

Now with the above derivation, we can proceed to evaluate CHF on a flat plate conveniently. We have seen in Fig. 8.2 that, as an approximation, we can consider the vapour jets to be equispaced because they are likely to form from the peaks of Taylor waves. Hence, instead of the whole heating surface, we can only consider a cell structure which is repeated as shown in Fig. 8.2.

Obviously, the area of this cell will be x^2 (it is a square with a side of length x). Let us assume that the diameter of the jet is D_j. Then the rate at which vapour is formed in a cell is

$$\dot{m} = \frac{\pi}{4} D_j^2 u_g \rho_g \qquad (8.7)$$

The heat associated with vapour formation is given by the latent heat required for this rate of vapour formation

$$\dot{Q} = \frac{\pi}{4} D_j^2 u_g \rho_g h_{fg} \qquad (8.8)$$

where h_{fg} = latent heat of evaporation.

Now, within a cell this heat comes from the supplied heat flux (q) over an area x^2, thus

$$\dot{Q} = qx^2 = \frac{\pi}{4} d_j^2 u_g \rho_g h_{fg} \qquad (8.9)$$

giving

$$q = \frac{\pi}{4} \left(\frac{d_j}{x} \right)^2 u_g \rho_g h_{fg} \tag{8.10}$$

It was first guessed that the jet diameter is $x/2$, and this was later also confirmed to be true. Hence, we have $d_j/x = 1/2$. This results in

$$q = \frac{\pi}{16} u_g \rho_g h_{fg} \tag{8.11}$$

Now, when the heat flux is critical heat flux (CHF), the vapour stem is just able to be sustained, which according to Eq. 8.6 is

$$u_g = \left(\frac{2\pi\sigma}{\rho_g \lambda_{KH}} \right)^{\frac{1}{2}} \tag{8.12}$$

Here, we make one assumption that the diagonal pitch of the vapour jets, x or Taylor wavelength for 1D wave, is equal to Kelvin–Helmholtz wavelength at the critical heat flux. This is a reasonable assumption looking at the instability structures. Thus,

$$\lambda_{KH} = x = \frac{\lambda_T}{\sqrt{2}} = 2\pi\sqrt{3} \left[\frac{\sigma}{(\rho_l - \rho_g)g} \right]^{\frac{1}{2}} \tag{8.13}$$

Substituting λ_{KH} from Eq. 8.13 and u_g from Eq. 8.12 in Eq. 8.11 gives, at critical condition,

$$q_{cr} = \frac{\pi}{16} \rho_g h_{fg} \left(\frac{2\pi\sigma}{\rho_g} \frac{1}{2\pi\sqrt{3}} \right)^{\frac{1}{2}} \left[\frac{(\rho_l - \rho_g)g}{\sigma} \right]^{\frac{1}{4}} \tag{8.14}$$

$$= \frac{\pi}{16 \times 3^{\frac{1}{4}}} \rho_g^{\frac{1}{2}} h_{fg} \left[\sigma(\rho_l - \rho_g)g \right]^{\frac{1}{4}} \tag{8.15}$$

Hence,

$$q_{cr} = 0.149 \rho_g h_{fg} \left[\frac{\sigma(\rho_l - \rho_g)g}{\rho_g^2} \right]^{\frac{1}{4}} \tag{8.16}$$

Kutateladze (1952) was the first to derive an expression which is close to the above expression, given by

$$q_{cr} = K_l \rho_g h_{fg} \left[\frac{\sigma(\rho_l - \rho_g)g}{\rho_g^2} \right]^{\frac{1}{4}} \tag{8.17}$$

He proposed the value of K_l to be 0.16. K_l is in reality Kutateladze number given by

$$K_l = u_g \rho_g^{\frac{1}{2}} \left[\sigma(\rho_l - \rho_g)g \right]^{\frac{-1}{4}} \tag{8.18}$$

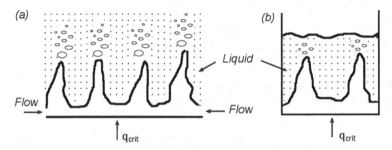

Fig. 8.3 **a** CHF with lateral vapour entry. **b** CHF without lateral vapour entry. (*Courtesy:* "Boiling Condensation and Gas-Liquid flow" by P. B. Whalley, with permission from Oxford University Press)

The constant value of 0.16 for this number is a topic which will be discussed along with flooding during condensation. However, Zuber (1959) derived the above equations in a little different way and got a slightly different value of the constant ($K_l = 0.131$). His equation can be written as

$$q_{cr} = 0.131 \rho_g h_{fg} \left[\frac{\sigma(\rho_l - \rho_g)g}{\rho_g^2} \right]^{\frac{1}{4}} \left[\frac{\rho_l}{\rho_l + \rho_g} \right] \tag{8.19}$$

The last term of the right-hand side approaches unity because $\rho_l \gg \rho_g$, giving an equation of the form of Eq. 8.17. This Eq. 8.19 closely follows experimental data on CHF.

However, the original equation we derived, Eq. 8.16, also performs quite well against experimental data if two conditions are fulfilled (Lienhard and Dhir 1973).

These conditions are

1. There is no liquid entry from the side unlike that is seen in Fig. 8.3a. In this case, the entire jet mass does not originate from applied heat flux and hence our derivation fails. Our derivation considers a situation depicted in Fig. 8.3b.
2. The heater length is large compared to the Taylor wavelength. This is because, our assumption of repeated jet structure is a statistical averaging which works only if there are a sufficient number of jets. Figure 8.4 shows how the data starts deviating from the equation as the heater length decreases.

Sun and Lienhard (1970) extended the above work to horizontal cylinders. It was found to be very similar for large cylinders

$$q_{cr} = 0.116 h_{fg} \rho_g^{\frac{1}{2}} \left[\sigma(\rho_l - \rho_g)g \right]^{\frac{1}{4}} \tag{8.20}$$

For small cylinders, they found the constant to be

$$K_l = 0.116 + 0.3 e^{-3.4 R'^{\frac{1}{2}}} \tag{8.21}$$

Fig. 8.4 CHF data and
correlation for a flat plate

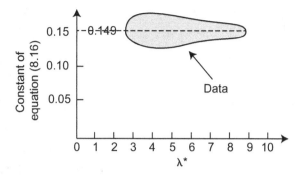

Fig. 8.5 CHF on a
horizontal cylinder

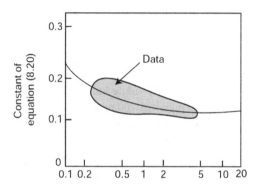

where $R' = R/L'$ and $L' = $ is Laplace constant

$$\left[\frac{\sigma}{(\rho_l - \rho_g)g} \right]^{\frac{1}{2}}$$

Figure 8.5 shows the match of this correlation with experimental data. It can be seen that with a larger diameter, the constant approaches 0.116.

8.1.2 Model for Vertical Surfaces

Around the same time as Zuber (1958), Chang (1957) proposed a wave- based model for CHF on a vertical surface. Later on, Chang (1962) developed a more general model with and without forced flow in the presence of liquid subcooling. He also considered CHF to be a problem of stability of a single bubble growing on a vertical surface as shown in Fig. 8.6.

In this model, the liquid turbulence is considered as the disturbing force, while surface tension is considered as the stabilizing force. Chang (1962) made several assumptions, such as

Fig. 8.6 Bubble model of
Chang (1962)

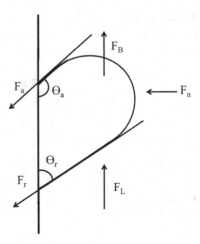

1. The analysis of a single bubble is representative of the average behaviour with
   respect to bubble dynamics, frequency of departure and nucleation site density.
2. Bubbles are spherical and they are at hydrodynamic equilibrium. In the case of
   subcooled boiling, they are also at thermodynamic equilibrium.

Under these assumptions, he carried out a force balance on the bubble attached to
the vertical wall. In reality, the bubble configuration is as shown in Fig. 8.6. As
the fluid flow tends to move the bubble upward, two different contact angles of
advancing (θ_a) and receding (θ_r) are formed. However, Chang approximated them as
$\theta_a \approx \theta_r \approx 90°$. The two surface tension forces of F_a and F_r can thus be approximated
as $F_a \approx F_r \approx F_s$. This gives the following estimate of forces:
Buoyancy Force

$$F_B = C_B r_b{}^3 (\rho_l - \rho_g) g$$

Surface tension force

$$F_S = C_S r_b \sigma$$

Tangential inertial force

$$F_t = C_t r_b{}^2 \rho_l u_{rel}{}^2$$

Normal inertial force

$$F_n = C_n r_b{}^2 \rho_l v_{rel}{}^2$$

where u_{rel}, v_{rel} are liquid velocity relative to the bubble and C_B, C_S, C_t, and C_n are
constants. Making a vertical force balance, we get at the point of departure

$$F_B + F_t = 0 \tag{8.22}$$

giving

$$C_B g(\rho_l - \rho_g)r_b^2 + C_t \rho_l u_{rel}^2 r_b = 0 \qquad (8.23)$$

From this, the departure size of the bubble can be calculated as

$$r_b = \left(\frac{C_t}{2C_B}\right)\left(\frac{\sigma u_{rel}^2}{\rho_l V_b^4}\right)\left\{\left[1 + 4\frac{C_S C_B}{C_t^2}\left(\frac{V_b}{u_{rel}}\right)^4\right]^{\frac{1}{2}} - 1\right\} \qquad (8.24)$$

where V_b is the detaching velocity of the bubble, given by

$$V_b = \left[\frac{g\sigma(\rho_l - \rho_g)}{\rho_l^2}\right]^{\frac{1}{4}} \qquad (8.25)$$

As stated earlier, the forces that determine the stability of a rising bubble on a vertical wall are the inertia and surface tension forces. The ratio of these forces is the Weber number given by

$$We = \frac{\rho_l v_{rel}^2 r_b}{\sigma} \qquad (8.26)$$

The critical velocity at which these forces act at CHF can be related to the critical Weber number, We_{cr}

$$v_{rel,cr} = \left(\frac{We_{cr}\sigma}{\rho_l r_b}\right)^{\frac{1}{2}} \qquad (8.27)$$

Using this velocity and radius, Chang (1962) derived the CHF on vertical surface, which is

$$q_{cr,vert} = 0.098 h_{fg} \rho_g^{\frac{1}{2}} \left[\sigma(\rho_l - \rho_g)g\right]^{\frac{1}{4}} \qquad (8.28)$$

Bernath (1960) showed that the ratio between the vertical and horizontal surface CHF is constant. Thus

$$\left(\frac{q_{cr,vert}}{q_{cr,hori}}\right) = 0.75 \qquad (8.29)$$

This gives critical heat flux on a horizontal surface as

$$q_{cr,hori} = 0.13 h_{fg} \rho_g^{\frac{1}{2}} \left[\sigma(\rho_l - \rho_g)g\right]^{\frac{1}{4}} \qquad (8.30)$$

It can be noted that this agrees completely with Eq. 8.19 which suggested (by Zuber, 1958) the constant to be 0.13. Lienhard and Dhir (1973) suggested the Kutateladze number K_l which is the constant in the CHF expression but is not a constant number. It varies with the liquid and also its pressure. Kutateladze and Malenkov (1974) showed that, in fact, the value of K_l depends on the velocity of sound, U_s, in the vapour. The dimensionless parameter on which this depends is given by

$$\frac{1}{U_s{}^*} = \frac{1}{U_s}\left[\frac{g\sigma}{\rho_l - \rho_g}\right]^{\frac{1}{4}} \tag{8.31}$$

It was found that, for $15 \times 10^{-5} < \frac{1}{U_s{}^*} < 70 \times 10^{-5}$, K_l value lies in the range

$$0.06 < K_l < 0.19$$

8.2 Microlayer-Based Thermal Model

The above models are more or less hydrodynamics-based models. A different approach looking at the liquid microlayer formation process was suggested by Katto and Yokoya (1968) and Haramura and Katto (1983). This model identifies the formation of large vapour lumps as a transient process arising out of the merging of smaller vapour bubbles. The formation and evaporation of microlayer or in other words drying out of microlayer is identified as a periodic process where the layer thickness decreases during the hovering period of the large bubble as shown in Fig. 8.7.

The merging of the small bubbles and its relation to microlayer are described in Fig. 8.3. It can be seen that the microlayer dries out at a rate which is determined by the heat flux, and the heat flux is maximum when the microlayer dries out completely at the final stage of hovering duration. Thus, a heat balance gives

$$\rho_l h_{fg}(1 - \alpha)\delta_c = q_c \tau_d \tag{8.32}$$

where α is the volumetric fraction of vapour in microlayer and δ_c is the microlayer thickness.

Using bubble dynamics, we can find out that the hovering period can be given as τ_d as

$$\tau_d = \left(\frac{3}{4\pi}\right)^{\frac{1}{5}}\left[\frac{4(\zeta\rho_l + \rho_g)}{g(\rho_l - \rho_g)}\right]^{\frac{3}{5}} v_1^{\frac{1}{5}} \tag{8.33}$$

where v_1 is the volumetric growth rate of vapour.

For horizontal plate $\zeta = 11/16$. If a vapour mass is assumed to be formed over an area of square shape for which each side is equal to the diagonal of Taylor wavelength,

Fig. 8.7 Periodic dryout of microlayer

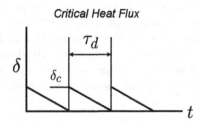

Fig. 8.8 Microlayer and its relation to bubble merging

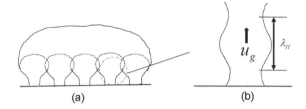

(a)   (b)

x (this is somewhat similar to what we have considered in Sect. 8.1), v_1 can be estimated as

$$v_1 = \lambda_T{}^2 \frac{q}{\rho_g h_{fg}}$$ (8.34)

putting $x = \lambda/\sqrt{2}$

$$2\pi\sqrt{3}\sqrt{\frac{\sigma}{\rho_l - \rho_g}g}$$

We get

$$v_1 = \frac{12\pi^2 \sigma q}{\rho_g g h_{fg}(\rho_l - \rho_g)}$$ (8.35)

Figure 8.8b shows that, as an approximation, the microlayer thickness can be assumed to be the maximum when it is $\frac{1}{4}$th of Kelvin–Helmholtz instability, giving

$$\delta_c = \frac{\lambda_{KH}}{4} = \frac{\pi}{2}\frac{\sigma}{\rho_g u_g{}^2}$$ (8.36)

This can be obtained from Eq. 8.12. Now, the heat balance in the vapour stem gives

$$u_g \alpha \rho_g h_{fg} = q$$ (8.37)

which results in

$$q_{cr} = K_l h_{fg} \rho_g^{\frac{1}{2}} \left[\sigma(\rho_l - \rho_g)g\right]^{\frac{1}{4}}$$ (8.38)

Thus even by the alternate route, we get a correlation for CHF which is very similar to Kutateladze or Zuber.

8.3 Factors Affecting Critical Heat Flux

Critical heat flux is affected by a large number of factors having moderate to large influence. CHF is primarily a function of the fluid as evident from the thermophysical properties involved in the CHF correlation Eq. 8.17. However, the following factors, which explicitly do not occur in the correlation, are also found to have considerable influence.

8.3.1 Liquid Subcooling

If the bulk liquid is subcooled, the CHF increases substantially. Subcooling of the bulk liquid may be a result of a large heat loss from the vessel where pool boiling takes place. However, since it is difficult to control such heat loss, it is difficult to control subcooling. This is why there is a scarcity of data for CHF of subcooled boiling. Ivey and Morris (1966) presented the following correlation based on their experiment on a horizontal wire:

$$\frac{q_{cr,sub}}{q_{cr,sat}} = 1 + \left(\frac{\rho_g}{\rho_l}\right)^{\frac{3}{4}} \frac{C p_l \Delta T_{sub}}{h_{fg}} \tag{8.39}$$

where $q_{cr,sub}$ and $q_{cr,sat}$ are CHF under subcooled and saturated conditions respectively, and ΔT_{sub} is the amount of subcooling of the bulk liquid. This equation suggests that even a small subcooling can give considerable enhancement of CHF. Elkassabgi and Lienhard (1988) extended the CHF study with liquid subcooling to a number of fluids such as R-113, iso-propanol, methanol and acetone. Although they found similar linear relationships as Ivey and Morris (1966) at a lower range of subcooling, their data showed that the CHF saturates to an asymptotic value ranging $40K < \Delta T < 60K$.

They correlated the lower ΔT range of data as

$$\frac{q_{cr,sub}}{q_{cr,sat}} = 1 + 4.28 \frac{\rho_l C_{p_l} \Delta T_{sub}}{\rho_g h_{fg}} \left[\frac{K_l(\rho_l - \rho_g)^{\frac{1}{4}} \rho_g^{\frac{1}{2}}}{\sigma^{\frac{3}{4}}}\right]^{\frac{1}{4}} \tag{8.40}$$

Shoji and Yoshihara (1991) also observed similar asymptotic behaviour although with a higher asymptotic limit. They observed the emission of a large number of small bubbles of micron sizes which is known as Micro-bubble Emission Boiling (MEB). The above equation can also be written as

$$\frac{q_{cr,sub}}{q_{cr,sat}} = 1 + 4.28 Ja Pe^{\frac{-1}{4}} \tag{8.41}$$

where

$$Pe = \text{Peclet number} = \left[\frac{\sigma^{\frac{3}{4}}}{\alpha(\rho_g - \rho_l)^{\frac{1}{4}} \rho_g^{\frac{1}{2}}}\right]$$

$$Ja = \text{Jacob number} = \frac{\rho_l C_{p_l} \Delta T_{sub}}{\rho_g h_{fg}}$$

The value of CHF for the asymptotic limit at high subcooling was suggested as

$$\frac{q_{cr,sub}}{\rho_g h_{fg}\sqrt{\dfrac{R_g T_{sat}}{2\pi}}} = 0.01 + 0.0047 e^{-1.11\times10^{-6}X} \tag{8.42}$$

where

$$X = \frac{R(R_g T_{sat})^{\frac{1}{2}}}{\alpha}$$

R_g = Ideal gas constant.

8.3.2 Effect of Pressure

Since CHF is a function of fluid properties, it is expected that pressure will have an influence on the value of CHF. The following are the trends of various thermophysical properties with pressure:

1. $h_{fg} \longrightarrow$ Continuously reduces with pressure till critical point.
2. $\sigma \longrightarrow$ Continuously fall with pressure.
3. $\rho_g \longrightarrow$ Continuously increases with pressure.
4. $(\rho_g - \rho_l) \longrightarrow$ Continuously fall with pressure.

As a consequence of the interplay of these properties, CHF first increases with reduced pressure, p^* $(= p/p_{critical})$ reaches a maximum and then decreases with p^* to reach a value of zero at the critical point.

This variation can be given by

$$\frac{q_g}{q_{g0}} = 2.8 p^{*0.4}(1 - p^*) \tag{8.43}$$

where q_{g0} is the CHF at $p^* = 0.1$.

It is interesting to note that for water this peak value of CHF is reached at $p = 70$ bar which is exactly the same as the value of CHF reached during flow boiling (to be discussed later). For $p^* < 0.1$, in place of Eq. 8.43, the following equation is suggested

$$\frac{q_g}{q_{g0}} = 1.05(p^{*0.2} + p^{*0.5}) \tag{8.44}$$

8.3.3 Effect of Surface Condition, Roughness and Wettability

We know that the heating surface has got a large effect on the boiling process. Both surface roughness and wettability of the surface influence CHF as well. Berenson (1960) carried out boiling experiments with n-pentane on surfaces of varying roughness and found that CHF increases with roughness. Liaw and Dhir (1986) and Haramura (1991) also showed that roughness increases CHF value.

Apart from roughness, surface wettability also affects CHF significantly. Usually, surface wettability is designated by the contact angle. The retreating contact angle on a surface is usually manipulated either by the coating of materials on the surface or oxidizing the surface itself. However, this also changes surface thermal resistance affecting the wall superheat. Liaw and Dhir (1986) demonstrated the effect of surface wettability to show that CHF decreases with retreating contact angle shown in Fig. 8.9.

This dependence can be given by

$$\frac{q_c}{\rho_g h_{fg} \left[\sigma g \frac{\rho_l - \rho_g}{\rho_g{}^2} \right]^{\frac{1}{4}}} = 0.1 e^{-\frac{\theta}{45°} + 0.055} \tag{8.45}$$

where $\theta =$ contact angle in degrees.

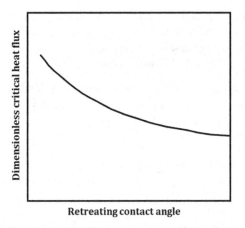

Fig. 8.9 Effect of wettability on CHF

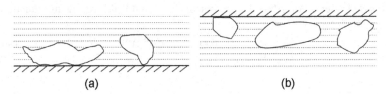

Fig. 8.10 Bubble on horizontal heated surfaces: **a** facing upward and **b** facing downward

8.3.4 Effect of Flow (Heater Orientation and Flow Velocity)

The manner in which a liquid goes past the heater influences the occurrence of critical heat flux. For a pure pool boiling where the liquid is initially at rest, the flow of liquid sets in due to natural convection. We know that natural convection heat transfer is critically dependent on the orientation of the plate because the cause of fluid flow is buoyancy here. Thus, all the phenomena which affect the CHF like bubble formation, departure, coalescence, film formation and instability are influenced by the orientation of the heater. Figure 8.10 shows two extreme configurations of heater orientation horizontal facing upward and downward. It is obvious that the bubbles will be readily carried away in the first configuration, while they will accumulate on the surface in the second.

The experiments on surfaces of varying orientation correlated as

$$\frac{q_{cr,\phi}}{q_{cr,0}} = \left[1 - \frac{\phi}{190°}\right]^{\frac{1}{2}} \tag{8.46}$$

for $0° \leqslant \phi \leqslant 165°$. This indicates that the CHF decreases with increasing inclination of the heater with the horizontal plane.

When instead of a stagnant pool, flow is imposed on a heater, the CHF occurs on the heater with film structure depending on heater dimension and flow velocity. The vapour forms a jet structure on a cylinder discussed earlier as shown in Fig. 8.11a. However, for wires of diameter less than 1 mm, with an increase in liquid velocity, a 2D sheet structure of vapour is formed as shown in Fig. 8.11b.

When velocity is sufficiently high to form such 2D structure, CHF is proportional to the imposed liquid velocity.

8.4 Film Boiling

The phenomenon beyond CHF is called film boiling because in this range of wall superheat, due to the high rate of vapour formation, a vapour film is formed on the heated surface which hinders the heat flow. As described in connection with the discussion on the Nukiyama curve, the first part of film boiling between CHF and Leidenfrost point is called the 'unstable' or 'transition' film boiling. Beyond the Leidenfrost point, it is called 'stable' or 'fully developed' film boiling. In the unstable part of film boiling since the film appears and disappears, at a given point of time only a part of the heating surface remains covered with vapour and the rest is wetted by the liquid. As the superheat increases, the ratio between the vapour and liquid-bathed part increases but till the Leidenfrost point some part remains wetted by the liquid. For fully developed film boiling, a continuous vapour film forms on the surface but due to large wall superheat, thermal radiation starts playing a major role. The zones of transition and fully developed film boiling are shown in Fig. 8.12.

Fig. 8.11 Film structure in
the presence of liquid
velocity

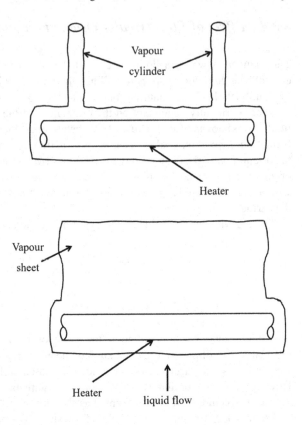

Technologically speaking, film boiling is technologically less important than
nucleate boiling. This is because nucleate boiling gives a larger amount of heat
transfer with a lower wall superheat. This is why high-temperature boiling equipment such as boilers and steam generators are designed to avoid film boiling. Film
boiling will raise the metal temperature in the equipment leading to failure under
pressure or simply through the melting of the metal. However, even for this purpose
one needs to know film boiling and its dependence on various parameters. Particularly
for cases where failure of the equipment can have a catastrophic effect like that in a
nuclear reactor, film boiling and its consequence need to be studied in detail. Apart
from this, there are a large number of applications in which film boiling takes place
naturally—cryogenic systems, liquid cooling of high-speed nozzles and quenching
of steel during heat treatment. In the following section, we will explain simple cases
of film boiling from the fundamentals of heat transfer.

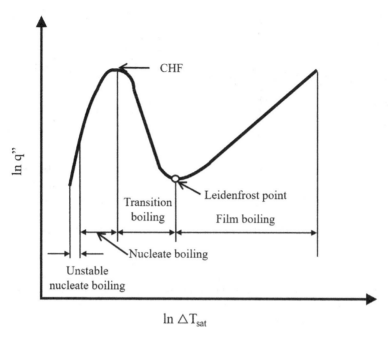

Fig. 8.12 Nukiyama curve and film boiling

8.4.1 Fully Developed Film Boiling

Bromley (1950) was the first one to analyse fully developed or stable film boiling. He took the idea of the formation of a thin film of one phase on a surface from the theory of film condensation of Nusselt (this will be discussed in full detail later). Nusselt theory assumes that the plate is covered by a thin, continuous, laminar, subcooled liquid film. Using an analogy, Bromley assumed a continuous, laminar, superheated vapour film on the heating surface inside the saturated liquid. He also used another major assumption of the Nusselt theory that the heat transfer through the vapour film takes place by pure conduction, i.e., the effect of heat convection by the film is neglected.

Following the same method as used for film condensation given in Chap. 11, we can set an equation for the vapour film for mass and momentum balance. These equations are similar to typical boundary layer equations. We must note that we are not solving the energy equation because convective heat transfer is neglected in the film. Then, what is the need to solve these equations? It has to be noted that Nusselt film condensation theory was developed for a liquid film falling on a vertical wall under gravity as shown in Fig. 8.13a. Similarly, Bromley developed the theory for a rising vapour film on a vertical wall as shown in Fig. 8.13b.

Since in both cases the film thickness changes along the wall and hence although the thermal conductivity may be the same throughout the film, the heat transfer will

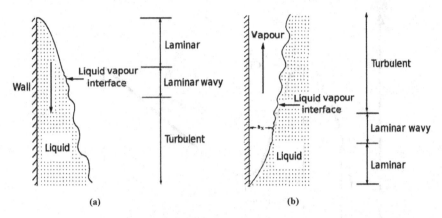

Fig. 8.13 Comparison of film condensation and film boiling. **a** Nusselt's falling liquid film and **b** Bromley's rising vapour film

vary due to film thickness. The continuity and momentum equation merely helps to evaluate thin film thickness at any location of the plate. Using the method described in Chap. 11, the film thickness can be given by

$$\delta_x = C \left[\frac{\mu_g \dot{M}}{\rho_g (\rho_l - \rho_g) g b} \right]^{\frac{1}{3}} \tag{8.47}$$

where $\dot{M}$ is the vapour mass flow rate and b is the width of the plate. We can assume that the liquid at the film interface flows at the same velocity as the vapour. This makes the solution similar to that of Nusselt theory of film condensation (this is described in detail in Chap. 11 and hence not repeated here). Under this condition, the constant C takes a value identical to that in the Nusselt theory of condensation, $C = \sqrt[3]{3} = 1.44$. It has to be remembered that in this equation the fluid (vapour) properties are to be evaluated at the mean film temperature of

$$T_f = \frac{T_w + T_{sat}}{2}$$

Now since we are considering heat transfer only by conduction (similar to Nusselt's condensation theory), the equivalent film boiling heat transfer coefficient can be given by

$$h_{f bx} = \frac{K_g}{\delta_x} \tag{8.48}$$

where $h_{f bx}$ is the local heat transfer coefficient due to conduction in the vapour film alone, K_g the thermal conductivity of vapour and δ_x is the local film thickness. However, we have discussed earlier that at the stable film boiling regime, the difference between wall temperature and liquid (saturated) temperature is large and hence radi-

Fig. 8.14 a Wavy
vapour–liquid interface, and
b Smooth vapour–liquid
interface

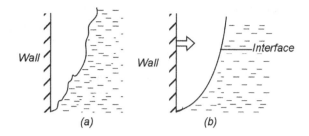

(a)   (b)

ation heat transfer between the liquid–vapour interface and the wall is significant. In
reality, the liquid–vapour interface is wavy as shown in Fig. 8.14a. However, if we
assume the interface to be smooth, it appears like Fig. 8.14b. Under this condition,
the local radiation heat flux between the wall and the vapour–liquid interface is given
by

$$q = \frac{\sigma(T_w{}^4 - T_{sat}{}^4)}{R_r} \tag{8.49}$$

The relation comes from the Stefan–Boltzmann Law of Radiation where σ is the
Stefan–Boltzmann constant ($\sigma = 5.67 \times 10^{-8} \text{W/K}^4$) and R_r is the radiation resistance between the two surfaces, namely the wall and the liquid interface. This resistance between two parallel plates (locally we can assume that the wall and the interface are parallel) can be calculated to be

$$R_r = \frac{1}{\epsilon_w} + \frac{1}{\epsilon_l} - 1 \tag{8.50}$$

If we do not consider the wall and the interface to be parallel or consider the waviness
of the film, we have to add the resistance due to the 'view factor' or 'shape factor'
between the wall and the interface. In the case of parallel infinite film assumption,
this is taken as unity ($F_{wl} = 1$).

Now using Eq. 8.50 and the definition of equivalent heat transfer coefficient due
to radiation alone, we can get from Eq. (8.49),

$$h_r = \frac{q}{T_w - T_{sat}} = \frac{\sigma}{\frac{1}{\epsilon_w} + \frac{1}{\epsilon_l} - 1} \frac{T_w{}^4 - T_{sat}{}^4}{T_w - T_{sat}} \tag{8.51}$$

where ϵ_w and ϵ_l are wall and liquid surface emissivities, respectively. The net heat
transfer coefficient through the film can be assumed to be a simple sum of the heat
transfer coefficients due to pure conduction and radiation. Thus

$$h_{net} = h_{f\,bx} + h_r = \frac{K_g}{\delta_x} + h_r \tag{8.52}$$

Fig. 8.15 Vapour control
volume for analysis

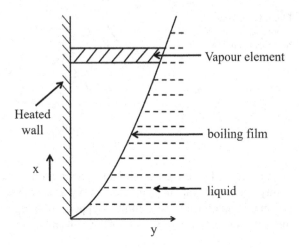

However, we cannot determine this net heat transfer easily since the local vapour
flow rate ($\dot{M}$) is not known. Please note that as more and more liquid evaporates,
the film thickness increases and the local vapour flow rate also increases with height
(x coordinate). As shown in Fig. 8.15 if we take a small vapour control volume, the
rate of change of vapour mass flow rate over the short length can be taken as $d\dot{M}$.
This is then, the liquid mass evaporated over a thin length absorbing latent heat of
vaporization h_{fg}. Over this small control volume, the heat transfer can be calculated
using the net heat transfer coefficient h_{net}.

Thus, the energy balance suggests that the heat transfer due to evaporation must
have come from the wall, giving

$$\Delta h d\dot{M} = h_{net}(T_w - T_{sat})b dx \qquad (8.53)$$

where Δh is the enthalpy difference between the vapour at film temperature T_f and
the saturation temperature T_{sat} (please note that Δh is not related of heat transfer
coefficient). Thus, we can write

$$\frac{(T_w - T_{sat})b}{\Delta h} \int_0^x dx = \int_0^{\dot{M}} \frac{d\dot{M}}{h_{fbx} + h_r} \qquad (8.54)$$

Please note that at the bottom of the plate, film thickness is zero giving $\dot{M} = 0$ at
$x = 0$.

Now we can introduce film thickness as a function of the mass flow rate of the
vapour from Eq. 8.47, as

$$\delta_x = C_1 \dot{M}^{\frac{1}{3}} \qquad (8.55)$$

where

$$C_1 = C \left[\frac{\mu_g}{\rho_g (\rho_l - \rho_g) g b} \right]^{\frac{1}{3}}$$

This gives

$$h_{fbx} = \frac{K_g}{C \dot{M}^{\frac{1}{3}}}$$

Substituting this in Eq. 8.53 and integrating both sides gives

$$\frac{(T_w - T_{sat}) bx}{\Delta h} \int_0^x dx = \frac{3C}{K_g b_1{}^4} \left[\frac{1}{3} b_1{}^3 \dot{M} - \frac{1}{2} b_1{}^2 \dot{M}^{\frac{1}{3}} - ln(1 + b_1 \dot{M}^{\frac{1}{3}}) \right] \quad (8.56)$$

where

$$b_1 = \frac{h_r C_1}{K_g}$$

Solving this transcendental equation, one can get the mass flow rate at a particular location and the local heat transfer equation from Eqs. 8.52 and 8.47. However, the task of finding an analytical equation from this for the average heat transfer equation is more complex. Roetzel (1979) accomplished this task brilliantly to suggest the final form of the equation, as

$$\frac{64}{81} z^4 = \frac{64}{81} z^3 y - \frac{8}{9} z^2 y^{\frac{2}{3}} + \frac{4}{3} z y^3 - ln \left(1 + \frac{4}{3} z y^3 \right) \quad (8.57)$$

where
$$z = \frac{h_r}{h_g}$$
$$y = \frac{h}{h_g}$$
h = average heat transfer coefficient in the presence of radiation
h_g = average heat transfer coefficient in the absence of radiation

$$\frac{4}{3} \left[\frac{\Delta h \rho_g (\rho_l - \rho_g) g \lambda_g{}^3}{16 \mu_g (T_w - T_{sat})} \frac{1}{x} \right]^{\frac{1}{4}}$$

Roetzel also showed that this complex equation can be approximated as

$$y = 1 + \frac{1}{5} z \left(4 + \frac{1}{1 - \frac{3}{z}} \right) \quad (8.58)$$

We can evaluate the value of y from a known value of z (i.e., h_r and $\bar{h}_g$). From y, the average heat transfer coefficient over the entire plate length x can be evaluated for film boiling. He also showed that for the entire range of z, Eq. 8.57 approximates the original exact solution of Eq. 8.56 within 1%. Before this, Bromley (1950) also suggested the implicit relation,

$$\bar{h} = \bar{h}_g \left(\frac{\bar{h}_g}{\bar{h}}\right)^{\frac{1}{2}} + h_r \qquad (8.59)$$

This is an approximate relation which is widely used in applications and often further approximated as

$$\bar{h} = \bar{h}_g + 0.75 h_r \qquad (8.60)$$

However, it must be said that the correlations of Roetzel (1979) given by Eqs. 8.56 and 8.57 are much more accurate. For boiling on plates of various other orientations, a large number of correlations are available. However, the following general correlation converts the vertical plate result (given above) to other orientations satisfactorily

$$\bar{h}_\theta = \bar{h}_{vertical} \sin \theta^{\frac{3}{8}} \quad \text{for } 30° < \theta < 180° \qquad (8.61)$$

where $\bar{h}_\theta$ is the film boiling heat transfer coefficient of a wall at an angle of θ with the horizontal and $\bar{h}_{vertical}$ is that for the vertical plate described above.

Examples

Problem 1: Water is boiled at atmospheric pressure by a horizontal polished copper heating element of diameter 5 mm and emissivity 0.05 immersed in water. If the surface temperature of the heating wire is 350 °C, determine the rate of heat transfer from the wire to the water per unit length of the wire.

The properties of water at the saturation temperature of 100 °C are $h_{fg} = 2257$ kJ/Kg, and density is 957 kg/m^3. The properties of the vapour at the film temperature are as follows. Vapour density is 0.441 kg/m^3. $C_{p_v} = 2000$ J/kgK. Viscosity is 1.73×10^{-5} kg/m.s. Thermal conductivity of the vapour is 0.0357 W/mK.

Solution: The excess temperature in this case is 250 °C. For water, CHF occurs at about an excess temperature of 30 °C. So, we can safely say that film boiling is taking place. The heat flux for film boiling on a horizontal cylinder is given by

$$\dot{q}_{film} = C_{film} \left[\frac{g k_g^3 \rho_v \left(\rho_l - \rho_g\right) \left[h_{fg} + 0.4 C_{p_g} \left(T_s - T_{sat}\right)\right]}{\mu_g D \left(T_s - T_{sat}\right)} \right]^{\frac{1}{4}} (T_s - T_{sat})$$

On substituting the values, we get

$$\dot{q}_{film} = 5.93 \times 10^4 \text{ W/m}^2$$

The radiation heat flux is given by

$$\dot{q}_{rad} = \epsilon\sigma(T_s^4 - T_w^4)$$

which gives radiation flux to be $157\,\text{W/m}^2\,\text{K}$.

The total heat flux is equal to $5.94 \times 10^4\,\text{W/m}^2\,\text{K}$, which gives the rate of heat transfer to be

$$Q_{total} = q_{total} \times Area$$
$$= 933\,W$$

Problem 2: Calculate the critical heat flux for the following fluids at 1 atm: mercury, ethanol and refrigerant R-12. Compare these results to the critical heat flux for water at 1 atm. The properties of the fluids are given in the following table.

	h_{fg} (kJ/kg)	ρ_g (kg/m³)	ρ_l (kg/m³)	$\sigma \times 10^3$ (N/m)	T_{sat} (K)
Mercury	301	3.90	12740	417	630
Ethanol	846	1044	757	17.7	351
R-12	165	6.32	1488	15.8	243
Water	2257	0.596	957.9	58.9	373

Solution: The critical heat flux can be estimated by the modified Zuber–Kutateladze correlation, which is

$$q_{cr} = 0.149\rho_g h_{fg} \left[\frac{\sigma(\rho_l - \rho_g)g}{\rho_g^2}\right]^{\frac{1}{4}}$$

Consider the numerical values for mercury

$$q''_{max} = 0.149 \times 301 \times 10^3 \times 3.90 \times \left[\frac{417 \times 10^{-3} \times 9.8 \times (12740 - 3.90)}{3.90^2}\right]^{\frac{1}{4}}$$

$$q''_{max} = 1.34\,\text{MW/m}^2$$

For the other fluids, the results are tabulated along with the ratio of the critical heat fluxes to that of water

	q''_{max} (MW/m²)	$q''_{max}/q''_{max,water}$
Mercury	1.34	1.06
Ethanol	0.512	0.41
R-12	0.241	0.19
Water	1.26	1.00

Exercises

1. What is the critical heat flux for boiling water at 1 atm on the surface of the moon, where the gravitational acceleration is one-sixth that of the earth?
2. Estimate the current at which a 1-mm diameter nickel wire burnout when submerged in water at atmospheric pressure. The electrical resistance of the wire is $0.129\omega/m$.
3. Advances in very large-scale integration (VLSI) of electronic devices on a chip are often restricted by the ability to cool the chip. For mainframe computers, an array of several hundred chips, each area of $26\,mm^2$, may be mounted on a ceramic substrate. A method of cooling the array is by immersion in a low boiling point fluid such as refrigerant R-113 at one atmosphere and 321 K. Calculate the critical heat flux and estimate the power dissipated by the chip when it is operating at 45% of CHF; what is the corresponding value of the chip temperature. The properties of saturated liquid are

$$\mu = 5.147 \times 10^4\,N.s/m^2$$

$$C_p = 983.8\,J/kgK$$

and

$$Pr = 7.183,$$

$$C_{sf} = 0.004,\ n = 1.7$$

4. The vacuum insulation around a spherical liquid helium vessel develops a leak and allows the heat transfer rate of $1kW/m^2$ to land on an external surface of the vessel. An amount of saturated liquid helium at atmospheric pressure boils at the bottom of the vessel. Calculate the excess temperature by assuming nucleate boiling with $C_{sf} = 0.02$ and $S = 1.7$. Compare the heat leak with the critical heat flux for nucleate boiling in the pool of liquid helium.

References

Bellman R, Pennington RH (1954) Effects of surface tension and viscosity on Taylor instability. Q Appl Math 12(2):151–62
Berenson PJ (1960) Transition boiling heat transfer. In: 4th National heat transfer conference, AIChE Preprint 18, Buffalo, NY
Bernath L (1960) A theory of local-boiling burnout and its application to existing data. Chem Eng Progr 56
Bromley LA (1950) Heat transfer in stable film boiling. Chem Eng Prog 46:221–227
Chang YP (1962) Some possible critical conditions in nucleate boiling ASME. J. Hear Trunsjk 85:89–100
Chang YP (1957) A theoretical analysis of heat transfer in natural convection and in boiling. Trans ASME J Heat Transf 79:1501–1513

Elkassabgi Y, Lienhard JH (1988) The peak pool boiling heat flux from horizontal cylinders in subcooled liquids. ASME J Heat Transfer 110:479–486

Haramura Y (1991) Temperature uniformity across the surface in transition boiling. Trans ASME J Heat Transf 113:980–984

Haramura Y, Katto Y (1983) New hydrodynamic model of critical heat flux, applicable widely to both pool and forced convection boiling on submerged bodies in saturated liquids. Int J Heat Mass Transf 26(3):389–399

Ivey HJ, Morris DJ (1966) Critical heat flux of saturation and subcooled pool boiling in water at atmospheric pressure. International Heat Transfer Conference Digital Library. Begel House Inc

Katto Y, Yokoya S (1968) Principal mechanism of boiling crisis in pool boiling. Int J Heat Mass Transf 11:993–1002

Kutateladze SS, Malenkov LG (1974) Heat transfer at boiling and barbotage similarity and dissimilarity. In: Proceedings of 5th international heat transfer conference, Tokyo, vol IV, p 1

Kutateladze SS (1952) Heat transfer in condensation and boiling, USAEC Rep. AEC-tr-3770

Liaw SP, Dhir VK(1986) Effect of surface wettability on transition boiling heat flux on vertical surface. In: Proceedings of international heat transfer conference, 2031-2036

Lienhard JH, Dhir VK (1973) Extended hydrodynamic theory of the peak and minimum pool boiling heat fluxes, NASA CR-2270, Contract No. NGL 18-001-035

Roetzel W (1979) Berechnung der Lei tung und Strahlung bei der Filmverdampfung and der ebenen Platte. Warme Stoffubertrag. 12:1–4

Shoji M, Yoshihara M (1991) Burnout heat flux of water on a thin wire. In Proc. of 28th National Heat Transfer Symposium of Japan, 1991

Sun KH, Lienhard JH (1970) The peak pool boiling heat flux on horizontal cylinders. Int J Heat Mass Transf 13(9):1425–39

Turner JS (1973) Buoyancy effects in fluids. Cambridge University Press

Zuber N (1958) On the stability of boiling heat transfer. Transactions ASME 80(3):711–4

Zuber N (1959) Hydrodynamic aspects of boiling heat transfer. AECU-4439

Chapter 9
Flow Boiling Heat Transfer

In Chap. 1, we have observed various regimes of two-phase flow when boiling takes place inside tubes. We have also defined various quantities which can characterize flow boiling such as void fraction, thermodynamic quality, actual quality and slip ratio. In Chap. 2, in addition to adiabatic flow, we have also demonstrated methods by which pressure drop in tubes can be estimated when the phase change takes place during flow. The various flow regimes and calculation methods for pressure drop clearly demonstrate that the analysis of flow boiling is much more complex and uncertainties involved are much greater in flow boiling compared to pool boiling. Consequently, the heat transfer mechanisms are also complex where various phenomena involving inertia, gravity and surface tension forces interact and due to rapid flow, the thermal equilibrium also is not attained on many occasions. Pool boiling presented in the previous chapter is dominated by natural convection, on the contrary flow boiling is predominantly forced convection. The nucleation and evaporation mechanisms here are complex and often indistinguishable. As a result, an engineering approach based on flow regimes rather than flow physics is usually adopted here.

9.1 Various Regimes of Flow Boiling

In Chap. 1, we have seen various regimes of flow boiling flow patterns such as bubbly flow, slug flow, churn flow, annular flow and drop or spray flow. The sketch of these flow patterns was shown for vertical upflow in Fig. 1.6. Here, we reproduce the same figure along with various heat transfer regimes of flow boiling as shown in Fig. 9.1. The various regimes of boiling observed are

1. Single phase convection in liquids.
2. Subcooled boiling.
3. Saturated nucleate boiling.

© The Author(s) 2023
S. K. Das and D. Chatterjee, *Vapor Liquid Two Phase Flow
and Phase Change*, https://doi.org/10.1007/978-3-031-20924-6_9

Fig. 9.1 Flow pattern and associated heat transfer regimes in flow boiling (*Courtesy:* "Heat Transfer in Condensation and Boiling" by Karl Stephan, with permission from Springer)

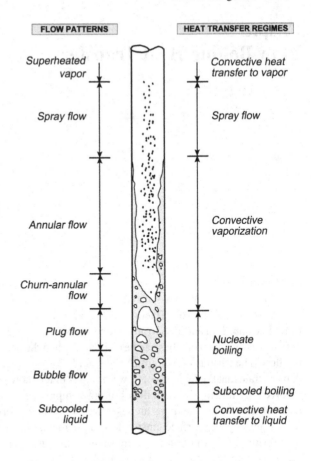

4. Convective boiling.
5. Spray or drop flow.
6. Single phase convection in vapour.

The subcooled boiling (we will discuss later) falls in the bubbly flow pattern with a low void fraction. Saturated nucleate boiling takes place in higher void fraction bubbly flow, slug/plug flow and churn or semi-annular flow region. Convective boiling almost coincides with an annular flow pattern. Heat transfer in the spray flow regime extends even beyond the thermodynamic quality of unity ($x_{th}^* > 1.0$).

These flow patterns and boiling regimes can be explained from the plot of average wall and fluid temperature as shown in Fig. 9.2. This figure brings out various features of flow boiling. First, in the single-phase liquid heat transfer region near the entry of the tube, we see that the wall and average fluid temperatures increase simultaneously at the same rate. We know this is a feature of fully developed single phase flow with constant wall heat flux (we have assumed here that the heat flux input to the wall of the tube is constant). The important fact to observe here is, even before the thermodynamic quality of the fluid reaches zero ($x_{th}^* = 0$), bubbles appear in

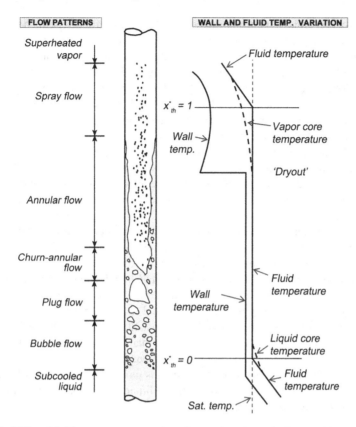

Fig. 9.2 Wall and fluid temperatures at various flow regimes in vertical upward flow (*Courtesy:* "Heat Transfer in Condensation and Boiling" by Karl Stephan, with permission from Springer)

the liquid and this region is called subcooled boiling. This can be explained in this way—although the average temperature of the fluid in this region is lower than the saturation temperature (at the local pressure), due to heat transfer from the wall, the liquid adjacent to the wall is superheated and the liquid at the core of the tube is largely subcooled (and the average fluid is at subcooled temperature). However, since the liquid near the wall is superheated, if the liquid superheat is enough to nucleate bubbles (as we have described in Chap. 4), the bubbles will be formed. It is called subcooled boiling because the average temperature of the cross section is still below saturated temperature but bubbles are being formed because of the lack of equilibrium across the cross section of the tube. As a result, although boiling has started at the wall, average fluid temperature continues to rise but the tube wall temperature becomes constant because bubbles are being formed near the wall and hence the fluid temperature near the wall does not rise. This continues until the average fluid temperature reaches saturation temperature (or average enthalpy of fluid in that cross section is equal to liquid saturation enthalpy h_f given in steam tables or property tables). It must be understood that although at this point the average fluid condition is saturated (liquid), its core is at a lower temperature and vapour bubbles present near the wall in such a proportion that the average enthalpy is that of saturated liquid enthalpy h_f. After this the saturated nucleate boiling takes place which is more or less like the nucleate pool boiling but the flow pattern changes due to the merging of bubbles and adherence of liquid to the wall, giving slug/plug and churn flow. The wall and fluid temperature remain constant here with the wall temperature a little above the fluid temperature.

As the fluid average condition becomes saturated, the only change that takes place is an increase in void fraction due to continuous vaporization. The churn flow is a stage in between slug and annular flows where irregular shapes of large vapour lumps form, thereby leading to annular flow where vapour remains at the core of the tube. In this region, the heat transfer mechanism gradually changes. Since the vapour at the core of the tube is approximately in a saturated state (of course with some dispersed liquid droplets), the temperature rise from the liquid–vapour interface to the wall gives the liquid near the wall only a small liquid superheat. This is because the liquid film at the wall is thin in annular flow. This small liquid superheat is not enough to form bubbles on the wall. Thus, as the liquid film becomes thinner, resulting in more evaporation, nucleate boiling gets more and more suppressed. However, the heat is provided at the same rate (because of constant heat flux) at the wall. Hence, the fluid has to find out an alternative way (other than wall nucleation) to evaporate. This mechanism is evaporation at the liquid–vapour interface. From another point of view, since the thermal conductivity of liquid is much higher than that of vapour, the mismatch in heat flux at the interface on the liquid and vapour side is large and this is bridged by evaporative heat flux:

$$k_v \frac{dT}{dr}\bigg|_{v,interface} = k_l \frac{dT}{dr}\bigg|_{l,interface} + \dot{m}_{evaporation} h_{lv} \qquad (9.1)$$

where k is thermal conductivity, r is radial coordinate, $\dot{m}_{evaporation}$ is evaporative mass flux and h_{lv} is the latent heat. Convective boiling is a very special mechanism typical to flow boiling only. In pool boiling, similar phenomenon is not observed and it continues till the wall remains wetted with liquid. However, with continuous evaporation at one point, the liquid film completely evaporates. This is called 'dry-out'. Since the liquid at this point completely evaporates at the wall, the wall faces pure vapour which gets superheated and the wall temperature jumps. In practical application, this state should be avoided. This is because, due to high temperature and pressure, the tube material may give in causing tube damage and leakage. The fluid temperature also increases here because of the superheating of vapour, although some liquid droplets remain suspended in it. The region beyond this is called spray or drop flow. This is just opposite to bubbly flow. Here, due to the large superheat in the vapour, the average enthalpy quickly crosses the saturation vapour state but due to nonequilibrium some liquid droplets also remain giving actual quality below unity ($x^*_{actual} < 1.0$). After some more heat transfer, these droplets evaporate and we get a pure single phase heat transfer if the tube material still sustains. It can be seen

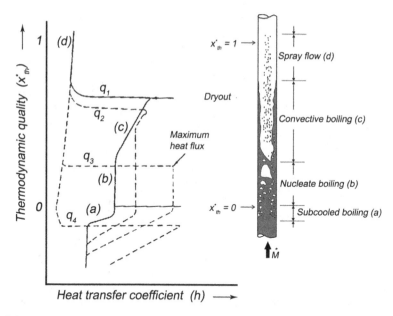

Fig. 9.3 Heat transfer coefficient in flow boiling (*Courtesy:* "Heat Transfer in Condensation and Boiling" by Karl Stephan, with permission from Springer)

from Fig. 9.2 that after dryout, the temperature difference between the fluid and wall increases indicating a fall in heat transfer coefficient (since heat flux supplied is constant). Thus, it makes sense to take a look at the heat transfer coefficient throughout the boiling process.

The picture depicted above is for medium heat flux. If the heat flux is increased, all the regimes of boiling depicted above do not occur in the way it is described above. This is shown in Fig. 9.3. Here, the heat transfer coefficient is plotted along the vertical tube. The solid line indicates the case of medium heat flux where all the boiling regimes are distinctly observable.

This is shown as curve q_1 where we find a constant heat transfer coefficient in the single phase liquid region. This is followed by an increase in the heat transfer coefficient in the subcooled boiling region leading to a constant value of the heat transfer coefficient due to a large temperature difference between the wall and fluid. This heat transfer coefficient then marginally increases because of an increase in vapour velocity due to evaporation of droplets in the spray flow region. However, as we move to higher heat flux (curve q_2), the subcooled boiling starts early due to the attainment of large liquid superheat at higher heat flux, and the subcooled boiling is extended further. The saturated boiling heat transfer is higher, and it also extends further squeezing the annular flow to a shorter length of the tube. The dryout occurs early here due to a larger heat flux. As we move to even higher heat flux (q_3), the annular flow is completely eliminated and the wall dryout occurs due to the crowding of bubbles near the wall (we will discuss it in the next chapter). Finally

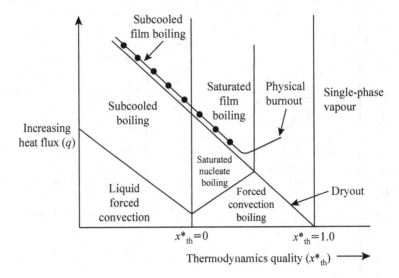

Fig. 9.4 Regions of flow boiling at different heat flux (Adapted from "Convective Boiling and Condensation" by John G. Collier and John R. Thome, with permission from Oxford University Press)

for extremely large heat flux (q_4), we may lose even saturated boiling regime and the wall dryout occurs during the subcooled boiling itself due to bubble crowding. These characteristics can be better understood by the heat flux versus quality plot provided by Collier and Thome (1994) as shown in Fig. 9.4.

9.2 Subcooled Flow Boiling

We have seen above that when the bulk enthalpy of the fluid is below the saturation enthalpy, boiling may still occur if the liquid near the wall is sufficiently superheated. We can also estimate the length over which this subcooled boiling can take place. Let us consider a case in which the liquid enters the tube in a subcooled state in which there are no bubbles (pure liquid), and the bulk temperature is below the saturation temperature (i.e., boiling point at the given pressure as shown in Fig. 9.5). Now the total heat transfer to the fluid over a length of dz from the entry can be given by the following expression when the heat flux q is constant,

$$q\pi Ddz = \dot{M}c_{p,l}\left[T_l(z) - T_{l,in}\right] \tag{9.2}$$

where D is the tube diameter, $\dot{M}$ is the inlet liquid flow rate, $c_{p,l}$ is the liquid specific heat, $T_l(z)$ is the temperature of the liquid at location z and $T_{l,in}$ is the liquid inlet temperature. Please note that this equation is valid only till the liquid does not boil,

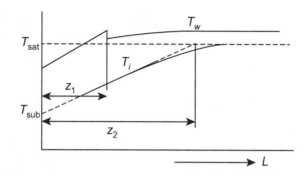

Fig. 9.5 Wall and bulk fluid temperature in subcooled and saturated flow boiling (*Courtesy:* "Heat Transfer in Condensation and Boiling" by Karl Stephan, with permission from Springer)

and its temperature increases due to sensible heating from the wall. Now, for sensible heating over a small length in this region, we can write heat transfer as

$$q\pi D dz = h\pi D dz \left[T_w(z) - T_l(z)\right] \tag{9.3}$$

where h is the local heat transfer coefficient and T_w the wall temperature. Combining Eqs. (9.2) and (9.3), we get

$$T_w = T_l(z) + \frac{q}{h} = T_{l,in} + q\left(\frac{\pi D z}{\dot{M} c_{p,l}} + \frac{1}{h}\right) \tag{9.4}$$

We know that, for a single phase fully developed flow with constant wall heat flux, the temperature of the bulk fluid increases linearly. Let us make an assumption that during subcooled flow boiling when the void fraction is small, this trend continues and the bulk fluid is primarily heated due to sensible heating at the wall. This is a reasonable assumption because of the small void fraction and the bulk enthalpy corresponding to the subcooled region. Now, for subcooled boiling to set in, a minimum liquid superheat is required (this is similar to nucleation during pool boiling as discussed earlier). Let us suppose that the minimum liquid superheat required for nucleation under the condition of the roughness (cavity size) is $\Delta T_{sup,ONB}$ or the liquid superheat required for the onset of nucleate boiling (ONB). At the wall, the liquid and the wall should be in equilibrium. Hence,

$$T_w - T_{sat} = \Delta T_{sup,ONB} \tag{9.5}$$

where T_{sat} is the saturation temperature at the tube pressure. Combining Eqs. (9.4) and (9.5), we get

$$T_w = T_{sat} + \Delta T_{sup,ONB} = T_{l,in} + q\left(\frac{D\pi z_1}{\dot{M} c_{p,l}} + \frac{1}{h}\right) \tag{9.6}$$

where z_1 is the distance at which subcooled boiling sets in. Hence,

$$z_1 = \left(\frac{\Delta T_{sub} + \Delta T_{sup,ONB}}{q} - \frac{1}{h} \right) \frac{\left(\dot{M} c_{p,l} \right)}{D\pi} \tag{9.7}$$

where ΔT_{sub} is the inlet subcooling of the liquid given by

$$\Delta T_{sub} = T_{sat} - T_{l,in}$$

Now, under the assumption of the same correlation extending up to the end of sub-cooled boiling (inception of saturated or fully developed boiling, FDB), we can say that the length, z_2, at which fully developed boiling starts is the point at which the bulk fluid attains a saturated temperature. In other words, $T_l(z) = T_{sat}$ in Eq. (9.2). This gives

$$q D\pi z_2 = \left(T_{sat} - T_{l,in} \right) \dot{M} c_{p,l} \tag{9.8}$$

$$z_2 = \frac{\Delta T_{sub} \dot{M} c_{p,l}}{q D\pi} \tag{9.9}$$

Subtracting Eq. (9.7) from (9.9) gives the length over which subcooled boiling takes place

$$\Delta z = z_2 - z_1 = \left(\frac{1}{h} - \frac{\Delta T_{sub,ONB}}{q} \right) \frac{\dot{M} c_{p,l}}{D\pi} = \left(\frac{1}{h} - \frac{\Delta T_{sub,ONB}}{q} \right) \frac{G c_{p,l} D}{4} \tag{9.10}$$

since $G = \frac{\dot{M}}{\pi/4 D^2}$. In Eq. (9.10), $\Delta T_{sub,ONB}$ is usually unknown quantity. We need to know it either from experimental data or our understanding and theoretical modelling of the phenomenon of onset of Nucleate Boiling (ONB). We will take up this next but mark that if the liquid enters the tube in a saturated state ($T_{l,in} = T_{sat}$), we get $q = h \left(\Delta T_{sub,ONB} \right)$ which gives $\Delta z = 0$. In other words, the subcooled boiling is absent here. From the above expression, we can also explain the phenomena we have seen earlier. For example, with the increase of heat flux (q), the region (Δz) of subcooled boiling increases.

9.2.1 Inception of Subcooled Flow Boiling

Before deriving the condition for the inception of subcooled flow boiling, let us first try to get the limits for it. It is obvious that no bubble can form if the wall temperature is below the saturation temperature. Hence we must have $T_w > T_{sat}$. Combining Eq. (9.6) with the above condition, we get

$$T_w = \left[T_{l,in} + q \left(\frac{D\pi z}{\dot{M} c_{p,l}} + \frac{1}{h} \right) \right] > T_{sat} \tag{9.11}$$

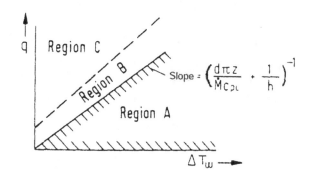

Fig. 9.6 Regions of subcooled pool boiling (*Courtesy:* "Heat Transfer in Condensation and Boiling" by Karl Stephan, with permission from Springer)

Substituting $\frac{G\pi D^2}{4} = \dot{M}$ and $\Delta T_{sub} = T_{l,in} - T_{sat}$, we get

$$T_w - T_{l,in} = q\left(\frac{4z}{GDc_{p,l}} + \frac{1}{h}\right) > \Delta T_{sub} \tag{9.12}$$

Here, z is the position at which the fluid temperature equals the saturation temperature in the limit and thereafter exceeds the saturation temperature making nucleation a possibility. Thus, if we plot the curve

$$q = \left(\frac{4z}{Gc_{p,l}D} + \frac{1}{h}\right)^{-1}\Delta T_{sub} \tag{9.13}$$

The region below it will show the regime in which nucleation is not possible. This is shown in Fig. 9.6 (Region A). The curve separating A and $(B + C)$ is a straight line with slope

$$\left(\frac{4z}{Gc_{p,l}D} + \frac{1}{h}\right)^{-1} = \left(\frac{d\pi z}{\dot{M}c_{p,l}} + \frac{1}{h}\right)^{-1}$$

and on this line, the wall temperature equals the saturation temperature of the fluid at the pipe pressure. Equation (9.10) also indicates that the factors which inhibit subcooling boiling are low heat flux, high fluid mass flux and large tube diameter. With these, the subcooled boiling shifts downstream and its inception is delayed.

However, subcooled boiling does not set up in its full extent immediately after the wall temperature exceeds saturation temperature. First, we only find single scattered nucleation sites (Region B, below the dotted line), and the heat transfer still remains primarily dominated by single phase convection at the wall. However, as the liquid near the wall gets more superheated, a fully developed subcooled boiling takes place and a large number of nucleation sites are activated (Region B, above the dotted line). As we have seen in the case of pool boiling, we must get a minimum amount of superheat for nucleation. In the case of flow boiling, liquid flow near the wall makes a significant change in the near-wall liquid temperature profile changing the condition for the inception of subcooled boiling. Let us now look at the quantitative

Fig. 9.7 Condition for
nucleation to take place at
wall first (*Courtesy:* "Boiling
Condensation and
Gas-Liquid flow" by P. B.
Whalley, with permission
from Oxford University
Press

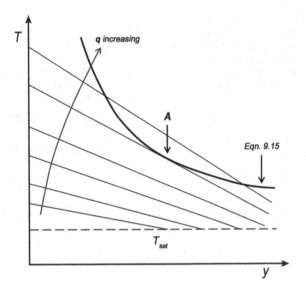

aspect of this condition for the inception of subcooled nucleate boiling. We have seen
that in pool boiling, the nucleation process is dominated by the temperature profile
in the fluid created by the transient conduction during the waiting period between
two consecutive bubble formations. In flow boiling, the situation is slightly different
since, due to the steady flow of liquid, a temperature profile exists near the wall from
which heat is transferred by forced convection. The layer of liquid just adjacent to
the wall is almost stagnant (no-slip condition) giving a pure conductive heat transfer
as shown in Fig. 9.7. It can be seen that in this region, the fluid temperature varies
linearly due to the domination of conduction in the near-wall no-slip region. Under
this condition, the temperature profile is given by Fourier's law of conduction as

$$q = -k_l \frac{T - T_w}{y} \tag{9.14}$$

where y is the distance in the liquid from the wall at which temperature is T and k_l
is the thermal conductivity of the liquid. Now, as shown in Chap. 4, the bubble tip
should be at a temperature higher than the liquid superheat required for nucleation
for the bubble to sustain, thus giving the following condition (for the bubble tip at a
distance y):

$$T - T_{sat} > \frac{2\sigma T_{sat}}{\rho_g y h_{fg}} \tag{9.15}$$

The plot in Eq. (9.15), assuming the two sides to be equal, divides the distance
temperature plot into two halves, which shows where nucleation is probable. Now
if we add to that Eq. (9.14) to show the temperature profiles at different heat fluxes,
we can superimpose the plots of these two equations as shown in Fig. 9.7. Davis and

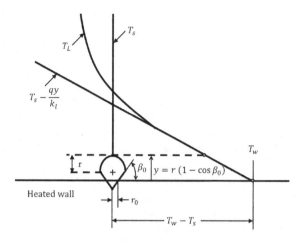

Fig. 9.8 Condition for nucleation in subcooled flow boiling (*Courtesy:* "Heat Transfer in Condensation and Boiling" by Karl Stephan, with permission from Springer)

Anderson (1966) presented a simple linearized theory for the criteria of subcooled boiling inception, which we present logically; the curve for Eq. (9.14) which touches the nucleation criteria (9.15) tangentially denotes the case in which the tip of the bubble in having just enough liquid superheat to continue to grow given by equation

$$T_w - \frac{qy}{k_l} = T_{sat} + \frac{2\sigma T_{sat}}{\rho_g y h_{fg}} \tag{9.16}$$

or

$$\Delta T_{\text{sup},ONB} = \frac{1}{y \left(\frac{q}{k_l} y^2 + \frac{2\sigma T_{sat}}{\rho_g y h_{fg}} \right)} \tag{9.17}$$

This is shown schematically in Fig. 9.8.

Now to derive an expression for finding out the condition for subcooled boiling, we also notice that for the condition of tangentiality of Eqs. (9.14) and (9.15), the slope of both curves must be equal. Thus,

$$-\frac{q}{k_l} = \frac{2\sigma T_{sat}}{\rho_g y^2 h_{fg}} \tag{9.18}$$

Substituting the value of y from Eq. (9.18) into Eq. (9.17) gives (meaning nucleation criteria is fulfilled at the bubble tip)

$$T_w - T_{sat} = \left(\frac{q h_{fg} \rho_g}{2\sigma T_{sat} k_l} \right)^{\frac{1}{2}} \left(\frac{4\sigma T_{sat}}{\rho_g h_{fg}} \right) \tag{9.19}$$

or

$$\left(\Delta T_{\text{sup},ONB} \right)^2 = \frac{8\sigma T_{sat} q}{\rho_g h_{fg} k_l} \tag{9.20}$$

Hence, the criterion for the inception of subcooled nucleate boiling is

$$\Delta T_{\text{sup},ONB} = T_{w,ONB} - T_{sat} = \left(\frac{8\sigma T_{sat}q}{\rho_g h_{fg} k_l}\right)^{\frac{1}{2}} \tag{9.21}$$

It is interesting to observe that if we consider pool boiling where the nucleation starts when the temperature reaches the liquid superheat condition given by Eq. (9.15), at a much lower cavity radius (for water at 1 bar and $\Delta T_{\text{sup},ONB} = 5K$, $r = 6.5\mu\text{m}$) the nucleation starts. But when we consider flow boiling, for water at 1 bar, Eq. (9.21) gives 4.38 K which is close to 5 K; the value of y calculated from Eq. (9.16) is $15\mu\text{m}$ which is more than double that for pool boiling. This is due to the temperature profile in the liquid near the wall under the steady-state condition. This simple theory was preceded by Bergles and Rohsenow (1964) who derived a dimensional equation by accounting for the variable slope of the saturation line $p_{sat} - T_{sat}$. According to them

$$\Delta T_{\text{sup},ONB} = \frac{5}{9}\left[\left(\frac{q}{1120}\right)^{0.463} p^{-0.535}\right]^{p^{0.0234}} \tag{9.22}$$

where q is in $^W/_{m^2}$, p in bar and ΔT in K. The range suggested for it was 1.03 bar $\leq p \leq 138$ bar.

It must be stated here that the approach given above is a very simplified version of the more detailed analysis by Hsu (1962) on pool boiling and suppression of saturated nucleate boiling. To ensure that the tip of the bubble does not condense and continue to grow, it was assumed that the tip of the bubble is in equilibrium with a superheated liquid which has just got enough liquid to superheat for equilibrium. For a spherical bubble, this temperature for spherical shape was derived in Chap. 4 as

$$\Delta T = T - T_{sat} = \frac{2\sigma T_{sat}}{\rho_v h_{lv} r}\left(1 + \Delta T\omega^*\right) \tag{9.23}$$

where

$$\omega^* = \frac{\rho_l - \rho_v}{\rho_l \sigma}\frac{d}{dT}\left(\frac{\rho_l \sigma}{\rho_l - \rho_v}\right) = \frac{d}{dT}\left[\ln\left(\frac{\rho_l \sigma}{\rho_l - \rho_v}\right)\right] \tag{9.24}$$

From the above equations, the liquid superheat can be estimated as

$$T(r) - T_{sat} = \frac{2\sigma T_{sat}}{\rho_v h_{lv} r}\left(\frac{1}{1 - \frac{2\sigma T_{sat}}{\rho_v h_{lv} r}\omega^*}\right) \tag{9.25}$$

Let us now consider the geometry for bubble growth given in Fig. 9.9. The height of the bubble and the cavity mouth diameter r_0 are given by

$$r_0 = r \sin \beta_0 \tag{9.26}$$

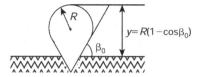

Fig. 9.9 Bubble geometry sitting on a cavity mouth (*Courtesy:* "Heat Transfer in Condensation and Boiling" by Karl Stephan, with permission from Springer)

Fig. 9.10 Bubble growth in subcooled liquid with T_w constant (*Courtesy:* "Heat Transfer in Condensation and Boiling" by Karl Stephan, with permission from Springer)

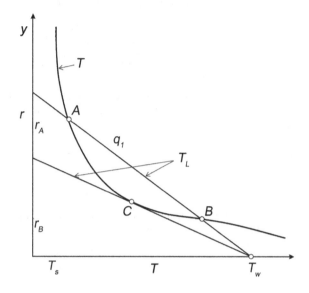

where r is the bubble radius and β_0 is the bubble contact angle with the surface. For a hemispherical bubble, $\beta_0 = \pi/2$ and $y = r = r_0$. Now, from this we can examine the inception nucleation in greater detail than the simple theory presented earlier. We recollect the constant heat lines for temperature versus distance from the wall in the liquid and the condition for nucleation (minimum liquid superheat) can be plotted in the same curve as shown in Fig. 9.10. The figure shows the minimum condition for nucleation in tangency of the heat flux. However, this figure is different from the previous Fig. 9.7 because in this figure it is assumed that the wall temperature is kept constant. How can we keep the wall temperature constant while heat flux at the wall is increased? It can only be done by increasing the strength of convection or increasing the mass flux. Thus, here unlike Fig. 9.7, as the heat flux increases the straight line representing liquid conduction (Eq. (9.14)) comes down. At a given heat flux, say q_1 the nucleation curve of Eq. (9.22) is intersected by the heat flux line (Eq. (9.14)) at two points A and B. For bubbles with $r > r_a$, the liquid superheat is insufficient at the tip to grow and with $r < r_b$ also it is the same case and hence bubbles are only possible for $r_b < r < r_a$. Now as the heat flux is increased to q_2 ($q_2 > q_1$) keeping the wall temperature constant, the high liquid superheat gets limited to a thinner layer in the liquid due to increase in strength of convection, and just nucleation with a single critical radius r_c (at point C) is possible. For a further increase in heat flux, nucleation becomes impossible due to increased strength of convection. This physical understanding will help us to understand the suppression of nucleation in annular forced convection boiling to be discussed later. Thus, following the same procedure of equating temperature and temperature gradient of the heat flux and liquid superheat curve and some manipulation with Eq. (9.25), the bubble radius for the more accurate analysis of Hsu can be found as

$$r_c = \frac{2\sigma T_{sat}}{\rho_v h_{lv}\left[\frac{d}{dT}\ln\left(\frac{\rho_l \sigma}{\rho_l - \rho_v}\right) + \left(\frac{k_l \rho_v h_{lv}}{2q\sigma T_{sat}}\right)^{1/2}\right]} \tag{9.27}$$

Equation (9.16) can be regarded as a simplified version as this equation although less accurate often Hsu's (1962) equation is written in a simplified form (Sato and Matsumura 1964),

$$\Delta T_{\text{sup},ONB} = \frac{4\sigma T_{sat} \upsilon_{lv} h_l}{k_l h_{lv}}\left[1 + \sqrt{\frac{k_l h_{lv}\Delta T_{sub}}{2\sigma T_{sat}\upsilon_{lv} h_l}}\right] \tag{9.28}$$

where h_l = liquid heat transfer coefficient and $\upsilon_{lv} = \upsilon_v - \upsilon_l$. Hsu (1962) presented a simplified form of this equation as

$$q_{ONB} = \frac{k_l h_{lv}\rho_v}{12.8\sigma T_{sat}}(T_w - T_{sat})^2 \tag{9.29}$$

Kandlikar and Spiesman (1997) presented a non-dimensional form of nucleation criterion which can be useful in assessing the subcooled boiling data. Kandlikar (1997) took a slightly different approach and assumed the bubble top temperature as that of the stagnation point of the thermal boundary layer. He suggested the correlation to be

$$q_{ONB} = \frac{k_l h_{lv}\rho_v}{C\sigma T_{sat}}(T_w - T_{sat})^2 \tag{9.30}$$

where $C = 9.2$. However, investigation on the subcooled boiling in a turbulent falling film by Marsh and Mudawar (1989) correlated the data for $C = 3.5$. Table 9.1 gives an overview of the various models of onset of Nucleate boiling (ONB) for subcooled flow boiling.

Table 9.1 Overview of models on ONB for subcooled flow boiling

Reference	Assumptions	Model development	ONB correlation proposed
Hsu and Graham (1961)	Nucleation starts when superheat criterion is satisfied at the distance of one bubble diameter from the wall	By solving transient conduction problems within a range of cavity sizes	$q''_{ONB} = \frac{k_l h_{lv} \rho_v (T_w - T_{sat})^2}{12.8\sigma T_{sat}}$
Sato and Matsumara (1962)	The nucleus obtains thermal energy indirectly from the surrounding liquid	Bubble radius is determined by solving the equation of thickness of superheated layer	$q''_{ONB} = \frac{k_l h_{lv} \rho_v (T_w - T_{sat})^2}{8\sigma T_{sat}}$
Marsh and Mudawar (1989)	The assumption of the linear temperature profile is inadequate	Developed for turbulent liquid falling film	$C = 3.5$
Kandlikar et al. (1997)	Liquid temperature at the bubble top equals that of the stagnation point in the thermal boundary layer	Location of stagnation point was calculated numerically	$q''_{ONB} = \frac{k_l h_{lv} \rho_v (T_w - T_{sat})^2}{9.2\sigma T_{sat}}$

9.2.2 Subcooled Boiling Heat Flux

Kandlikar (1998) explained in detail the various sub-regions of subcooled flow boiling as given in Fig. 9.11.

These regimes can be listed as

1. The onset of nucleate boiling (ONB): As discussed in detail in the previous section, this is the beginning of bubble formation by nucleation at the wall.

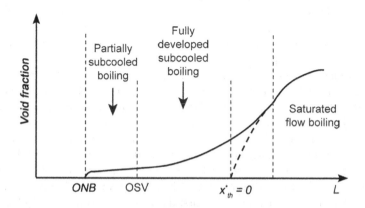

Fig. 9.11 Various regimes of subcooled boiling (*Courtesy:* OECD Halden Reactor Project Report)

2. Partial subcooled boiling: This is also known as highly subcooled boiling with very few bubbles and heat transfer dominated by liquid convection.
3. Fully developed subcooled boiling (FDB): This is also known as slightly sub-cooled boiling where although subcooled, nucleate boiling dominates. The bound-ary between partial subcooled boiling and fully developed subcooled boiling is known as Net Vapour Generation (NVG) or Onset of Significant Void (OSV).
4. Saturated flow boiling: This is the regime in which the thermodynamics quality can be between 0 and 1 ($0 < x^* \leq 1$). However since it is followed by annular flow, usually $x^* < 0.5$.

9.2.2.1 Partial Subcooled Boiling

In the partial subcooled boiling region, only a few bubbles are formed as only a few nucleation sites are activated. Since the wall area available for single phase convection heat transfer is large, convection dominates, while a small part of the heat transfer comes from phase change due to nucleation. As the wall temperature is increased, the convective heat transfer increases during subcooled boiling. This is because, for single phase heat transfer from Newton's law of cooling:

$$q_{SPL} = h_{SP} \left(T_w - T_{bulk} \right) \tag{9.31}$$

Here, q_{SPL} is heat flux due to single phase liquid convection, h_{SP} is the single phase heat transfer coefficient, T_w is wall temperature and T_{bulk} is the bulk liquid temperature. In Fig. 9.12, q_{SPL} is plotted against wall temperature for constant liquid subcooling. It can be seen that as wall temperature increases, in the single phase region the heat flux increases linearly with wall temperature till the wall temperature is equal to the saturation temperature. Then the convective heat flux q_{SPL} becomes non-linear, attains a maximum value and finally reduces to zero.

This trend for heat flux can be explained by keeping in mind that in partial boiling only fewer bubbles are formed, and hence the heat transfer is mostly due to single phase convection with little area coverage by the nucleation sites containing bubbles. As the wall temperature increases, the number of bubbles increases and consequently the share of heat transfer by nucleate boiling increases. However, more tube surface covered by bubbles means the area for liquid convection decreases and hence the single phase heat flux decreases. As the nucleation starts with the increase of wall temperature initially due to increase in temperature difference between wall and fluid, the heat transfer increases and the single phase convection increases increasing the convective heat flux. Thus, with the increase of wall temperature two competing mechanisms take place in the liquid—one, heat flux decreases due to less area avail-able for convection, and other, heat flux increases due to increase in ΔT between the wall and fluid. In the first part of subcooled boiling the second mechanism dom-inates, then the two mechanisms balance each other to give a maximum heat flux of liquid phase convection and finally the first mechanism starts dominating bringing down the heat flux. As the wall temperature is further increased, the wall surface gets

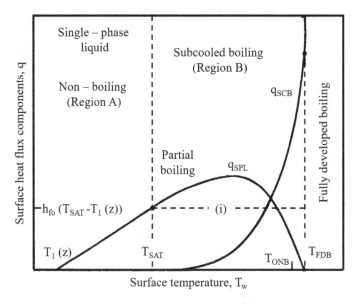

Fig. 9.12 Heat flux during subcooled boiling (*Courtesy:* OECD Halden Reactor Project Report)

completely covered with nucleation sites and single phase convection almost gets eliminated bringing q_{SPL} to zero. After this, the fully developed subcooled boiling takes place.

Bowring (1962) suggested that during the entire subcooled boiling, the total heat flux is the sum of single phase liquid heat flux, q_{SPL}, and the nucleate boiling heat flux, q_{SCB}, for subcooled boiling, giving

$$q = q_{SPL} + q_{SCB} \tag{9.32}$$

The subcooled boiling heat flux due to nucleation is a strong non-linear function of wall temperature as shown in Fig. 9.12. This essentially follows the pattern of the heat flux–wall temperature relationship of nucleate pool boiling (recall, the inception of subcooled boiling was also predicted by Hsu's model which is an adaptation of pool boiling inception). One such simple correlation can be of the power law type,

$$\Delta T_w = T_w - T_{sat} = \psi q_{SCB}^n \tag{9.33}$$

where ψ is a function of fluid properties as well as the surface interaction parameter C_{sf} discussed in the chapter on pool boiling. The value of the exponent n usually ranges from 0.25 to 0.5. It must be understood that although q_{SCB} takes a positive value for positive wall superheat ($\Delta T_w > 0$), the nucleation does not start immediately after the wall temperature exceeds saturation temperature. As we have seen in the previous section, ONB occurs only when a certain amount of superheat at the wall is reached. Figure 9.13 shows it clearly.

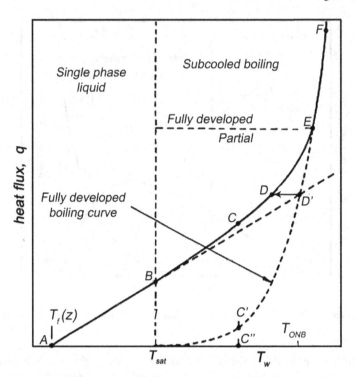

Fig. 9.13 Subcooled boiling curve (*Courtesy:* OECD Halden Reactor Project Report)

Here, the main curve $ABCDEF$ has three regions—AB shows single phase heat transfer, $BCDE$ partial subcooled boiling and EF fully developed subcooled boiling (FDB). Now, since ONB takes place only at a wall superheat corresponding to point D' on the fully developed boiling curve (Eq. (9.33), drawn in the dotted line), till the point D', the heat flux will follow the line ABD'. As the first nucleation takes place, the surface temperature drops to D (Bertoletti et al., 1964). With a further increase of heat flux, the wall temperature increases along DEF. Thus, the difference between the q and q_{SCB} in Fig. 9.13 gives the convective liquid heat flux q_{SPL} which are separately plotted in Fig. 9.12. In Fig. 9.13, these two fluxes as a component of the total flux are shown. Now since the curve for q_{SPL} given in Fig. 9.12 cannot be obtained directly because the area covered by bubble nucleation which is not available for single phase convection is not known, we resort to various indirect methods to evaluate it. The following methods are suggested. Before describing these methods, it may be clarified that the difference between the curves for total heat flux and nucleate boiling heat flux as given in Fig. 9.13 gives the convective heat transfer. At the point F, the two cures approach each other indicating the disappearance of the single phase component of heat flux because the single phase heat flux drops to zero as shown in Fig. 9.12. The methods described below use these physical interpretations.

Fig. 9.14 Convective heat
flux and its approximation
(*Courtesy:* "Heat Transfer in
Condensation and Boiling"
by Karl Stephan, with
permission from Springer)

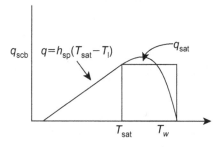

1. Griffith et al. (1958) suggested a simplification of the q_{SPL} curve as shown in
   Fig. 9.12 which was further extended by Bowring (1962). According to this
   method, it is assumed that as long as wall temperature remains below saturated
   fluid temperature, the heat flux increases linearly and beyond saturation point it
   remains constant as long as fully developed boiling does not occur. This means

$$q_{SCB} = \begin{cases} h_{sp}[T_{sat} - T_l(z)] & \text{for } T_w \leq T_{sat} \\ q_{sat} & \text{for } T_{sat} < T_w < T_{FDB} \\ 0 & \text{for } T_w \geq T_{FDB} \end{cases} \qquad (9.34)$$

This essentially means that the curve for q_{SPL} is approximated as shown in
Fig. 9.14.
Forster and Grief (1959) also noted that in Fig. 9.13:

$$q_E = 1.4q_D \qquad (9.35)$$

Bowring (1962) used this correlation to derive an expression for local subcooling
at the point of fully developed boiling as

$$\Delta T_{sub}(z)_{FDB} = \frac{q}{1.4h_{SPL}} - \psi\left(\frac{q}{1.4}\right)^n \qquad (9.36)$$

2. Rohsenow (1953) had suggested a slightly different approach. He suggested that
   the single phase convection is decided by simple Newton's law of cooling in the
   form:

$$q_{SPL} = h_{SPL}\left[T_w - T_f(z)\right] \qquad (9.37)$$

The heat transfer coefficient h_{SPL} was calculated using a modified Dittus–Boelter
equation in the form

$$Nu = \frac{h_{SPL}d}{k_l} = 0.019\,Re_l^{0.8}\,Pr_l^{0.4} \qquad (9.38)$$

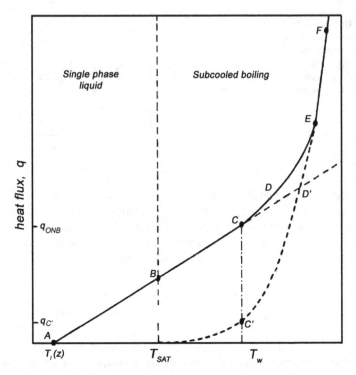

Fig. 9.15 Bergles and Rohsenow's (1964) method (*Courtesy:* OECD Halden Reactor Project Report)

where Nu stands for single phase Nusselt number and Re_l is the liquid Reynolds number in the tube. Further, he determined the subcooled nucleate boiling heat flux as $q_{SCB} = q_{total} - q_{SPL}$. Finally, he correlated the heat flux q_{SCB} with his famous nucleate pool boiling correlation given in Chap. 7. He found a good agreement of q_{SCB} thus found with experimental data.

3. Another method was suggested by Bergles and Rohsenow (1964). Here, they superimposed the fully developed subcooled nucleate boiling curve on the single phase liquid convection line as shown in Fig. 9.15. They calculated the q_{ONB} by using their own correlation (Eq. (9.19)). This is point C on the curve. The superposition is done only between C and F since before that although the fully developed flux has a positive value, nucleation does not start and hence it does not make sense to superimpose the curve. This method enables rapid calculation for q_{SPL} from fewer data points.

9.2.2.2 Fully Developed Subcooled Boiling

As we have seen in the previous section, the influence of single phase convection gradually decreases as the partial subcooled boiling progresses. Finally, it reaches a point where the contribution of convection is negligible and the entire heat flux is due to the wall nucleation. At this condition, the heat transfer is independent of subcooling and more like pure nucleate boiling where wall temperature depends on heat flux at the wall and the pressure of the fluid. However, there is one distinctive feature of fully developed subcooled boiling. It is less influenced by surface conditions (like surface roughness) compared to that in pool boiling. This is because, due to flow the requirement of heat flux is larger here compared to pool boiling (remember, only at high heat flux subcooled boiling occurs, as shown in Fig. 9.4), and hence a larger number of nucleation sites are activated making surface roughness plays a reduced role in the process.

The understanding of fully developed subcooled boiling is still not very complete. On the whole, it is seen as an extension of nucleate pool boiling. Mc Adams et al. (1949) presented data for both partial and fully developed subcooled boiling and have shown that for partial subcooled boiling, both inlet subcooling and fluid velocity influence the heat flux. However, once the fully developed subcooled boiling occurs, the curves of the same velocity and inlet subcooling merge. There are a large number of correlations for fully developed subcooled boiling. Thome et al. (1965) proposed

$$\Delta T_{sat} = 22.65 q^{0.5} e^{-p/87} \tag{9.39}$$

where q is in MW/m^2, p in bar and ΔT_{sat} in K. Similarly Jens and Lottes (1951) proposed

$$\Delta T_{sat} = 25 q^{0.25} e^{-p/62} \tag{9.40}$$

with the same units as in Eq. (9.39). The actual mechanism of fully developed subcooled boiling is still debated. In spite of its closeness to nucleate pool boiling, one must appreciate that the bulk fluid is still at an average enthalpy below the equilibrium saturation enthalpy (that we get in the steam table, for steam–water system). One of the mechanisms of alternate cold and hot liquid movement was suggested by Brown (1967). He suggested that the mechanism of thermocapillary works as shown in Fig. 9.16. This mechanism originates from the movement of the interfacial fluid between the bubble and liquid due to temperature inequality on the bubble surface which brings about a variation in surface tension value. This effect is called the 'Marangoni effect'. This drags the liquid around the bubble from the hot to cold region and creates a wake region for the bubble. The bubble boundary layer thus plays an important role in the process.

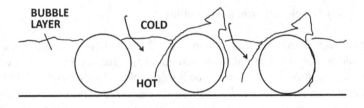

Fig. 9.16 Thermocapillary flow in subcooled boiling

9.3   Saturated Flow Boiling

As we have seen in the previous section, fully developed subcooled boiling is followed by saturated boiling. Theoretically, saturated boiling takes place when the mean liquid temperature reaches the saturation temperature at the given pressure; this means that thermodynamic quality is zero ($x^* = 0$). This essentially means that the average enthalpy of fluid reaches the enthalpy of the saturated fluid as given by the thermodynamic equilibrium property (as in the steam table, for example, for water). However, even under this condition there may be suppression of nucleate boiling rather than enhancement. This is because, in reality the core of the liquid in the tube may be subcooled and near wall, liquid may be superheated giving the average to be saturated. When the better heat transfer by the two-phase flow removes heat efficiently from the wall and there is a balance between heat transfer and the supplied wall heat flux, further nucleation giving saturated heat transfer will not occur. This will occur only when supplied wall heat flux is larger than what the two-phase mixture can remove. Assuming the bubble seating at the mouth of the cavity to be hemispherical, Bergles and Rohsenow (1964) derived the expression for minimum superheat required for nucleation of subcooled liquid as

$$\Delta T_{W,sat} = \Delta T_W - \Delta T_{sat} = 2\left(\frac{2q T_{sat}\sigma}{\rho_g h_{fg} k_l}\right)^{\frac{1}{2}} \tag{9.41}$$

Defining heat transfer coefficient for two-phase flow as

$$h_{tp} = \frac{q}{T_W - T_{sat}} \tag{9.42}$$

we get the condition for saturated boiling heat flux as

$$q \geq \frac{8\sigma T_{sat}}{\rho_g h_{fg} k_l}\left(h_{tp}\right)^2 \tag{9.43}$$

The two-phase heat transfer coefficient, h_{tp}, increases with mass flux (additional vapour formation is not included in it) as well as with vapour content. As a result the $T_{W,sat}$ decreases, and the heat flux being insufficient according to Eq. (9.43), the nucleation gets suppressed. This results in the conversion of subcooled boiling to convective boiling or eliminating it altogether. For calculating h_{tp}, to see whether condition (9.43) is satisfied, Dengler and Adams (1956) suggested the equation

$$\frac{h_{tp}}{h_{liquid}} = 3.5 \left(\frac{1}{X_{tt}}\right)^{0.5} \tag{9.44}$$

where h_{liquid} is the heat transfer coefficient of the pure liquid and X_{tt}, the Martinelli parameter. One must understand that the above estimate of h_{tp} is approximate and can be used only for estimating the condition for nucleation in saturated boiling given by Eq. (9.43). The more precise estimation of two-phase heat transfer in saturated boiling is given below.

The first important thing to note is that the mechanism of fully developed subcooled boiling and saturated flow boiling is essentially the same. However, the major difference between these two regimes where one follows the other is that in subcooled boiling, the local subcooling is the most important parameter whereas in saturated flow boiling the thermodynamic quality is the corresponding quantity. With constant heat flux, for a circular tube, the thermodynamic quality can be calculated as follows. Let us consider the portion of the tube with saturated boiling alone up to a distance z. Making heat balance as shown in Fig. 9.5,

$$\dot{m} \left[x^* (z) - x^* (sat)\right] h_{fg} = q\pi D (z - z_{sc}) \tag{9.45}$$

Now we know from the definition of mass flux and saturation point that $\dot{m} = G\left(\pi \frac{D^2}{4}\right)$ and $x^*(sat) = 0$. This gives from Eq. (9.25)

$$x^* (z) = \frac{4q}{DGh_{fg}} (z - z_c) \tag{9.46}$$

Now, on a scale of thermodynamic quality, the regions for subcooled saturated and convective boiling can be shown as given in Fig. 9.4.

Coming to the estimation of saturated boiling heat transfer, we use a simple additive correlation indicating that the effective heat transfer coefficient for two-phase flow is a simple sum of convective (h_{conv}) and nucleate boiling (h_{NB}) heat transfer coefficient, as

$$h_{tp} = h_{NB} + h_{conv} \tag{9.47}$$

The simplest method is to determine the convective heat transfer coefficient using the Dittus–Boelter equation assuming the flow to be turbulent:

$$h_{conv} = 0.023\,\mathrm{Re}_l^{0.8}\,\mathrm{Pr}_l^{0.4}\left(\frac{k_l}{D}\right) \qquad (9.48)$$

where

$$\mathrm{Re}_l = \frac{\dot{m}_l D}{\mu_l} = \frac{\dot{m}\,(1-x^*)\,D}{\mu_l}$$

and Pr_l is the Prandtl number of the liquid. Although the contribution of convective heat transfer in a saturated boiling regime is much smaller than either subcooled or convective (to be discussed later) boiling, Chawla (1967) has suggested that a better match of experimental data with a direct correlation of

$$\frac{h_{tp}}{h_{NB}} = 29\mathrm{Re}^{-0.3}\,Fr_l^{0.2} = RHS \qquad (9.49)$$

for $1.0 \leq RHS \leq 1.2$ where Fr_l is the liquid Froude number is given by

$$Fr_l = \frac{\dot{m}^2(1-x^*)^2}{\rho_l^2 g D} \qquad (9.50)$$

The nucleate boiling heat transfer coefficient, h_{NB}, can be calculated from the pool boiling correlations in the nucleate boiling regime as given in Chap. 6.

Pujol (1968) showed that the heat transfer coefficient for downflow saturated boiling is smaller than upflow, thus

$$h_{down} = 0.75 h_{up} \qquad (9.51)$$

For the horizontal tube, the situation is complex because the upper wall is poorly wetted as we have seen in Chap. 1. Thus, the heat transfer coefficient can vary circumferentially. It has been found that tube wall conductivity plays an important role in deciding this circumferential variance.

9.4   Convective Flow Boiling

Following saturated boiling in the form of slug or churn flow, gradually the vapour slugs become larger and the liquid is thrown to the wall giving rise to semi-annular flow first and then annular flow. This region is known as 'convective boiling' or more accurately 'forced convective evaporation'. The distinctive feature of this region is gradually decreasing nucleation eliminating it almost completely downstream. As a consequence of that, the phase changes take place here at the vapour–liquid interface by evaporation (and not by nucleate boiling).

The above feature can be explained in the following way. We can consider that the vapour at the core of the tube is in saturated state and as a result at the saturation temperature. However, as the liquid film at the wall becomes thinner and thinner from a semi-annular to annular regime, the space in the radial direction within the liquid becomes smaller and hence the wall superheat becomes smaller and smaller because the temperature gradient at the wall is the same due to constant heat flux as shown in Fig. 9.4. Thus, due to the gradual decrease in wall superheat, the nucleation sites get deactivated and nucleate boiling gets suppressed. However, as the same heat flux propagates through the liquid film, when it reaches the liquid–vapour interface, there is a mismatch between heat flux in the liquid and vapour side. As a result, evaporation takes place at the liquid–vapour interface to compensate for the suppression of wall nucleation. Thus, two mechanisms work simultaneously for convective boiling:

1. Suppression of saturated nucleate boiling.
2. Enhancement of wall liquid cooling.

Chen (1963) was the first one to suggest a method in which the above mechanisms were combined to present a correlation which calculates the convective heat transfer coefficient quite successfully. His suggestion was that we add the convective and nucleate boiling heat transfer coefficients to get the effective two-phase boiling heat transfer coefficient. However, the nucleate boiling heat transfer coefficient is taken with a suppression factor ($S \leq 1.0$) multiplied by it and the liquid convection heat transfer coefficient is taken with an enhancement factor ($F \geq 1.0$) multiplied to it, thus

$$h_{CB} = S h_{NB} + F h_{FC} \qquad (9.52)$$

where h_{CB} is heat transfer coefficient for convective boiling, h_{NB} is heat transfer coefficient for nucleate boiling, h_{FC} is heat transfer coefficient for forced convection, S is suppression factor for nucleate boiling and F is enhancement factor for forced convection. He suggested that the h_{NB} is calculated from the Forster–Zuber (1955) correlation in the form

$$h_{NB} = \frac{0.00122 \Delta T_{sat}{}^{0.24} \Delta p_{sat}{}^{0.75} c_{pl}{}^{0.45} \rho_l{}^{0.49} k_l{}^{0.79}}{\sigma^{0.5} h_{lv}{}^{0.24} \mu_l{}^{0.29} \rho_v{}^{0.24}} \qquad (9.53)$$

It can be shown that the above equation is dimensionally consistent and hence, if the SI unit is used for all the parameters on the right-hand side, the unit of h_{NB} will come out to be W/m^{2}K. In this equation like the previous explanation, the properties with l and v subscripts stand for liquid and vapour, respectively. ΔT_{sat} and Δp_{sat} are the corresponding temperature and pressure difference obtained from the vapour pressure curve.

Coming to the convective heat transfer coefficient, h_{FC}, we take only the liquid mass rate to calculate the turbulent heat transfer coefficient since the wall is cooled here by liquid alone in the absence of nucleation. Thus, using the well-known Dittus–Boelter equation, we have

$$Nu_{FC} = 0.023 Re_l{}^{0.8} Pr_l{}^{0.4} \qquad (9.54)$$

Fig. 9.17 Convection enhancement factor, F

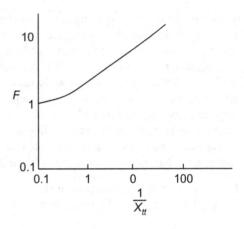

where Re_l is liquid Reynolds number, Pr_l is liquid Prandtl number and

$$Nu_{FC} = \frac{h_{FC} D}{k_l}$$

is the forced convection Nusselt number. The next task is to find out the enhancement factor, F, and suppression factor, S. This is actually the major proposition of Chen's (1963) work. He demonstrated that F is a function of the Martinelli parameter, X. For turbulent flow in both the phases $X = X_{tt}$. Under this condition, the correlation between F and X_{tt} can be given by

$$F = \begin{cases} 1 & \text{for } \frac{1}{X_{tt}} \leq 0.1 \\ 2.35\left(\frac{1}{X_{tt}} + 0.213\right)^{0.736} & \text{for } \frac{1}{X_{tt}} > 0.1 \end{cases} \qquad (9.55)$$

This can be put in the form of a curve as shown in Fig. 9.17. It can be seen that F is greater than unity ($F > 1.0$). Next, we take up the evaluation of the suppression factor, S, suggested by Chen. To find this, we need to calculate a two-phase Reynolds number.

$$Re_{TP} = Re_l F^{1.25} \qquad (9.56)$$

The suppression factor, S, is a function of Re_{TP} which is shown in Fig. 9.18.
Thus, with the combination of the above correlations for h_{FC} and h_{NB} along with Figs. 9.17 and 9.18, the convective boiling heat transfer coefficient can be evaluated from Eq. (9.52).

Apart from the above method of Chen, there are a large number of experimental correlations available in the literature. Some of them are simply modifications of Chen's correlation such as that by Gungor and Winterson (1986). They found better

Fig. 9.18 Nucleate boiling
suppression factor, S

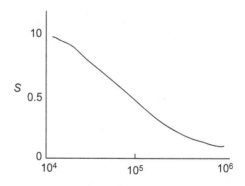

correlation for S and F for a large number of data by introducing Bond number
$Bo = (q/\dot{m}h_{fg})$ as

$$S = \left(1 + 1.15 \times 10^{-6} F^2 Re_l^{1.17}\right)^{-1}$$
$$F = 1 + 2.4 \times 10^4 Bo^{1.16} + 1.37 X_{tt}^{-0.86} \tag{9.57}$$

A large number of completely empirical correlations are also available.

Horizontal Tubes: The flow pattern in horizontal tubes with boiling is considerably
different from that in a vertical tube. Here, the bubbles and slug stay in the upper
part of the tube and the dryout on the upper wall occurs much before that in the
bottom wall of the tube. Figure 9.18 shows a detailed longitudinal picture of boiling
inside a horizontal tube. In such a case, the angle wetted by the liquid at a particular
cross section ψ is of special importance. The circumferential average heat transfer
coefficient is given by

$$\bar{h} = \frac{\varphi}{\pi}\bar{h}_b + \left(1 - \frac{\varphi}{\pi}\right)\bar{h}_{dr} \tag{9.58}$$

where $\bar{h}_b$ is heat transfer coefficient of the wetted arc and $\bar{h}_{dr}$ is heat transfer coefficient of the dried arc.

Based on the experiment, Shah (1976) suggested the following equation for the
calculation of the heat transfer coefficient

$$\frac{h(Z)}{h_{lo}} = (1 - x^*)^{0.8}\left[1 + \left(\frac{2}{C_o}\right)^{1.6}\right]^{0.5} \tag{9.59}$$

The Froude number is given by

$$Fr_{lo} = \frac{\dot{m}^2}{g d_h \varsigma_L^2} < 0.04$$

$$C_o = \left(\frac{1 - x^*}{x^*}\right)^{0.8} \left(\frac{\rho_G}{\rho_L}\right)^{0.5} \left(\frac{0.04}{Fr_{lo}}\right)^{0.288}$$

For $Fr_{lo} > 0.04$, the last term is taken as unity

$$h_{lo} = 0.023 Re_{lo}^{0.8} Pr_L^{0.4} \frac{k_L}{d_h} \quad \text{and} \quad Re_{lo} = \frac{\dot{m} d_h}{\mu_L} \tag{9.60}$$

However, the most accurate and advanced correlation for convective flow boiling is the one from Kandlikar (1990) applicable to both vertical and horizontal flows. He utilized the single-phase, liquid-only heat transfer coefficient to frame the nucleate boiling and convective boiling components of the turbulent regime given by

$$h_{tp,nb} = \left[0.6683 Co^{-0.2} f\,(Fr_{lo}) + 1058.0 Bo^{0.7} F_f\right](1 - x)^{0.8} h_{lo} \tag{9.61}$$

$$h_{tp,cb} = \left[1.136 Co^{-0.9} f\,(Fr_{lo}) + 667.2 Bo^{0.7} F_f\right](1 - x)^{0.8} h_{lo} \tag{9.62}$$

$$f\,(Fr_{lo}) = \begin{cases} 1 & \text{for } Fr_{lo} \geq 0.04 \\ (25 Fr_{lo})^{0.3} & \text{for } Fr_{lo} < 0.4 \end{cases} \tag{9.63}$$

$$Co = \left(\frac{1 - x}{x}\right)^{0.8} \left(\frac{\rho_v}{\rho_l}\right)^{0.5} \tag{9.64}$$

From the well-known Gnielinski correlation (1976),

$$h_{lo} = \frac{(f/8)\,Re_{lo} Pr_l\,(\lambda_l/D)}{1 + 12.7(f/8)^{1/2}\left(Pr_l^{2/3} - 1\right)} \tag{9.65}$$

From Petukhov and Popov (1963)

$$h_{lo} = \frac{(f/8)\,(Re_{lo} - 1000)\,Pr_l\,(\lambda_l/D)}{1 + 12.7(f/8)^{1/2}\left(Pr_l^{2/3} - 1\right)} \tag{9.66}$$

$$f = (0.79 \ln Re_{lo} - 1.64)^{-2} \tag{9.67}$$

The values of the fluid surface parameter, F_f, are recommended in Table 9.2. $F_f = 1$ for stainless steel tubes for all fluids.

Table 9.2 Recommended F_f values in Kandlikar's flow boiling correlation

Fluid	F_f	Fluid	F_f
Water	1.00	R-134a	1.63
R-11	1.30	R-152a	1.10
R-12	1.50	R-31/R-132	3.30
R-13B1	1.31	R-141b	1.80
R-22	2.20	R-124	1.00
R-113	1.30	Kerosene	0.488
R-114	1.24		

Examples

Problem 1: Water at 3 bar flows inside a vertical tube of 18 mm diameter under subcooled flow boiling conditions where the tube wall temperature is 10 °C above the saturation temperature. Estimate the heat transfer per meter length of the tube.

Solution: For forced convection subcooled boiling inside vertical tubes, the following relation is recommended by Jacob (1957):

$$h = 2.54\,(\Delta T)^3\,e^{p/1.551}\,(\mathrm{W/m^2 - K})$$

where p is pressure in $\mathrm{MN/m^2}$, and $\Delta T = T_w - T_{sat}$. Substituting these values, $h = 3082\ \mathrm{W/m^2}$-K. Heat transfer, $q = hA\,(T_w - T_{sat})$, where $A = \pi DL$. So, $q = 1743$ W/m.

Problem 2: In a forced convective boiling experiment, saturated water is flowing at 200 kg/h through a vertical copper tube of 1.6 cm inside diameter at atmospheric pressure. The tube wall temperature is held at 120 °C by condensing steam. Calculate the heat transfer coefficient at a point at which the quality is 20%. An annular flow pattern can be assumed. Saturation pressure corresponding to 120 °C is 2 atm.

Solution: The properties (from steam tables) corresponding to 1 atm are

$\rho_l = 958\ \mathrm{kg/m^3}$, $\rho_v = 0.6\ \mathrm{kg/m^3}$, $\mu_l = 28.3 \times 10^{-5}$ Pa-s, $\mu_v = 12.06 \times 10^{-6}$ Pa-s, $h_{fg} = 2257\mathrm{kJ/kg}$, $Pr_l = 1.75$, $k_l = 0.681$ W/mK, $C_{p_l} = 4.218$ kJ/kg-K, $T_{sat} = 100°$.

Total mass flow rate, $m = 200/3600 = 0.055$ kg/s.

$$Re_l = \frac{4m\,(1-x)}{\pi D\mu_l} = 12372$$

Martinelli parameter,

$$X_{tt} = \left(\frac{1-x}{x}\right)^{0.9} \left(\frac{\mu_l}{\mu_v}\right)^{0.1} \left(\frac{\rho_v}{\rho_l}\right)^{0.5} = 0.12$$

Hence, correction factor $F = 10$.

$$h_{convective} = 0.023 \left(\frac{k_l}{D}\right) Re_l^{0.8} Pr_l^{0.4} F = 23.01 \text{ kW/m}^2\text{K}$$

Two-phase Reynolds number,

$$Re_{TP} = Re_l F^{1.25} = 2.2 \times 10^5$$

Hence, suppression factor, $S = 0.18$.

$$h_{\text{nucleate boiling}} = 0.00122 \left[\frac{k_l^{0.79} e_{pl}^{0.45} \rho_l^{0.49}}{\sigma^{0.5} \mu_l^{0.29} h_{fg}^{0.24} \rho_v^{0.24}}\right] \Delta T^{0.24} \Delta P^{0.75} S$$

$$= 3.443 kW/m^2 K$$

$$h_{\text{tw phase}} = h_{\text{convective}} + h_{\text{nucleate boiling}} = 26.453 \text{ kW/m}^2\text{K}$$

Problem 3: Water with a mass flux of 500 kg/m^{2}s and at a pressure of 10 bar and 100 °C enters a heated channel of 20 mm inner diameter. Determine for a uniform wall heat flux of 200 kW/m^2 (a) the length over which single phase convection takes places and (b) the length over which subcooled boiling takes place. Single phase (liquid phase) heat transfer coefficient = 5 kW/m^{2}K. Wall superheat immediately before the inception of subcooled boiling is 5 K.

Solution: x_1 is the length over which single phase convection takes place. x_2 is the distance from the entrance at which the subcooled boiling stops and the nucleate boiling starts. From energy balance,

$$q \times \pi D \times x = \dot{m} \times C_{pl} \times [T_l - T_1] = h \times \pi D \times x [T_w - T_l]$$

where T_w, T_l are wall temperature and liquid temperature respectively at any distance 'x' from the entrance, q is wall heat flux and T_1 is the inlet temperature.

Wall superheat just before the inception of subcooled boiling = 5 K.

$$T_w - T_s = \Delta T_{sup}$$

From energy balance,

$$T_{l,x_1} - T_1 = \frac{q\pi D \times x_1}{\dot{m}C_{pl}}$$

$$T_{w,x_1} - T_{1,x_1} = \frac{q}{h}$$

$$\Delta T_{sup,x_1} + T_s - T_1 = \frac{q}{h} + \frac{q\pi D \times x_1}{\dot{m}C_{pl}} \quad (\because T_w - T_s = \Delta T_{sup})$$

At the entrance, $\Delta T_{sub} = T_s - T_1$.

$$\Delta T_{sup,x_1} + \Delta T_{sub,e} = q\left[\frac{\pi D x_1}{\dot{m}C_{pl}} + \frac{1}{h}\right]$$

therefore

$$x_1 = \frac{\dot{m}C_{pl}}{\pi D}\left[\frac{\Delta T_{sup,x_1} + \Delta T_{sub,e}}{q} - \frac{1}{h}\right]$$

$\dot{m}$ = mass flux × area = 0.157 kg/s. The saturation temperature corresponding to 10 bar is 179.91 °C and $T_1 = 100$°C. Therefore

$$\Delta T_{sub,e} = 79.91°C$$

C_{pl} at 10 bar is 4.4 kJ/kg-K. Therefore

$$x_1 = 2.47m$$

The length over which the single phase convection takes place = 2.47 m. Subcooled boiling takes place till the mean liquid temperature is equal to saturation temperature. So, from the energy equation,

$$q\pi D x_2 = \dot{m}C_{pl}(T_s - T_1)$$

So, $x_2 = 4.39$ m and the length over which subcooled boiling takes place $= x_2 - x_1 = $ 1.92 m.

Exercises

1. Estimate the wall superheat and heat transfer coefficient when water at 5 bar flows inside a vertical tube of 20 mm diameter under subcooled boiling conditions if the heat transfer per meter length of the tube is 2000 W.

2. Saturated water at 220 kg/h flows through a copper tube of 1.5 cm inside diameter at atmospheric pressure. The tube wall is maintained at a constant temperature of 135 °C. Assuming an annular flow pattern, calculate the local heat transfer coefficients for qualities 0.1, 0.2, 0.3, 0.4, and 0.5. Plot local heat transfer coefficient against quality.
3. Water with a mass flux of 350 kg/m²s and at a pressure of 5 bar and 100 °C enters a heated channel of inner diameter 20 mm. Determine the length over which subcooled boiling takes place if the heat flux is 100 kW/m². Wall superheat immediately before the inception of subcooled boiling is 8 K. Liquid phase (single phase) heat transfer coefficient can be determined from the Dittus–Boelter equation corresponding to inlet conditions.

References

Bergles AE, Rohsenow WM (1963) The determination of forced convection surface boiling heat transfer. Paper 63-HT-22. VIth National Heat Transfer Conference. Am. Soc. Mech. Eng' Amer. Soc. Chem. Eng., Boston

Bergles AE, Rohsenow WM (1964) The determination of forced-convection surface boiling heat transfer. Trans Am Soc Mech Eng Ser C J Heat Transf 86:365–372

Bowring RW (1962) Physical model based on bubble detachment and calculation of steam voidage in the subcooled region of a heated channel. OECD Halden Reactor Project Report HPR-10

Brown WT (1967) A study of flow surface boiling. Phd Thesis, Mech. Eng. Dept., M.I.T

Chawla JM (1967) Warmeubergang and Druckabfall in WaagerechtenRohrenbei der Stromung von VerdaropfendenKaltemitteln. VDI-Forschungsh. 523

Chen JC (1963) A correlation for boiling heat transfer to saturated fluids in convective flow. ASME 63-HT-64. In: Presented at 6th National heat transfer conference, Boston

Collier JG, Thome JR (1994) Convective boiling and condensation. Clarendon Press, Oxford

Davis EJ, Anderson GH (1966) The incipience of nucleate boiling in forced convection flow. Am Inst Chem Eng J 12:774–780

Dengler CE, Addams JN (1956) Heat Transfer mechanism for vaporization of water in a vertical tube. Chem Eng Progr Symp Se 52:95–103

Forster HK, Grief R (1959) Heat transfer to a boiling liquid mechanisms and correlations. Trans ASME J Heat Transf Ser C 81

Griffith P, Clark JA, Rohsenow WM (1958) Void Volumes in subcooled boiling systems. In: Paper 58-HT-19 presented at ASME – AICHE heat transfer conference, also Technical Report No. 12 (MIT)

Gungor KE, Winterson RHS (1986) A general correlation for flow boiling in tubes and in annuli. Int T Heat Mass Transf 29:351–358

Hsu VV (1962) On the size of range of active nucleation cavities on a heating surface. Trans Am Soc Mech Eng Ser C J Heat Transf 84:207–216

Hsu YY, Graham RW (1961) An analytical and experimental study of the thermal boundary layer and ebullition cycle in nucleate boiling. National Aeronautics and Space Administration, Washington, D.C

Jens WH, Lottes PA (1951) Analysis of heat transfer burnout, pressure drop and density data for high pressure water. Rep. ANL - 462

Kandlikar SG (1990) A general correlation for saturated two-Phase flow boiling heat transfer inside horizontal and vertical tubes. ASME J Heat Transf 112(1):219–228

Kandlikar SG (1991) Development of a flow boiling map for subcooled and saturated flow boiling of different fluids inside circular tubes. ASME J Heat Transf 113:190–200

Kandlikar SG (1997) Bubble nucleation and growth charachteristics in subcooled flow boiling of water. HTD – Vol 342, National Heat Transfer Conference, volume 4, ASME

Kandlikar SG (1998) Heat Transfer characteristics in partial boiling, fully developed boiling, and significant void flow regimes of subcooled flow boiling. J Heat Transf 120:395–401

Kandlikar SG (2022) A general correlation from saturated two phase flow boiling heat transfer inside horizontal and vertical tubes. ASME J Heat Transf 112:219–228

Kandlikar SG, Spiesman PH (1998) Effect of surface finish on flow boiling heat transfer. American Society of Mechanical Engineers, Heat Transfer Division (Publication)

Marsh WJ, Mudawar I (1989) Predicting the onset of nucleate boiling in way free-falling turbulent liquid films. Int J Heat Mass Transf 32(2):361–378

McAdams WH, Kennel WE, Mindsen CSL, Carl R, Picronell PM, Dew JE (1949) Heat Transfer at high rates to water with surface boiling. Ind Eng Chem 41(9):1945–1953

Petukhov BS, Popov VN (1963) Theoretical calculation of heat exchange and frictional resistance in turbulent flow in tubes of an incompressible fluid with thermophysical properties. Teplofizika vysokikh temperatur 1(1):85–101

Pujol L (1968) Boiling heat transfer in vertical upflow and downflow tubes. Ph.D. Thesis at Lehigh University, Pennsylvania

Thom JRS, Walkes WM, Fallon TA, Reising GFS (1965) Boiling in subcooled water during flow up heated tubes or annuli. Symp Inst Mech Eng Paper 6. London

Rohsenow WM (1952) A method of correlation Heat Transfer Data from surface Boiling Liquids. Trans ASME 74

Rohsenow WM (1953) Heat transfer with evaporation. Heat transfer-a symposium held at the University of Michigan during the summer of 1952. University of Michigan Press, Michigan, pp 101–150

Sato T, Matsumura H (1964) On the conditions of incipient subcooled-boiling with forced convection. Bull JSME 7(26):392–398

Sato T, Matsumara H (2022) On the conditions of Incipient subcooled Boiling with forced convection. Bulletin of TSME 07(26):392–398

Shah MM (1976) A new correlation for heat transfer during boiling flow through pipes. ASHRAE Trans 82(2):66–86

Chapter 10
Flow Boiling Crisis and Post Dryout Heat Transfer

10.1 Critical Heat Flux in Flow Boiling

In the previous chapter, we have discussed various modes of flow boiling such as subcooled, saturated or convective boiling. We also indicated that the critical heat flux (CHF) occurs during any of these regimes of boiling depending on the applied heat flux. For example, CHF can occur even during subcooled boiling if the heat flux is sufficiently large. On the other hand, it is likely to occur during convective boiling (annular flow) for moderate heat flux due to the complete dryout of the annular film. This brings out the question—what is CHF in flow boiling? We have seen that for pool boiling CHF indicates a transition from nucleate to film boiling whereas in flow boiling unique phenomenon like film dryout occurs. Hence, it is important to identify what is really meant by CHF in flow boiling.

10.1.1 Two Phenomena in Flow Boiling, 'DNB' and 'Dryout'

The above question becomes even more relevant because of the existence of a large number of terms that are used to indicate these phenomena. Generally, it can be said that the occurrence of a sudden rise of tube wall temperature in flow boiling is what is referred to as CHF or dryout or departure from nucleate boiling (DNB). However, all the above terms along with 'burnout' (meaning literally, physical tube burnout) are often used interchangeably meaning the same phenomenon while in reality difference exists between them. The reason for this is the fact that in flow boiling, the sudden rise of heating surface temperature can happen due to a number of reasons. This is not the case with pool boiling where the sudden rise of heater wall temperature happens only due to film formation on the wall which prevents the liquid to wet the wall. In the case of flow boiling, two distinctly different phenomena may lead to this sudden rise in wall temperature. These are as follows:

S. K. Das and D. Chatterjee, *Vapor Liquid Two Phase Flow and Phase Change*, https://doi.org/10.1007/978-3-031-20924-6_10

Fig. 10.1 Crowding of
bubbles and DNB

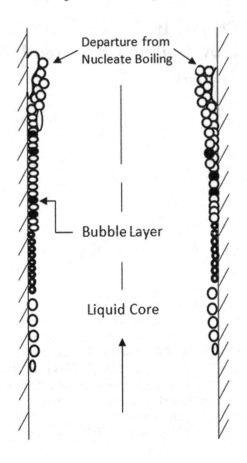

a. At low quality (i.e., in subcooled or saturated boiling), the mechanism for wall
   temperature rise is similar to that in pool boiling. However, in pool boiling the
   film of vapour blankets the surface depriving the wall of liquid wetting. As a
   result, the wall temperature shoots up to compensate for the poor heat transfer
   coefficient of vapour at the wall. In flow boiling, bubbles still remain separated
   and they do not form a vapour film at the wall. On the contrary, when the rate
   of bubble formation is higher than the rate at which they are removed, there is
   a crowding of bubbles at the wall. This also deprives the liquid flow to the wall
   and forms a stagnant long bubble at the wall which cannot move away from the
   wall due to bubble crowding. This is shown in Fig. 10.1. It is generally known
   as Departure from Nucleate Boiling (DNB).
   This can occur either in a subcooled or saturated boiling state depending on
   wall heat flux. There are also other phenomena that may lead to the rise in wall
   temperature and DNB at low quality such as due to high heat flux the liquid
   cannot re-wet the surface or proximity of the vapour slug to the wall which
   creates a dry patch in the wall and leads to the wall temperature rise. This is

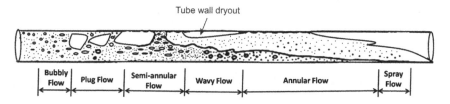

Fig. 10.2 Occurrence of DNB for slug proximity to wall

shown in Fig. 10.2 which shows that the possibility of such DNB is more in the horizontal flow where the slug tends to stick towards the upper part of the tube due to buoyancy.

However, this can also happen due to the wavy nature of the liquid film causing vapour to touch the wall at some location as shown in Fig. 10.3.

However, for subcooled and saturated flow regions, the crowding of bubbles is the major DNB mechanism. It shows hysteresis-like pool boiling, as shown in Fig. 10.4.

b. At high quality (i.e., during convective boiling), the mechanism of wall temperature rise is different. It is due to drying out of the wall as the annular liquid film completely evaporates. This is called dryout. This is shown in Fig. 10.5 where the liquid layer dries out but small liquid particles remain suspended in the vapour. During dryout, the sudden change in wall temperature is completely reversible and it does not show any hysteresis as seen in DNB or pool boiling. The dryout curve is shown in Fig. 10.6.

It is seen clearly from the above discussions that the mechanisms for wall temperature rise (or a sudden decrease in wall heat transfer coefficient) are different in low- and high-quality regions. Both of these are referred to as critical heat flux (CHF) in flow boiling. However, sometimes the term burnout is also used for them although these conditions may not lead to physical burnout. In the former Soviet Union (USSR), the terms *Boiling Crisis of First Kind* for DNB and *Boiling Crisis of Second Kind* for dryout were used. Probably boiling crisis is a more generic and accurate term for these but it is not very much in use today.

The two types of boiling crisis, namely DNB and dryout, are shown in Fig. 10.7. It shows that, at lower quality, DNB occurs which is almost linear with vapour quality; on the other hand, in the dryout region, CHF decreases more drastically with quality. When the vapour content of the two vapour phase flows is high, the CHF (dryout) can occur at relatively lower heat flux. Under this condition, in the downstream region, liquid droplets suspended in the vapour core move to the wall and get partially evaporated giving rise to spray cooling. This is called 'deposition controlled dryout'. The consequence of this is a wall cooling causing CHF to decrease less drastically with quality. It can be noted that the deposition controlled dryout region shows a lower gradient of CHF decrease than the normal dryout region.

Fig. 10.3 DNB due to wavy
annular flow

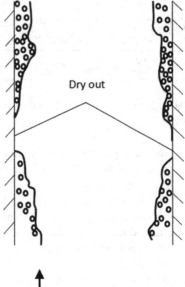

Fig. 10.4 DNB at low
quality

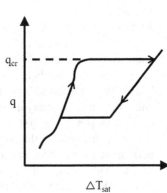

Fig. 10.5 Dryout in vertical
tube flow

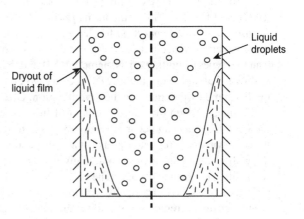

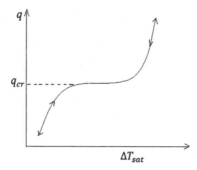

Fig. 10.6 Dryout curve for high quality

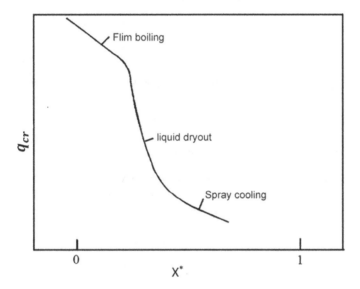

Fig. 10.7 Dependence of CHF on local quality (*Courtesy:* "Heat Transfer in Condensation and Boiling" by Karl Stephan, with permission from Springer)

10.1.2 Limits of CHF in Flow Boiling

It is intuitive that critical heat flux will depend on parameters like mass flux (or mass velocity, G), tube diameter (D), inlet subcooling ($\Delta T_{sub} = T_{sat} - T_{in}$), system pressure ($P_{sys}$) and tube length ($L$). Other parameters such as the thermal conductivity of tube material may also have a significant contribution because they can decide the uniformity of the tube wall temperature. Based on simple energy balance, one can calculate the upper and lower limits of CHF during flow boiling.

The lower limit or the minimum value of CHF can be found out assuming that it occurs when almost the entire liquid in a cross section is subcooled. Hence, the total heat input, in the form of a uniform heat flux, is used to increase the average fluid temperature from T_{in} at the inlet to $T_l(z)$. Thus,

$$q\pi Dz = \dot{m}C_{pl}[T_l(z) - T_{in}]   \tag{10.1}$$

Now in subcooled boiling, both the wall and fluid temperatures are expected to rise linearly at the same rate and hence the heat input can also be expressed as

$$q\pi Dz = h_{sc}\pi Dz[T_w(z) - T_l(z)]   \tag{10.2}$$

where h_{sc} is the subcooled boiling heat transfer coefficient. Equations (10.1) and (10.2) can be combined to eliminate $T_l(z)$,

$$q = \frac{T_w(z) - T_{in}}{\frac{D\pi z}{\dot{m}C_{pl}} + \frac{1}{h_{sc}}}   \tag{10.3}$$

using $G = \frac{4\dot{m}}{\pi D^2}$, we get the critical heat flux, when the wall is above saturation temperature, $T_w - T_{in} = T_{sat} - T_{in} = \Delta T_{sub}$

$$q_{cr} \geq \frac{\Delta T_{sub}}{\frac{4z}{GC_{pl}D} + \frac{1}{h_{sc}}}   \tag{10.4}$$

Now, for the upper limit of CHF, the entire fluid must reach the saturated equilibrium condition, indicating that the entire liquid evaporates. Again making energy balance for this

$$q\pi Dz = \dot{m}(h_g - h_{in})$$
$$= \dot{m}(h_f + h_{fg} - h_{in})   \tag{10.5}$$

where h_f, h_g and h_{fg} are saturated liquid enthalpy, saturated vapour enthalpy and enthalpy of evaporation ($h_{fg} = h_g - h_f$). Now, if we consider heat transfer only upto the saturated condition is reached,

$$\dot{m}(h_g - h_{in}) = \dot{m}C_{pl}(T_{sat} - T_{in})   \tag{10.6}$$

Now the CHF has to be lower than this. By putting $\dot{m} = \frac{G\pi D^2}{4}$, we get the limits given by Eqs. (10.4) and (10.6) which can be plotted as shown in Fig. 10.8 where the zone in which CHF lies is shown as a marked region. Here, ΔT_{sub} is shown from right to left to indicate that the higher the subcooling, the lower the temperature at entry. The axis is taken from right to left to indicate that higher subcooling means lower inlet temperature.

Fig. 10.8 The limits of CHF in flow boiling (*Courtesy:* "Convective Boiling and Condensation" by John G. Collier and John R. Thome, with permission from Oxford University Press)

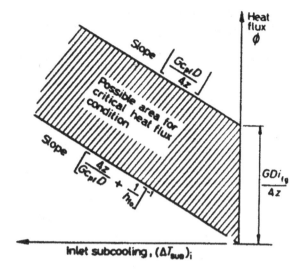

Stephan (1992) has demonstrated that for inlet subcooling of 50 k and mass velocity of $1000 \, \text{kg/m}^2\text{s}$, water boiling inside a 1m long tube of 25mm diameter gives the following limit of CHF for an assumed $h_{sc} = 1000$ W/m^2 k

$$3.62 \times 10^5 \, \text{W/m}^2 < q_{cri} < 1.4 \times 10^7 \, \text{W/m}^2$$

10.2 Parametric Variation of CHF in Flow Boiling

Starting from the 1960s, the progress of nuclear reactor technology and installation of large reactors throughout the world necessitated better understanding of in-tube boiling and associated CHF conditions. Since the majority of the reactors are Boiling Water Reactors (BWR), the occurrence of CHF conditions is a major safety concern to such reactors. Consequently, a large number of studies on CHF were carried out between the 1960s and 1980s after which there was a slump due to worldwide antinuclear protests. However, the data and understanding of CHF generated during the above period gave a clear idea of the variation of CHF during flow boiling due to parametric change. Most of these studies were carried out for vertical upflow conditions with electrical heating providing constant wall heat flux.

Previously, we mentioned that critical heat flux can be a function of several parameters. Instinctively we mentioned that

$$q_{crit} = f(G, \Delta T_{sub}, P, D, z) \tag{10.7}$$

The hundreds of experiments carried out and available in open literature as well as in the classified data of various nuclear establishments around the world reported the

Fig. 10.9 CHF as a function
of inlet subcooling

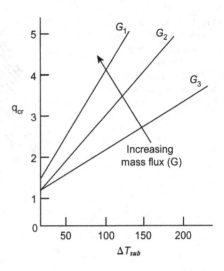

dependence of CHF (q_{crit}) on G, P, D and z. They also reported dependence of q_{crit}
on local quality (x^*). However, the local quality and local enthalpy, $h(z)$, are not
independent quantities but they are correlated to inlet subcooling and applied heat
flux as

$$h(z) = h_f + \frac{4qz}{GD} - \Delta T_{sub} C_{pl} \tag{10.8}$$

$$x^*(z) = \frac{1}{h_{fg}[\frac{4qz}{GD} - C_{pl}\Delta T_{sub}]} \tag{10.9}$$

Hence, critical heat flux (CHF) can be expressed as a function of either enthalpy
$h(z)$ or quality $x^*(z)$ along with other parameters, as

$$q_{crit} = f_1(G, h(z), P, D, z) \tag{10.10}$$
$$= f_2(G, x^*(z), P, D, z) \tag{10.11}$$

Let us now examine the nature of the dependence of CHF on the above factors. When
we keep other parameters like diameter (D), tube length (z), pressure (P) and mass
velocity (G) constant, the CHF increases linearly with inlet subcooling as shown
in Fig. 10.9. The figure also contains the plots of CHF for different values of mass
velocity G. It can be seen that the slope of the curve increases with mass velocity. For
this reason, many correlations use linear function between CHF and inlet subcooling.
Figure 10.10 shows the variation of CHF with mass flux exclusively keeping other
parameters constant.

When plotted against mass velocity (G) alone for fixed diameter, length and
inlet subcooling, the variation of CHF increases non-linearly (Fig. 10.10). The CHF
decreases with tube length for a given diameter, mass velocity and inlet subcooling

Fig. 10.10 CHF changes
with mass flux (*Courtesy:*
"Boiling Condensation and
Gas-Liquid flow" by P.B.
Whalley, with permission
from Oxford University
Press)

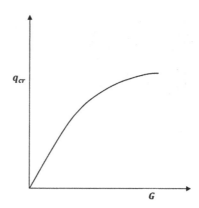

Fig. 10.11 CHF and P_{crit}
variation against length
(*Courtesy:* "Boiling
Condensation and
Gas-Liquid flow" by P.B.
Whalley, with permission
from Oxford University
Press)

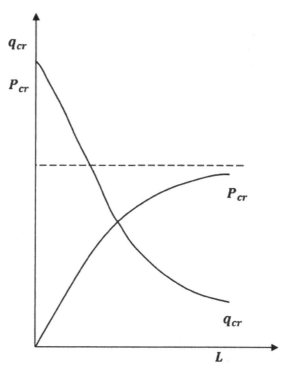

as shown in Fig. 10.11. However, it may be important also to plot the critical power,
P_{crit}, which asymptotically approaches a value that is sufficient to produce a quality
of unity by evaporating all the liquid also shown in the same figure. This power is
given by

$$P_{crit} = \pi D z q_{crit} \tag{10.12}$$

Fig. 10.12 The change of
CHF with mass velocity in
flow boiling (*Courtesy:*
"Boiling Condensation and
Gas-Liquid flow" by P.B.
Whalley, with permission
from Oxford University
Press)

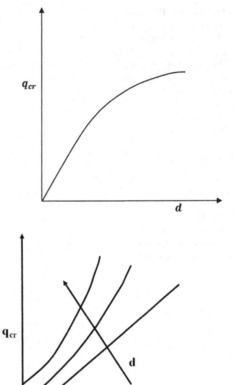

Fig. 10.13 CHF against
subcooling at different
diameters

The effect of tube diameter is also that of increasing CHF. The nature of this increase
depends on the inlet subcooling. Usually, CHF increases with the diameter as shown
in Fig. 10.12 for a fixed inlet subcooling, tube length and mass velocity.

However, the influence of diameter can also be understood by plotting CHF against
inlet subcooling for various tube diameters as given in Fig. 10.13 according to Lee
(1966). It can be observed that the nature of variation of CHF is dependent on
diameter. For lower values of z/D, the CHF varies linearly but at higher values,
non-linear increase is observed.

Usually, it is said that the influence of system pressure is not much on CHF.
However, this is an unqualified statement. Usually, the dependence on the pressure
is very complex. While for fixed inlet subcooling, the CHF increases with increasing
pressure in the low-pressure region, it passes through a maximum at around 30 bar
and then decreases with pressure as shown in Fig. 10.14. Yin et al. (1988) observed
a second maximum and it was found that this maximum was more pronounced.

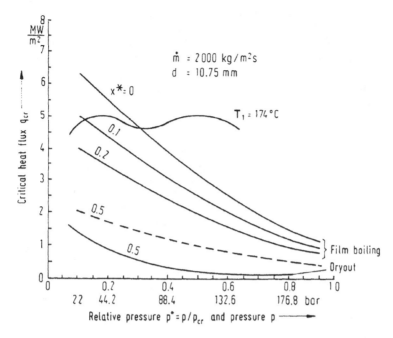

Fig. 10.14 Effect of pressure on CHF (*Courtesy:* "Heat Transfer in Condensation and Boiling" by Karl Stephan, with permission from Springer)

However, this is very much dependent on vapour quality. At lower quality, the CHF is known to decrease with pressure where the primary mechanism is that of film boiling.

The above discussions conclusively proved that Critical Heat Flux (CHF) in flow boiling is largely influenced by inlet subcooling. Again according to Eq. (10.2), the local quality (x^*) is a function of inlet subcooling ΔT_{sub}. Hence, it is important to examine the effect of local quality on critical heat flux. Let us first consider the influence of quality on CHF with mass velocity as a parameter. We recall from Fig. 10.9 that CHF depends more or less linearly on inlet subcooling and Eq. (10.2) shows that inlet subcooling in turn influences quality linearly. Hence, a linear dependence of CHF on exit quality is expected for a given tube diameter, tube length and mass velocity. While this is approximately correct, as can be seen from Fig. 10.15, the dependence on mass flux (or mass velocity G) seems to be different in subcooled and saturation regimes. While in the subcooled regime (negative quality, $x^* < 0$), higher mass flux gave higher CHF, in the saturated region at a low quality, a cross over of the curves is observed after which lower mass velocity gives higher heat flux, for a given exit quality. Thus, although correlated, in the actual application (keeping inlet subcooling constant and exit quality constant), there are two different conditions giving different trends of CHF.

Fig. 10.15 CHF dependence
on quality at different mass
fluxes (*Courtesy:* "Heat
Transfer in Condensation
and Boiling" by Karl
Stephan, with permission
from Springer)

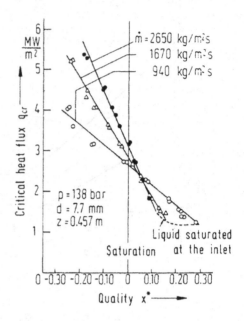

A similar effect can be observed when we examine the effect of tube length (L) on CHF. We recall that in Fig. 10.11 we have seen that with tube length CHF decreases sharply. However, when plotted against the exit quality of steam, surprisingly tube length appears to have very little effect.

Therefore, the question arises of which is the right way of looking at the phenomenon. Let us elaborate on the question a little bit more. Is it logical to look at CHF keeping the inlet subcooling constant and compare its variation with parameters like diameter, mass velocity and tube length? Or is it prudent to compare CHF with these parameters keeping the exit quality to be constant? These two points of view have two explanations.

(a) Let us first say that the total length of the tube comprises of two parts, one due to subcooled boiling and the other due to saturated boiling.

$$Z = Z_{sc} + Z_{sat} \tag{10.13}$$

$$Z_{sc} = \frac{DGC_{pl}\Delta T_{sub}}{4q_{cr}} \tag{10.14}$$

On one hand, Fig. 10.16 shows the variation of critical heat flux with exit quality. Figure 10.16 suggests that critical heat flux is an integral process since it incorporates the history of the flow related to the entire boiling length.

(b) The other curve (Fig. 10.15) suggests that critical heat flux is a completely local phenomenon. When the local values of q and x correspond to critical values, CHF occurs.

At the first sight, the above viewpoint appears to be completely contrary to each other. However, it can be readily found that these two viewpoints are actually complementary to each other. Let us consider the heat balance over the boiling length (assuming vapour generated in subcooled heat transfer is small) z_{sat}. The heat flux in this length supplies the total latent heat of evaporation,

$$\pi D Z_{sat} q_{cr} = \dot{m} x_{cr} h_{fg}$$

or

$$Z_{sat} = \frac{G D h_{fg}}{q_{fg}} x_{cr}$$

where

$$\dot{m} = \frac{\pi}{4} D^2 G$$

Thus, the above equation clearly demonstrates that q_{crit} as a function of quality (x) contains no less understanding than as a function of length (Z). However, the correlations available of CHF, based on x, are called 'local condition correlations' while those based on z are known as 'integral correlations'.

10.2.1 Correlations for CHF in Flow Boiling

As stated above, the correlations for CHF in flow boiling are of two types, 'local condition correlations' and 'integral correlations'. Often one type of correlation can be transformed into the other type by using the correlation between local quality (x^*) and the tube length (z). One question is of course relevant whether these correlations are for 'DNB' or for 'Dryout'. More pointedly, the question can be "is it possible to have a correlation which works both for DNB and dryout?" Logically, the answer should be 'No' and separate sets of correlations need to be framed for DNB or dryout. Unfortunately, very few equations are capable of delineating the two types of dryouts. Most of the correlations are intuitive and semi-empirical and hence tend to cover the entire range of CHF, be it DNB or be it dryout. Most of the 'local condition correlations' are based on the 'local condition hypothesis' of Barnett (1963). This hypothesis states that the CHF is solely a local phenomenon and a function of only the local quality (x^*) at point of overheating.

In other words,

$$q_{cr} = f[x^*(Z), G, D, P] \tag{10.15}$$

Figure 10.16 supports this hypothesis because the data points for experiments with different tube lengths collapse to one curve which shows the dependence of CHF on

Fig. 10.16 CHF dependence
on exit quality for different
tube diameters (*Courtesy:*
"Heat Transfer in
Condensation and Boiling"
by Karl Stephan, with
permission from Springer)

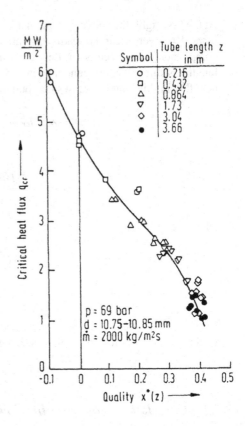

exit mass quality alone. The most commonly used correlation based on this hypothesis was that by Macbeth (1964) and Thompson (1964). This empirical correlation in a simplified form can be written as

$$q_{cr} = Ah_{fg}G^{\frac{1}{2}}.(1 - x^*) \tag{10.16}$$

where

$$A \approx 0.25\frac{\text{kg}^{\frac{1}{2}}}{\text{ms}^{\frac{1}{2}}}$$

This local correlation can evolve interesting insight into the process of CHF as will be shown later. Now, we show how Barnett and Macbeth with their assumption developed the local condition correlation. Their primary assumption was that the CHF varies linearly with inlet subcooling enthalpy

$$q_{cr} = A + B\Delta h_{sub} \tag{10.17}$$

where

$$\Delta h_{sub} = h_f - h_{in}$$

From the parametric dependence presented in the last section, it is obvious that A and B must be functions of D, G, P and z. Considering the subcooled flow to be a single phase

$$\Delta h_{sub} = C_{pl}(\Delta T_{sub}) \tag{10.18}$$

Now recalling Eq. (10.13), the total pipe length is given by

$$Z = Z_{sc} + Z_{sat}$$

$$\dot{m}[\Delta h_{sub} + x^*(z)h_{fg}] = q\pi Dz \tag{10.19}$$

putting in

$$\dot{m} = \frac{G\pi D^2}{4}$$
$$x^*(z) + \Delta h_{sub} = \frac{4qz}{GD} \tag{10.20}$$

From Eqs. (10.20) and (10.17), by eliminating Δh_{sub},

$$q_{cr} = \frac{A - Bx^*(z)h_{fg}}{1 - \frac{4Bz}{GD}} \tag{10.21}$$

This equation indicates a nearly linear relationship between CHF and exit quality (x^*) as shown in Fig. 10.16. It can be seen in this figure that this linearity holds good for low quality (i.e., for DNB); however, for high exit quality approaching dryout, it becomes inaccurate.

We can recast the above equation as

$$q_{cr} = \frac{A' - \frac{GDh_{fg}x(z)}{4}}{B' + z} \tag{10.22}$$

where

$$A' = \frac{AGD}{4B} = AB'$$
$$B' = \frac{GD}{4B}$$

Using these, Eq. (10.17) yields

$$q_{cr} = \frac{A' + \frac{GDh_{sub}}{4}}{B'} \tag{10.23}$$

Now, although Eqs. (10.22) and (10.23) give the same quantity (q_{cr}), Eq. (10.23) is dependent on independent variables and hence easier to use.

Bowring (1972) provided a better method to use the Macbeth correlation by providing the correlations for A' and B'

$$A' = \frac{2.317[\frac{GDh_{fg}}{4}]F_1}{(1 + 0.0143F_2D^{\frac{1}{2}}G)} \tag{10.24}$$

$$B' = \frac{0.077F_3GD}{(1 + 0.347F_4(\frac{G}{1356})^n)} \tag{10.25}$$

where $n = 2 - 0.00725P$, P being the system pressure in bar.

This correlation is valid for the range

$$2\,\text{bar} < P < 190\,\text{bar}$$
$$2\,\text{mm} < D < 190\,\text{mm}$$
$$15\,\text{m} < z < 3.7\,\text{m}$$
$$136\frac{\text{kg}}{\text{m}^2\,\text{s}} < G < 18600\frac{\text{kg}}{\text{m}^2\,\text{s}} \tag{10.26}$$

All the above correlations are based on data for steam–water system. Ahmad (1973) provided an excellent method for how the scaling law can be applied. He identified five dimensionless scaling factors,

$$\frac{z}{D}, \frac{\rho_l}{\rho_g}, \frac{\Delta h_{sub}}{h_{fg}}, \frac{q_{crit}}{Gh_{fg}} \text{ and } \psi$$

He proposed matching these parameters for the given fluid to steam–water system in the following way in which the equivalent steam–water system is chosen.

(i) z/D is made same for the system. They can be done by scaling up or down both z and D or keeping them the same for both the systems.

(ii) $\frac{\rho_l}{\rho_g}$ is made same for the fluid and water. This can be done by changing the pressure for steam–water system to reach a pressure at which the density ratio is the same as that of the given fluid.

(iii) $\frac{\Delta h_{sub}}{h_{fg}}$ is made the same for the fluid and water by changing the inlet temperature of the water, giving $T_{water,in}$.

(iv) If ψ is made the same for the fluid and water system, it leads to determine the equivalent mass velocity of the water.

(v) Now with the changed conditions of steam–water system as above, (q_{crit}) is calculated from the Macbeth–Bowring correlation.

Katto and Ohme (1984) provided a similar method for calculating CHF which uses four dimensionless parameters.

10.2.2 *Local Condition and the Location for CHF Initiation*

The 'local condition correlation' presented in the previous section can be useful to identify the location at which CHF will occur in the tube. Let us consider the case of heating the tube uniformly (i.e., the tube has a constant uniform wall heat flux over its entire length). The entire analysis is shown in a series of figure sketches as shown in Fig. 10.17 as argued by (Whalley, 1987). Figure 10.17b shows the approximate linear variation of CHF with exit condition as given by local condition correlation (this is eventually Fig. 10.15, resketched). Figure 10.17b shows the variation of quality with the length of various values of heat flux (say $q1$, $q2$ and $q3$). From these two figures, we can eliminate quality (one axis for Fig. 10.17a, the other axis for Fig. 10.17b), giving the plot of CHF with length, Fig. 10.17c for the heat fluxes $q1$, $q2$ and $q3$. Now the actually imposed heat fluxes can be as shown in Fig. 10.17c (shown with dotted lines) where three cases of CHF are compared with actually imposed heat flux. If we consider CHF level $q1$ and imposed flux, the curves do not intersect throughout the length of the tube. Even at the exit CHF remains higher than the imposed flux

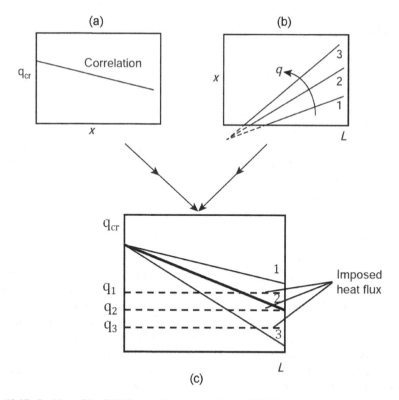

Fig. 10.17 Position of the CHF for a uniformly heated tube. CHF first occurs at the end of the tube (*Courtesy:* "Boiling Condensation and Gas-Liquid flow" by P.B. Whalley, with permission from Oxford University Press)

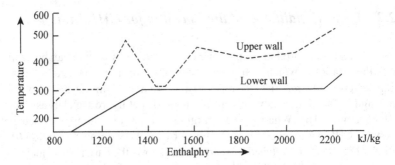

Fig. 10.18 Wall temperature for boiling in a horizontal tube (*Courtesy:* "Heat Transfer in Condensation and Boiling" by Karl Stephan, with permission from Springer)

and hence the CHF does not occur in this case. For the case where the CHF curve $q2$ touches the imposed flux only at the tube end of the tube (near the exit), it causes CHF there. Going further the heat fluxes $q3$, CHF occurs near the exit over a wider length. In general, for uniformly heated tubes, CHF occurs near the exit and then propagates backwards.

For horizontal tubes, the controlling parameter is the Froude number

$$F_n = x^*_{crit}\dot{m}\left[\frac{1}{\rho_l(\rho_g - \rho_l)gd}\right]^{\frac{1}{2}} \tag{10.27}$$

where x^*_{crit} corresponds to the inception of the boiling crisis. For $F_n > 7$, inertial force is large enough to avoid any stratification. For $F_n < 3$, stratification is pronounced. The partial wetting of the wall creates a circumferential temperature distribution in the wall enhancing circumferential heat transfer giving a complex temperature profile like the one given in Fig. 10.17.

10.3 Post Dryout Heat Transfer or Spray Cooling

The heat transfer mechanism occurs when the liquid film in the convective boiling dries out completely or in other words after CHF due to dryout occurs. This region is not simply vapour flow inside the tube but, as discussed during the CHF description, heat transfer with liquid droplets in vapour and consequent cooling of the wall. The mechanism is known by various names such as 'post dryout heat transfer', 'spray cooling' or 'liquid deficient region'. From the application point of view, this type of heat transfer is important in devices such as once through steam generators and refrigeration devices. It is probably most important with respect to boiling water nuclear reactors or even in pressurized heavy water reactors where dryout may occur due to loss of coolant accident (LOCA) or severe accident. This type of cooling is often employed during the cooling of a hot surface at a pre-assigned rate to get a particular microstructure on the surface and as a result a sought property as in the case of cooling of steel bars in steel plant by spraying small water droplets.

Unlike CHF in low quality which is due to film boiling (or bubble crowding), in the high-quality region, the liquid dries up, and in the last phase of drying, the film gets torn and dispersed into the vapour in the form of tiny liquid droplets. Subsequently, the heat is transferred to the vapour at the wall which gets superheated and which in turn transfers the heat to the droplets which gradually evaporate. This phase can be seen as just opposite to bubbly flow where the liquid is the continuous phase and bubbles are the dispersed phase. Here vapour is the continuous phase and droplet is the dispersed phase. For this similarity in both these cases, the homogeneous model works well for calculating pressure drop. However, from a heat transfer point of view, it has some unique features to note. Firstly, the liquid droplets in their suspended state offer large surface area or evaporation due to their tiny size leading to the rapid change in quality. Secondly, the droplets, due to motion, do penetrate the vapour boundary layer at the wall and remove heat additionally from the wall apart from cooling by vapour. This enhances the heat transfer coefficient at the wall. It is interesting to note that while film boiling provides a heat transfer coefficient of the order of 250–1500 W/m^2K, spray cooling provides a heat transfer coefficient of several times higher, typically 5000–40000 W/m^2K. In reality, during spray cooling, several mechanisms of heat transfer take place simultaneously, such as the following:

(a) Heat transfer by vapour convection at the wall.
(b) Heat transfer from the superheated vapour to the saturated droplets suspended at the core.
(c) Heat transfer due to the impact of the liquid droplets which penetrates the vapour boundary layer at the wall.
(d) Heat transfer by radiation from the wall to the vapour which acts as a participating medium and to the liquid droplets.

Out of these, the first two mechanisms make major contributions and the entire phenomenon can be explained in terms of these mechanisms.

Here two extreme cases may be thought of. In the case of total thermal inequilibrium which occurs at low mass flux and pressure, the wall heats up the vapour to higher superheat. Due to nonequilibrium, this heat does not go to the dispersed liquid droplet at the core and hence both wall and vapour temperature increases linearly. However, the actual quality remains the same but the thermodynamic quality increases due to heat transfer. The sketch of wall, vapour and liquid temperature is shown in Fig. 10.19 for this case.

For the case of complete equilibrium between vapour and liquid, the actual quality coincides with thermodynamic quality and the wall temperature rises suddenly and then decreases due to an increase in the vapour flow caused by evaporation. The liquid and vapour temperature remains the same till all the liquid evaporate and then the vapour temperature starts rising. This is shown in Fig. 10.20 which occurs at high pressure and mass flux. The reality lies in between these two extreme cases and the corresponding temperatures are shown in Fig. 10.21.

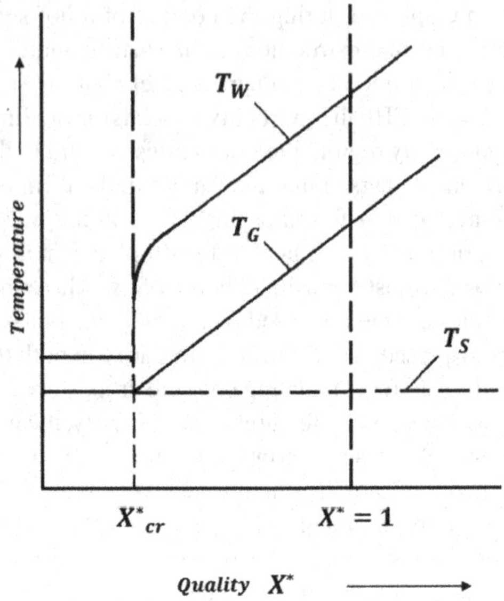

Fig. 10.19 Post dryout heat
transfer with complete lack
of thermodynamic
equilibrium (*Courtesy:*
"Heat Transfer in
Condensation and Boiling"
by Karl Stephan, with
permission from Springer)

Fig. 10.20 Post dryout heat
transfer with complete
thermodynamic equilibrium
(*Courtesy:* "Heat Transfer in
Condensation and Boiling"
by Karl Stephan, with
permission from Springer)

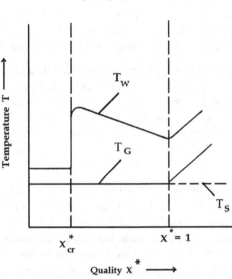

This figure can be explained by the heat balance over the length as to correlate
wall, vapour and liquid temperature to the thermodynamic quality as

$$x_{th}^* = \frac{1}{h_{fg}} \left(\frac{q\pi Dz}{\dot{m}} + h_{in} - h_f \right) \qquad (10.28)$$

Till the CHF occurs, there is a thermal equilibrium between the liquid and vapour after
which the loss of equilibrium becomes a function of the amount of wall superheat or

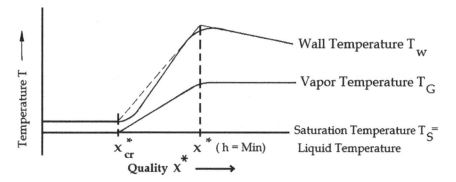

Fig. 10.21 The real picture of post dryout heat transfer (*Courtesy:* "Heat Transfer in Condensation and Boiling" by Karl Stephan, with permission from Springer)

in other words on the pressure and mass flux. When the wall superheat is small, the entire heat gets to the vapour and a complete inequilibrium is established. However, downstream equilibrium is partially restored and the liquid evaporates bringing the wall temperature down and presents further superheating of vapour. In terms of heat transfer coefficient, it can be told that at CHF first the heat transfer coefficient falls rapidly reaching a minimum and then it slowly increases due to spray cooling where the droplet evaporation mechanism takes over.

However, as we have seen during defining thermodynamics quality in a previous chapter, due to the nonequilibrium of liquid and vapour after dryout, actual quality is lower than thermodynamic quality. The actual quality can be calculated as follows, from the heat balance:

$$\dot{m}h_{in} - q\pi Dz = \dot{m}x^*_{actual}h_G + \dot{m}\left(1 - x^*_{actual}\right)h_f$$

or

$$x^*_{actual} = \frac{1}{h_G - h_f}\left(\frac{q\pi Dz}{\dot{m}} + h_{in} - h_f\right) \qquad (10.29)$$

Please compare this equation to Eq. (10.28). The only difference being the term that indicates the vapour enthalpy. Vapour enthalpy in reality will be in the superheated range because the vapour is not in equilibrium with the liquid. If there is equilibrium,

$$h_G = h_g$$

and hence,

$$h_g - h_f = h_{fg}$$

giving

$$x^*_{actual} = x^*_{th}$$

A number of methods are suggested for the calculation of post dryout heat transfer. A simple one is by (Groeneveld and Moeck, 1969) which uses

$$Nu = a Re^{n_1} Pr_G^{n_2} Y^{n_3} \qquad (10.30)$$

where

$$Nu = \frac{h_{dryout} d}{k_G}$$

$$Re = \frac{\dot{m} d}{\mu_G} \left[x^* + \frac{(1 - x^*) \rho_G}{\rho_L} \right]$$

Pr_G = vapour Prandtl number at wall temperature

$$Y = 1 - 0.1 \left[\left(\frac{\rho_G}{\rho_L} - 1 \right) (1 - x^*) \right]^{0.4}$$

Suggested values of constant and exponents are

$$a = 0.327 \times 10^{-2},$$
$$n_1 = 0.901,$$
$$n_2 = 1.32,$$
$$n_3 = -1.5$$

Another method specifically considered nonequilibrium and suggested the total heat flux to be a sum of the heat flux to the vapour and the liquid, thus

$$q = q_g + q_l \qquad (10.31)$$

here q_g raises the superheat of the vapour and q_l evaporates the liquid. Various authors have provided correlations for the ratio of q_l and total flux q as

$$\epsilon = \frac{q_l}{q} \qquad (10.32)$$

Once ϵ is known, wall and vapour temperatures can be calculated using thermodynamic relations. Various correlations were suggested, such as that by Plummer (1977)

Table 10.1 Constant of Plummer (1977), (Eq. 10.33)

Fluid	C_1	C_2
N_2	0.082	0.29
H_2O	0.07	0.4
R12	0.078	0.255
R113	0.078	0.13

$$\epsilon = c_1 ln \left[G \left(\frac{D_h}{\rho_g \sigma} \right)^{\frac{1}{2}} (1 - x_{CHF})^5 \right] + c_2 \qquad (10.33)$$

where D_h =Hydraulic diameter.

The values of the constants are given in Table 10.1.

Examples

Problem 1: Calculate the critical heat flux for the water for the following conditions:

Pressure $(P) = 60\,bar$
Dia of tube $= 2.5\,cm$
Length of tube $(Z) = 2.5\,m$
$\dot{m} = 800\,kg/m^2s$
subcooling $= 20\,°C$

Solution: By using the Thomson–Macbeth correlations,

$$q_G = \frac{A \frac{\Delta h_v md}{4} + \frac{md\bar{c}_{pl}\Delta t_u}{4}}{Z + B \frac{\Delta h_v md}{4}}$$

$$n = 2.0 - 0.00725\,P \text{ (P in bar)}$$
$$= 2.0 - 0.0072 \times 60$$
$$= 1.565$$

$$B = 0.308 \frac{\frac{F_3(P)}{\Delta h_v}}{1 + 0.347 F_4(p)[\frac{m}{1356}]^n}$$

$$= 0.308 \frac{\frac{0.89}{157 \times 10^3}}{1 + 0.347 \times 0.701[\frac{800}{1356}]^1 .565}$$

$F3(P)$, $F4(P)$ are obtained from pressure function charts; therefore,

$$B = 1.575 \times 10^{-7}$$

$$A = 2.314 \frac{F_1(P)}{1 + 0.0143 F_2(P) d^{\frac{1}{2}} m}$$

$$= 2.314 \frac{1.043}{1 + 0.0143 \times 0.934 (0.025)^{\frac{1}{2}} \times 800}$$

$$= 0.9$$

$$q_G = \frac{A \Delta h_v + \bar{C}_{pl} \Delta t_u}{\frac{Z4}{md} + B \Delta h_v}$$

$$= \frac{0.9 \times 1571 \times 10^3 + 4500 \times 20}{\frac{2.5 \times 4}{800 \times 0.025} + 1.575 \times 10^3 \times 1517 \times 10^3}$$

Hence, critical heat flux $= 2.102 \times 10^6 \frac{W}{m^2}$.

Problem 2: Calculate the critical heat flux for the fluid specified below, while flowing through a 3cm dia pipe (horizontal) with a mass flux of 800 kg/m²-s and length in 3.5 m.

subcooling $= 25\,\text{kg/m}$, $\rho_l = 1205\,\text{kg/m}^3$, $\rho_g = 9.57\,\text{kg/m}^3$,

$K_l = 113\,\text{W/mK}$, $K_g = 7.7\,\text{W/mK}$, $h_l = 5.07\,\text{kJ/kg}$,

$\Delta h_g l = 86.1\,\text{kJ/kg}$, $\sigma = 4.78 \times 10^{-3}\,\text{N/m}$,

$C_p l = 1.87\,\text{kJ/kgK}$, $C_p s = 1.31\,\text{kJ/kgK}$,

$\mu_l = 1.27 \times 10^{-4}$, $\mu_g = 4.63 \times 10^{-6}$,

$T_{sat} = 27.1\,\text{K}$, $P_{sat} = 101.3\,\text{kPa}$

Solution:

$$\phi_H = \frac{mD}{\mu_l} \left(\frac{\mu_l^2}{\sigma D \rho_l} \right)^{-1.57} \left(\frac{(\rho_l - \rho_g)/g D^2}{\sigma} \right)^{-1.05} \left(\frac{\mu_l}{\mu_g} \right)^{6.41}$$

$$\phi_H = \frac{800 \times 0.03}{1.27 \times 10^{-4}} \left(\frac{(1.27 \times 10^{-4})^2}{4.78 \times 10^{-3} \times 0.03 \times 1205} \right)^{1.57}$$

$$\times \left(\frac{(1205 - 9.57)/9.81 \times 0.03^2}{4.78 \times 10^{-3}} \right)^{-1.05} \left(\frac{1.27 \times 10^{-4}}{4.63 \times 10^{-6}} \right)^{6.41}$$

$$= 3.44 \times 10^{20}$$

Boiling Number

$$N_B = \frac{q_c}{mh_{lg}} = 5.75 \times \phi_H^{-0.34} \frac{L}{D}^{-0.51} \left[\frac{\rho_l - \rho_g}{\rho_g}\right]^{1.27} \left[1 + \frac{\Delta h_{sub}}{h_{lg}}\right]^{1.64}$$

$$= 5.75 \times \left(3.44 \times 10^{20}\right)^{-0.34} \frac{3.5}{0.03}^{-0.51} \left[\frac{1205 - 9.57}{9.57}\right]^{1.27}$$

$$\times \left[1 + \frac{25}{86.1}\right]^{1.64}$$

$$= 5.75 \times 1.04 \times 10^{-7} \times 0.08 \times 460 \times 1.6$$

$$= 3.7 \times 10^{-5}$$

therefore

$$q_c = 3.7 \times 10^{-5} \times m \times h_{lg}$$
$$= 3.7 \times 10^{-5} \times 800 \times 86.1 \times 10^3$$
$$= kW/m^2$$

Problem 3: Calculate critical heat flux for following subcooled water while flowing in round tubes.

Pressure $(P) = 140\,bar$
Dia of tube $= 10\,cm$
Mass flow rate $= 2000\,kg/m^2$
subcooling $= 26\,K$

Solution: Properties at 140 bar:

$$\rho_l = 620\,kg/m^3, \quad \rho_v = 87\,kg/m^3, \quad K = 0.5\,W/mK,$$
$$h_{fg} = 1071\,kJ/kg, \quad h_f = 1571.5\,kJ/kg, \quad T_{sat} = 336.6°C$$
$$C_p = 5.9\,kJ/kgK, \quad \mu_l = 8.1 \times 10^{-5}$$

Assume an initially $q = 5 \times 10^6\,W/m^2$

$$\Psi_0 = 230 \left[\frac{5 \times 10^6}{2000 \times 1071 \times 10^3}\right]^{0.5} = 11.11$$

$$h_{FC} = 0.023 \left(\frac{Gd}{\mu_l}\right)^{0.8} Pr_L^{0.4} \left(\frac{K}{d}\right)$$

$$= 23412.2\,W/m^2 K$$

$$T_w - TL = \frac{(\Psi_0 - 1)(T_{sat} - T_L) + \left(\frac{q}{h_{FC}}\right)}{\Psi_0}$$

$$= 42.88$$

$$q_B = q - h_{FC}(T_w - TL)$$

$$= 5 \times 10^6 - 23412.2 \times (42.88)$$

$$= 3.996 \times 10^6$$

Initial thickness of liquid sublayer, δ

$$\delta = \frac{\pi (0.0584)^2}{2} \left(\frac{\rho_v}{\rho_L}\right)^{0.4} \left(1 + \frac{\rho_v}{\rho_L}\right) \frac{\sigma}{\rho_v} \left(\frac{\rho_v H_{fg}}{q_B}\right)$$

$$\delta = 2.43 \times 10^{-4}$$

$$x_0 = \frac{i_L - i_{sat}}{H_{fg}}$$

x_0 is the thermodynamic equilibrium quality, i_L is the enthalpy of subcooled liquid and i_{sat} is the enthalpy of saturated liquid.

$$x_0 = \frac{(1571.5 - (6 \times 26)) - 1571.5}{1071}$$

$$x_0 = -0.146$$

$$x_{e,N} = \frac{-154 \times 5 \times 10^6}{1071 \times 10^3 \times 2000}$$

$$x_{e,N} = -0.36$$

therefore $x_{e,N} < x_0$

$$x = \frac{x_0 - x_{e,N} \exp\left(\frac{x_0}{x_{e,N}} - 1\right)}{1 - x_{e,N} \exp\left(\frac{x_0}{x_{e,N}} - 1\right)}$$

$$x = 0.044$$

$$\alpha = \frac{x}{\left[x + (1-x)\frac{\rho_v}{\rho_L}\right]} = 0.245$$

$$\rho = \frac{1}{\left[\frac{x}{\rho_v} + \frac{(1-x)}{\rho_L}\right]} = 484$$

$$\mu_v = 23 \times 10^{-5}$$

$$\mu = \mu_v \alpha + \mu_L (1 - \alpha)(1 + 2.5\alpha) = 1.042 \times 10^{-7}$$

$$R_e = \frac{Gd}{\mu}$$

$$= \frac{2000 \times 0.01}{1.042 \times 10^{-4}} = 1.92 \times 10^5$$

$$\frac{1}{\sqrt{f}} = 2.0 \log_{10}\left(R_e\sqrt{f}\right) - 0.8$$

$$\therefore f = 0.0155$$

$$\tau_w = f\frac{G^2}{8\delta} = 16$$

$$y_{\delta+} = \delta\sqrt{\frac{\tau}{\mu}} = 6.076$$

$$U_{\delta+} = 5 + 5m\left(\frac{y_{\delta+}}{5}\right) = 5.975$$

$$K = \frac{242\left[1 + k\left(0.355 - \alpha\right)\right]}{\left[0.0147 + \left(\frac{\rho_v}{\rho_L}\right)^{0.733}\right]\left[1 + 90.3\left(\frac{\rho_v}{\rho_L}\right)^{368}\right]} \times R_e^{-0.8} = 0.08$$

therefore

$$U_B = KU_\delta = 0.08 \times 1.09 = 0.0872$$

$$L_B = \frac{2 \times \pi\left(\rho_v + \rho_L\right)}{\rho_v\rho_L U_B^2} = 0.15$$

$$T = \frac{L_B}{U_B} = 1.74$$

$$\therefore q^1 = \frac{\delta\rho_L H_{fg}}{\tau} = \frac{2.43 \times 10^{-4} \times 620.7 \times 1071 \times 10^3}{1.74} = 0.93 \times 10^5 \,\text{W/m}^2$$

We have to iterate like this by assuming q upto $q = q_1$, then q is called critical heat flux.

Problem 4: A subcooled water is flowing in a vertical tube upwards calculate critical heat flux for the following conditions.

Pressure $P = 1500$ Psi
Dia of tube $D_e = 25$ inches
$\frac{L}{D_e} = 150$
$G = 4 \times 10^6$ lb/h ft^2
ΔT subcooling $= 50\,°$F

Solution: Properties at 1500 Psi:

$$\rho_l = 42.5 \, \text{lb/ft}^3$$
$$\rho_g = 3.67 \, \text{lb/ft}^3$$
$$h_{sat} = 340 \, \text{Btu/lb}$$
$$h_{fg} = 555.7 \, \text{Btu/lb}$$

Enthalpy at 50°F subcooled = 290 Btu/lb. For subcooled region Westinghouse APD correlation.

$$q_{cri} = \left(0.23 \times 10^6 + 0.094G\right)\left(3 + 0.01\Delta T_{sub}\right)\left(0.435 + 1.23e^{-0.0093\frac{L}{D_e}}\right)$$
$$\times \left(1.7 - 1.4e^{-a}\right)$$

$$a = 0.532 \left(\frac{\rho_l}{\rho_v}\right)^{\frac{1}{3}}\left(\frac{h_{sat} - h_i}{h_{fg}}\right) = 0.532 \left(\frac{42.5}{3.67}\right)^{\frac{1}{3}}\left(\frac{50}{555.7}\right) = 0.198$$

$$q_{cri} = \left(0.23 \times 10^6 + 0.094 \times 4 \times 10^6\right)\left(3 + 0.01 \times 50\right)$$
$$\times \left(0.435 + 1.23e^{-0.0093 \times 150}\right)\left(1.7 - 1.4e^{-0.198}\right)$$
$$= 8.06 \times 10^5 \, \text{Btu/h ft}^2$$

Problem 5: Calculate the flooding critical heat flux while water flows down through the channel and has the following conditions.

Pressure $P = 200 \, \text{Psi}$
Channel c/s $= 0.2 \times 0.4 \, \text{inch}$
Heating length $= 20 \, \text{inch}$
Mass flow rate (W) $= 10^5 \, \text{lb ft}^2\text{h}^{-1}$

Solution: Properties at 200 Psi:

$$\rho_l = 54.32 \, \text{lb/ft}^3$$
$$\rho_g = 0.444 \, \text{lb/ft}^3$$
$$t = 195°C$$
$$h_{fg} = 843.4 \, \text{Btu/lb}$$

$A_C = 0.2 \times 0.4$, $A_h = 0.4 \times 20$. Therefore, $C = 0.63$ for two-side heating section.

$$q_{CRF} = \frac{c^2 A_c H_{fg}\sqrt{2\rho_g g \Delta\rho w}}{A_h\left[1 + \frac{\rho_g}{\rho_L}^{\frac{1}{4}}\right]^2}$$

$$q_{CRF} = \frac{0.63^2 \times 5.33 \times 10^{-4} \times 843.4 \times \sqrt{2 \times 0.444 \times 4.17 \times 10^8 \times 54 \times 10^5}}{0.0555\left[1 + \frac{0.444}{54.32}^{\frac{1}{4}}\right]^2}$$

$$= 8.8 \times 10^7 \, \text{Btu/h ft}^2$$

Exercises

1. Calculate critical heat flux for the following conditions:

   Diameter of tube = 2.0mm
   Pressure = 4 bar
   Mass flux = 500 kg/m^{2}s
   Liquid = Liquid CO_2
   Orientation = Vertical

   Find out also the maximum and minimum possible critical heat fluxes.
2. Find the critical heat flux for the following conditions.
   Heat is given from the inner heated tube (in the annulus), water is flowing upwards, pressure is 80 bar, mass flux is 1500 kg/m^{2}s, $d = 12$ mm and $d_0 = 8$m.
3. Find the critical heat flux for following conditions (small diameter tubes) fluid is water

   Mass flux = 50 kg/m^{2}s
   Dia of tube = 1.45 mm [less than the Laplace constant]
   Pressure = 10 bar
   Orientation = Vertical

   Find also the possible maximum and minimum heat fluxes.
4. Water is flowing through helical coiled tubes, find the critical heat flux for the following conditions.

   Mass flow rate = 600 kg/m^{2}s
   Pressure = 8 MPa
   Inside dia = 16 mm
   Thickness = 2 mm
   Coil dia = 405 mm
   Pitch = 33 mm coil length = 4500 mm

References

Ahmad SY (1973) Fluid to fluid modelling of critical heat flux: a compensated distortion model. Int J Heat Mass Transf 16:641–661
Alekseev GV (1964) Burnout heat fluxes under forced flow. In: IIIrd International Conference on peaceful uses of atomic energy, Geneva, A/Conf. P. 28/P/327a
Barnett PG (1963) The scaling of forced convection boiling heat transfer; United Kingdom
Bowring RW (1972) A simple but accurate round tube uniform heat flux, dryout correlation over the pressure range 0,7-17 MN/m2 (100-2500 psia). AEEW-Rep. 789
Groeneveld DC (1973) Post-dryout heat transfer at reactor operating conditions, pp 321–350
Groeneveld DC, Moeck EO (1969) An investigation of heat transfer in the liquid deficient regime. No. AECL-3281 (Rev.). Atomic Energy of Canada Ltd., Chalk River Nuclear Labs, Chalk River, Ontario

Katto Y, Ohno H (1984) An improved version of the generalized correlation of critical heat flux for the forced convective boiling in uniformly heated vertical tubes. Int J Heat Mass Transf 27(9):1641–1648

Lee DH (1966) Burnout in a channel with nonuniform circumferential heat flux. AEEW-Rep. 477

Macbeth RV (1963) Burn-out analysis. Part 4. Application of a local condition hypothesis to world data for uniformly heated round tubes and rectangular channels. AEEW-Rep. 267

Macbeth RV (1964) Burn-out analysis. Part 5. Examination of published world data for rod bundles. No. AEEW-R-358. United Kingdom Atomic Energy Authority. Reactor Group. Atomic Energy Establishment, Winfrith, Dorset, England (United Kingdom)

Plummer DN (1974) Post critical heat transfer to flowing liquid in a vertical tube. PhD thesis, Mass. Inst. Technology

Plummer LN (1977) Defining reactions and mass transfer in part of the Floridan aquifer. Water Resour Res 13.5:801–812

Stephan K (1992) Heat transfer in condensation and boiling. Springer

Thompson B, Macbeth RV (1964) Boiling water heat transfer burnout in uniformly heated round tubes: a compilation of world data with accurate correlations; United Kingdom

Whalley PB Handbook of phase change: boiling and condensation Oxford University Press

Whalley PB (1987) Boiling, condensation, and gas-liquid flow. United States

Yin ST, Liu TJ, Huang YD, Tain RM (1988) An investigation of the limiting quality phenomenon of critical heat flux. In: Veziroglu TS (ed) Particulate phenomena and multiphase transport, vol 2. Hemisphere Publishing Corporation, Washington, pp 157–173

Chapter 11
Condensation: Nusselt Theory and External Condensation

Condensation is the process of changing vapour into liquid. If the temperature of a surface in contact with the vapour is below saturation temperature which is able to bring the vapour temperature below saturation temperature, condensation takes place. The liquid which is formed gets subcooled by giving more heat to the surface and hence more condensation takes place on this liquid. Thus, condensation is always associated with mass transfer.

There are different resistances to heat transfer during condensation-vapour phase resistance (convection and diffusion), interface resistance and liquid phase resistance. The liquid phase resistance is dominant in most of the practical cases. In the liquid phase, the resistance is due to both convection and conduction. Usually, condensation is associated with desuperheating and subcooling as well. If the vapour is in a superheated state, it must first cool to the saturated state by sensible heat transfer and then condense. Also since the wall remains below the saturation temperature, the condensed liquid gets subcooled. Condensation is associated with mass transfer where one phase diffuses into the other. Often condensation is associated with non-condensible gases such as air in steam. This also brings the role of mass transfer in the process.

Condensation is a natural phenomenon. The formation of dew is a common example. Cloud formation and rain are also associated with condensation. Condensation is extremely important in power and process industry. It has applications in separation, process cooling, steam heating, refrigeration and power plant operation applications. In the present chapter, condensation is described with the classical theory from Wilhelm Nusselt as well as the subsequent developments that emerged after him. In a large number of applications, condensations of mixtures of vapours take place. We will consider this phenomenon in a separate chapter.

© The Author(s) 2023
S. K. Das and D. Chatterjee, *Vapor Liquid Two Phase Flow
and Phase Change*, https://doi.org/10.1007/978-3-031-20924-6_11

11.1   Different Types of Condensation

There are different types of condensation. Film condensation and dropwise condensation are the most commonly investigated phenomena. The two major types of condensation are shown in Fig. 11.1. The film condensation is associated with the formation of the film on the cold surface while dropwise condensation forms droplets that remain separated.

   These are somewhat analogous to film boiling and nucleate boiling. However, unlike boiling, film condensation is of more practical importance because it is difficult to achieve sustained dropwise condensation. This is because although in all cases only drops are formed in the beginning (please check your experience of how you find the condensed water on the lid of the vessel in which you boil water), with an increase of condensation, eventually the film is formed. The formation of the film also depends on the wetting behaviour of the surface, i.e., whether the surface is hydrophilic or hydrophobic. This is characterized by the angle that the droplet makes at the contact region of the surface, known as the 'contact angle'. The following convention of terminology is usually adopted:

- Contact angle <90°, Hydrophilic surface.
- 90° < Contact angle <150°, Hydrophobic surface.
- Contact angle >150°, Superhydrophobic surface.

This is shown in Fig. 11.2. The above wetting characteristics depend on a number of parameters such as surface roughness, surface texture and liquid properties such as surface tension.

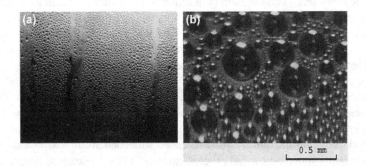

Fig. 11.1 Two major types of condensation. **a** Film condensation and **b** Dropwise condensation

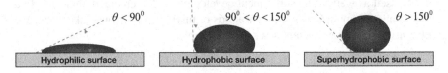

Fig. 11.2 Liquid drops on surfaces. **a** Hydrophilic, **b** Hydrophobic and **c** Superhydrophobic

Apart from these, there can be misty flow analogous to dispersed flow in post dryout regions in boiling. Another important area of condensation study is the direct contact condensation. Here the cooling fluid is directly injected into the vapour in the form of spray which gives a large contact area and intimate mixing.

11.2 Film Condensation

Film condensation is the most common type of condensation where the condensed liquid forms a film on the cold surface which separates the vapour from the surface and impedes heat transfer. The liquid, thus formed, flows downwards under gravity and fresh vapour is condensed. However, the film which wets the surface is a major hindrance to heat transfer which can only be removed in a more efficient way by dropwise or direct contact condensation, but these have their own problems (to be discussed in subsequent sections). Even though the heat transfer process during condensation has got a series of resistances such as the vapour phase resistance (convection and diffusion), interface resistance and liquid phase resistance, the liquid phase resistance is the controlling one under most of the practical cases.

11.2.1 Nusselt's Theory of Condensation

Nusselt (1916) presented an apparently simple analysis for laminar film condensation on a vertical wall. It is astonishing to observe that this theory, even though based on a number of simplifying assumptions, gives a remarkably accurate estimation of heat transfer during condensation and is hence used as the basis of design for most of the condensing equipment till today. Only minor modifications have been suggested to this theory over the last century. The theory assumes the following:

1. The vapour is at rest and does not offer any shear stress at the liquid–vapour interface.
2. The flow of the liquid is only under gravity in the presence of viscosity and not by the interfacial shear.
3. The acceleration of the liquid film is neglected and hence inertia term does not appear in the momentum equation.
4. Heat transfer across the film is assumed to be by pure conduction, i.e., convection effects are neglected.
5. The fluid properties are assumed to be constant.
6. Vapour is assumed to be in the saturated state without any superheating.
7. The liquid is considered to be in thermal equilibrium with vapour at the interface, i.e., subcooling is eliminated.

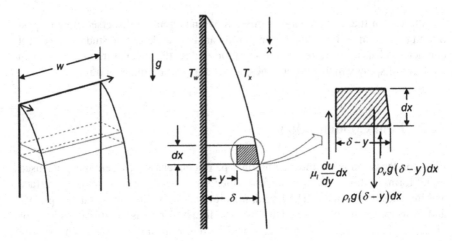

Fig. 11.3 Condensing film on a vertical wall

Under these assumptions, the momentum equation for a liquid element within the film, which grows in thickness down the wall due to condensation (as shown in Fig. 11.3), can be given by balancing the wall viscous stress against net force due to gravity and buoyancy as

$$\mu_l \frac{d^2u}{dy^2} = -(\rho_l - \rho_v)\, g \qquad\qquad (11.1)$$

The boundary conditions can be obtained from no slip of liquid at the wall and no shear at liquid–vapour interface as

$$u = 0 \quad \text{at} \quad y = 0 \qquad\qquad (11.2)$$

$$\frac{du}{dy} = 0 \quad \text{at} \quad y = \delta \qquad\qquad (11.3)$$

Equation (11.1) can be solved with these boundary conditions to give

$$u = \frac{g(\rho_l - \rho_v)}{\mu_l}\left(\delta y - \frac{y^2}{2}\right) \qquad\qquad (11.4)$$

Now, according to the assumption of pure conduction in the film, a linear temperature gradient will be obtained in the film and the heat flux will be given by

$$q = \frac{k_l \Delta T}{\delta} \qquad\qquad (11.5)$$

where ΔT is the temperature difference across the film at a particular location. This heat flux can be equated to the heat required for condensation which is the increase in mass flow rate in the film times the latent heat.

Mass flow rate per unit width

$$M = \int_0^\delta \rho_l u \, dy = \rho_l \frac{(\rho_l - \rho_v)\,g}{\mu_l} \frac{\delta^3}{3} \tag{11.6}$$

or,

$$\frac{dM}{dx} = \frac{\rho_l\,(\rho_l - \rho_v)\,g}{\mu_l} \delta^2 \frac{d\delta}{dx} \tag{11.7}$$

Now, equating the heat conducted to the latent heat transfer,

$$q = h_{fg} \frac{dM}{dx} = \frac{k\,(\Delta T)}{\delta} \tag{11.8}$$

Thus, putting these equations together,

$$\int_0^\infty \delta^3 d\delta = \frac{\mu_l k_l \Delta T}{\rho_l \Delta \rho g h_{fg}} \int_0^x dx \tag{11.9}$$

Thus, the final equation from which the film thickness can be obtained is

$$\delta = \left[\frac{4\mu_l k_l \Delta T x}{\rho_l (\rho_l - \rho_v) g h_{fg}} \right]^{1/4} \tag{11.10}$$

Hence, from Eq. (11.5), we get

$$q = \frac{k_l \Delta T_{sat}}{\delta} = h_x \Delta T_{sat} \tag{11.11}$$

where $\Delta T_{sat} = T_w - T_v = T_w - T_{sat}$ (because vapour is assumed to be saturated). The local heat transfer coefficient comes out as

$$h_x = \left[\frac{\rho_l (\rho_l - \rho_v) g h_{fg} k_l^3}{4\mu_l \Delta T_{sat} x} \right]^{1/4} \tag{11.12}$$

The mean heat transfer coefficient can be obtained as

$$\begin{aligned} h &= \tfrac{1}{L} \int_o^L h_x dx \\ &= \tfrac{2\sqrt{2}}{3} \left[\frac{\rho_L (\rho_l - \rho_v) g h_{fg} k_l^3}{\mu_l (T_w - T_{sat}) L} \right]^{1/4} \end{aligned} \tag{11.13}$$

Thus, the mean Nusselt number is given by

$$Nu = 0.943 \left[\frac{\rho_l(\rho_l - \rho_v)gh_{fg}L^3}{\mu_l k_l \Delta T_{sat}} \right]^{1/4} \tag{11.14}$$

This equation can also be written as

$$Nu = 0.943 \left[\frac{Gr_l}{J} \right]^{1/4} \tag{11.15}$$

where

$$Gr_l = \frac{\rho_l(\rho_l - \rho_v)gL^3}{\mu_l^2}$$

and

$$J = \frac{k_l \Delta T}{\mu_l h_{fg}}$$

This is similar to the expression for natural convection. For constant heat flux boundary conditions, the vertical plate equations are given

$$Nu = 0.943 \left[\frac{\rho_l(\rho_l - \rho_v)g\, h_{fg}L^3}{\mu_l k_l \Delta \bar{T}} \right]^{1/4} \tag{11.16}$$

where

$$\Delta \bar{T} = \frac{q}{k_l L} \int_o^L \delta dx$$

Horizontal Tube: Extension of Nusselt theory can be made for condensation on a horizontal tube which is a geometry of considerable practical importance. Most of the surface condensers are made of bundles of horizontal tubes. In the case of a horizontal tube, it can be seen as a wall of varying angle with horizontal (0 to π for half of the tube) as shown in Fig. 11.4. Thus, the governing equation turns out to be

$$\mu_l \frac{d^2u}{dy^2} + g \sin\theta \, (\rho_l - \rho_v) = 0 \tag{11.17}$$

where u is the tangential velocity to the cylindrical coordinate system coinciding with the geometry of the cylinder and y is the distance measured radially outward from the tube surface. This equation with zero velocity at the tube surface and zero shear stress assumed by Nusselt theory at the condensate surface gives the velocity of the condensate film

$$u = \frac{g \sin\theta}{\mu_l} \Delta\rho \left(\delta y - \frac{y^2}{2} \right) \tag{11.18}$$

Fig. 11.4 Condensing film
on a horizontal tube

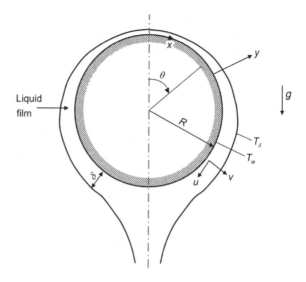

Balance of radial conduction across the condensate film and latent heat required for
evaporation yields

$$\frac{\delta}{R}\frac{d}{d\theta}\left\{\int_0^\delta u\,dy\right\} = \frac{k_l \Delta T}{\rho h_{fg}} \tag{11.19}$$

where R is the tube radius. Combining the above equations, we get the following
differential equation for the film thickness :

$$\delta \frac{d}{d\theta}\left(\delta^3 \sin\theta\right) = \frac{3}{2}\frac{\mu_l k_l d\,\Delta T}{\rho\,\Delta\rho g h_{fg}} \tag{11.20}$$

where d is the tube diameter. Putting

$$z = \delta^4 \cdot \frac{\rho\,\Delta\rho g h_{fg}}{\mu_l k_l d\,\Delta T} \tag{11.21}$$

For an isothermal surface when ΔT is constant, the film equation may be recast as

$$\sin\theta\frac{dz}{d\theta} + \frac{4}{3}z\cos\theta - 2 = 0 \tag{11.22}$$

By symmetry, $\frac{dz}{d\theta} = 0$ at $\theta = 0$ gives

$$z_{\theta=0} = \frac{3}{2} \tag{11.23}$$

So that the film thickness at the top of the tube is

$$\delta_{\theta=0} = \left(\frac{3}{2}\frac{\mu_l k_l d \Delta T}{\rho \Delta \rho g h_{fg}}\right)^{1/4} \tag{11.24}$$

The solution of Eq. (13.18), subject to the condition that z is finite at $\theta = 0$ or by symmetry $\frac{dz}{d\theta} = 0$, is

$$z = \frac{2}{\sin^{4/3}\theta} \int_0^\theta \sin^{1/3}\theta d\theta \tag{11.25}$$

Hence, the local heat flux $q = k_L \Delta T / \delta$ may be written as

$$q = \left\{\frac{\rho \Delta \rho g h_{fg} k_l{}^3 \Delta T^3}{\mu_l d}\right\}^{1/4} \left\{\frac{2}{\sin^{4/3}\theta} \int_0^\theta \sin^{1/3}\theta d\theta\right\}^{-1/4} \tag{11.26}$$

The mean heat flux up to angle θ is then given by

$$\bar{q}_\theta = \left\{\frac{\rho \Delta \rho g h_{fg} k_v{}^3 \Delta T^3}{\mu_l d}\right\}^{1/4} \psi(\theta) \tag{11.27}$$

where

$$\psi(\theta) = \frac{1}{2^{1/4}}\frac{1}{\theta} \int_0^\theta \frac{\sin^{1/3}\theta}{\left(\int_0^\theta \sin^{1/3}\theta d\theta\right)^{1/4}} d\theta \tag{11.28}$$

and

$$\int_0^\theta \frac{\sin^{1/3}\theta}{\left(\int_0^\theta \sin^{1/3}\theta d\theta\right)^{1/4}} d\theta = \frac{4}{3}\left\{\int_0^\theta \sin^{1/3}\theta d\theta\right\}^{3/4} \tag{11.29}$$

resulting

$$\bar{q}_\theta = \left\{\frac{\rho \Delta \rho g h_{fg} k_l{}^3 \Delta T^3}{\mu_l d}\right\}^{1/4} \frac{4}{2^{1/4}3} \cdot \frac{1}{\theta}\left\{\int_0^\theta \sin^{1/3}\theta d\theta\right\}^{3/4} \tag{11.30}$$

When we consider the films on both the sides of the cylinder to be symmetrical, we can only consider one half of it such that $\theta = \pi$, giving the mean flux for the tube, as

$$\overline{q} = \frac{4}{2^{1/4}3\pi}\left\{\frac{\rho\Delta\rho gh_{fg}k_l{}^3\Delta T^3}{\mu_l d}\right\}^{1/4}\left\{\int_0^\pi \sin^{1/3}\theta d\theta\right\}^{3/4} \qquad (11.31)$$

and

$$\int_0^\pi \sin^{1/3}\theta d\theta = 2^{7/3}\pi^2/\{\Gamma\,(1/3)\}^3 \qquad (11.32)$$

where Γ is the gamma function. This yields

$$\overline{q} = (8/3)\,(2\pi)^{1/2}\{\Gamma\,(1/3)\}^{-9/4}\left\{\frac{\rho\Delta\rho gh_{fg}k_l{}^3\Delta T^3}{\mu_l d}\right\}^{1/4}$$
$$= 0.728018\ldots\left\{\frac{\rho\Delta\rho gh_{fg}k_l{}^3\Delta T^3}{\mu_l d}\right\}^{1/4} \qquad (11.33)$$

$$h = \frac{\overline{q}}{\Delta T} = 0.728018\ldots\left\{\frac{\rho\Delta\rho gh_{fg}k_l{}^3}{\mu_l d\Delta T}\right\}^{1/4} \qquad (11.34)$$

$$\overline{Nu} = \frac{hd}{k_l} = 0.728018\ldots\left\{\frac{\rho\Delta\rho gh_{fg}d^3}{\mu_l k_l\Delta T}\right\}^{1/4} \qquad (11.35)$$

or simply

$$Nu = 0.728\left[\frac{\rho_l(\rho_l - \rho_v)gh_{fg}d^3}{\mu_l k_l\Delta T_{sat}}\right]^{1/4} \qquad (11.36)$$

In dimensionless form

$$\overline{Nu} = 0.728018\ldots\{Gr_d/J\}^{1/4} \qquad (11.37)$$

where

$$Gr_d = \frac{\rho\Delta\rho gd^3}{\mu_l{}^2} \qquad (11.38)$$

Finally, it may be noted that the film thickness becomes infinite at the bottom of the tube, and the assumption $\delta = R$ is not valid anymore. However, as the film thickness towards the lower side of the tube becomes large giving a poor rate of heat transfer, the effect of inaccurate heat transfer coefficient is not significant for the entire tube.

For the horizontal tube with uniform surface heat flux, Fujii et al. (1972) suggested

$$\overline{Nu} = 0.695\left\{\frac{\rho\Delta\rho gh_{fg}d^3}{\mu_l k_l\overline{\Delta T}_d}\right\}^{1/4} \qquad (11.39)$$

with

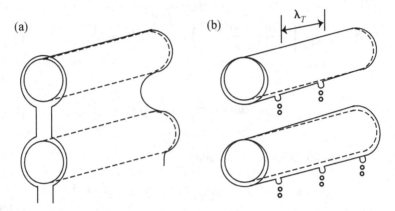

Fig. 11.5 Film condensation on a bank of tubes. **a** Sheet and **b** Drips

$$\overline{\Delta T}_d = \frac{1}{\pi} \int_0^\pi \Delta T d\theta \qquad (11.40)$$

where d is the diameter of the tube. This means that for the horizontal tube only the constant and the length scale (d in place of L) change. Nusselt predicted the constant as 0.725 due to a mistake in the calculation (some of the texts still follow the mistake and use 0.725).

Laminar film condensation on other geometries is also important such as that on inclined elliptical tubes, studied by Fieg and Roetzel (1944).

Bank of Tubes: For bank of tubes, the heat transfer coefficient reduces because the liquid from the top tubes falls on the tubes below making the film thickness on them more as shown in Fig. 11.5. Actually, the liquid drips at distinct points separated by a length equal to the wavelength of Taylor instability as shown in Fig. 11.5.

$$\lambda_T = 2\pi \sqrt{3} \sqrt{\frac{\sigma}{(\rho_l - \rho_v)g}} \qquad (11.41)$$

Due to this, the usual suggested mean heat transfer coefficient given by

$$h_{bundle} = \frac{h_{horizontal}}{n^{1/4}} \qquad (11.42)$$

where n is the number of tubes in a vertical row is not acceptable.

A more accurate correlation given by Kern is

$$h_{bundle} = \frac{h_{horizontal}}{n^{1/6}} \qquad (11.43)$$

The average heat transfer coefficient of the nth tube in a row is given by Kern as

$$\frac{h_n}{h_{horizontal}} = n^{5/6} - (n-1)^{5/6} \text{ for } n \geq 10 \tag{11.44}$$

and

$$h_{bundle} = \frac{1}{n} \sum_{k=1}^{n} h_k \tag{11.45}$$

11.2.2 Deviations from Nusselt's Theory

As is obvious from the preceding discussion, Nusselt's theory is based on a number of simplifying assumptions. However, it produces surprisingly accurate predictions even with such apparently unrealistic assumptions. The reason why Nusselt's equation predicts such an accurate result is this, in the following way, it cancels the errors induced by various assumptions so that the errors tend to compensate for each other.

1. The vapour shear stress at the liquid–vapour interface and inertia due to liquid film acceleration retard the liquid film. So neglecting them overestimates heat transfer.
2. Convection heat transfer in the film and subcooling of liquid enhance heat transfer. Hence, neglecting these effects underpredicts heat transfer.

Thus, the two sets of effects compensate for each other, and Nusselt's theory, which neglects these effects, predicts heat transfer considerably accurately. However, although this compensation occurs in most of the practical situations, there are instances in which considerable deviations from the Nusselt theory are observed. To understand the deviations from this theory, we have to formulate the condensation problem in the most general way. To do this, we must consider a vapour boundary layer over the liquid film which is caused by the vapour velocity and interfacial shear. Figure 11.6 shows liquid and vapour layers in motion.

The general equation for the two layers including inertia, convection and interfacial drag is given as follows:

Condensate Film: Continuity

$$\frac{\partial u}{\partial x} + \frac{\partial v}{\partial y} = 0 \tag{11.46}$$

Momentum

$$\rho_l \left(u \frac{\partial u}{\partial x} + v \frac{\partial u}{\partial y} \right) = \mu_l \frac{\partial^2 u}{\partial y^2} - \frac{\partial p}{\partial x} + \rho_l g \tag{11.47}$$

Energy

$$\rho_l C_{pl} \left(u \frac{\partial T}{\partial x} + v \frac{\partial T}{\partial y} \right) = k_l \frac{\partial^2 T}{\partial y^2} \tag{11.48}$$

Fig. 11.6 Liquid film and
vapour boundary layer

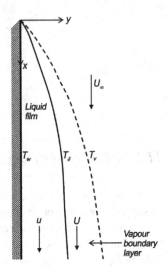

Vapour Boundary Layer: Continuity

$$\frac{\partial U}{\partial x} + \frac{\partial V}{\partial y} = 0 \tag{11.49}$$

Momentum

$$\rho_v \left(U \frac{\partial u}{\partial x} + V \frac{\partial u}{\partial y} \right) = \mu_v \frac{\partial^2 U}{\partial y^2} - \frac{\partial p}{\partial x} + \rho_v g \tag{11.50}$$

Here, u and v are liquid velocities, U and V are vapour velocities and vapour is
considered to be saturated while properties are considered to be constant. Under
these conditions, boundary conditions are given by

At $y = 0$

$$u = v = 0 \quad \text{and} \quad T = T_w \text{ (for constant wall temperature)}$$

$$\frac{\partial T}{\partial y} = \text{constant (for constant wall heat flux)}$$

At $y = \delta$ Conservation of mass

$$\rho_l \left(v - u \frac{d\delta}{dx} \right) = \rho_v \left(V - U \frac{d\delta}{dx} \right) \tag{11.51}$$

Streamwise continuity

$$u = U \tag{11.52}$$

Continuity of shear stress

$$\mu_l \frac{\partial u}{\partial y} = \mu_v \frac{\partial U}{\partial y} \tag{11.53}$$

For negligible interfacial thermal resistance

$$T = T_s = T_v = \text{constant} \tag{11.54}$$

At $y \to \infty$

Natural convection

$$U \to 0 \tag{11.55}$$

Forced convection

$$U \to U_\infty \text{ for flat plate } U \to U_\infty \sin \phi \text{ for flow on a tube} \tag{11.56}$$

Since δ is not known, we have to close the problem from mass and energy conversion as

$$k_l \left(\frac{\partial T}{\partial y} \right)_\delta = \rho_l h_{fg} \frac{d}{dx} \left[\int_o^\delta u \, dy \right] \tag{11.57}$$

$$k_l \left(\frac{\partial T}{\partial y} \right)_o = \rho_l \frac{d}{dx} \left[\int_o^\delta \left(h_{fg} + C_{pl} \left(T_\delta - T \right) \right) u \, dy \right] \tag{11.58}$$

It can be readily observed that dropping the terms as per assumptions made by Nusselt ($U = V = 0$, $\frac{\partial p}{\partial x} = 0$, etc.) reduces these equations to the ones given in Nusselt's theory. Thus, these equations represent more general cases. The solutions of these equations are complex and can only be achieved by numerical methods as will be discussed later.

Effect of Subcooling: Due to subcooling of the condensate if the additional energy required is accounted for, we get an expression for effective latent heat which can be used in the Nusselt correlation to bring out the subcooling effect. This is given by

$$h^*_{fg} = h_{fg} + \frac{3}{8} C_{pl} (T_{sat} - T_w) \tag{11.59}$$

However, Rohsenow (1956) suggested the following effective latent heat (or enthalpy of evaporation) which not only considers subcooling but also slight curvature in liquid temperature profile (due to convection),

$$h^*_{fg} = h_{fg} + 0.68 \, C_{pl} (T_{sat} - T_w) \tag{11.60}$$

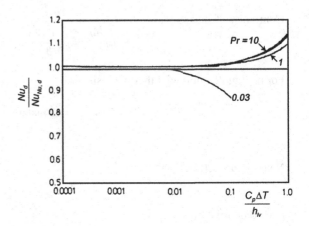

Fig. 11.7 Film condensation
including inertia and
convection effects

Effect of Vapour Superheat: In a way similar to above, the effect of vapour super-heat can also be incorporated within the enthalpy of evaporation. The resulting effective enthalpy of evaporation including both superheat and subcooling effects is given by

$$h^*_{fg} = C_{pv}(T_v - T_s) + h_{fg} + 0.68\,C_{pl}(T_{sat} - T_w) \tag{11.61}$$

where C_{pv} is the specific heat of the vapour.

Effects of Inertia and Convection: Sparrow and Gregg (1959) were first to consider the effect of inertia and heat convection. However, they neglected interfacial shear and hence the vapour equations were not necessary. Using similarity transformation, they showed that

$$\frac{Nu}{Nu_{Nu}} = f\left[\frac{C_{pl}\Delta T}{h_{fg}}, \text{Pr}\right] \tag{11.62}$$

where Nu_{Nu} is the Nusselt number predicted by Nusselt's theory. This function is shown in Fig. 11.7. It can be seen that at a low value of $C_{pl}\Delta T/h_{fg}$, the solution coincides with Nusselt's solution. However, at higher values of this parameter, the result deviates from Nusselt's solution considerably. It can be seen that the deviation is more severe for low Prandtl number fluid. At a higher Prandtl number, the deviation is due to the dominant effect of convection while at a lower Prandtl number (liquid metals), the inertial effect dominates. Usually, $C_{pl}\Delta T/h_{fg}$ is not too high and hence at a moderate Prandtl number, Nusselt's solution suffices.

Effect of Interfacial Shear: Chen (1961) was the first one to investigate the effect of interfacial shear taking the vapour boundary layer into consideration. He suggested two types of parameters for two extreme limits of the Prandtl number as

$$\frac{Nu}{Nu_{Nu}} = f_1\left(\frac{C_{pl}\Delta T}{h_{fg}}\right) \quad \text{as Pr} \to \infty \tag{11.63}$$

$$\frac{Nu}{Nu_{Nu}} = f_2 \left(\frac{k_l \Delta T}{\mu h_{fg}}\right) \quad \text{as } Pr \to 0 \tag{11.64}$$

These two functions can be written explicitly as

$$\frac{Nu}{Nu_{Nu}} = \left[\frac{1 + 0.68H + \frac{0.02}{Pr}H^2}{1 + \frac{0.85}{Pr}H - \frac{0.15}{Pr}H^2}\right]^{1/4} = \left[\frac{1 + 0.68\,Pr\,J + 0.02\,Pr\,J^2}{1 + 0.85J - 0.15\,Pr\,J^2}\right]^{1/4}$$

$$\tag{11.65}$$

where $J = \frac{k_l \Delta T}{\mu_l h_{fg}}$ and $H = Pr\,J = \frac{C_{pl}\Delta T}{h_{fg}}$. These equations were found to be within 1% accuracy of the complete numerical simulation. However, the exact solution was presented by Koh (1962).

Variable Surface Temperature: Roetzel (1992) brought out the effect of variable wall temperature on laminar film condensation. For this case, one has to use the area average wall temperature in the Nusselt formula. He has proven that the heat flux through the wall can be estimated correctly using the mean wall temperature.

Deviations in Tube Bundles: Chen (1961) suggested that for tube bundles of n vertical rows, the effect of subcooling can be incorporated directly in the heat transfer coefficient as

$$\bar{h}_n = \left[1 + 0.2\frac{C_{pl}(T_{sat} - T_w)}{h_{fg}}(n-1)\right] n^{-1/4} h_{hor} \tag{11.66}$$

where h_{hor} is the heat transfer coefficient of a single horizontal tube. In a practical condenser, the effect of the flow of the vapour is taken into consideration by incorporating a vapour shear heat transfer coefficient, h_τ, as

$$h_{hor} = \left[\frac{h_\tau^2}{2} + \left(\frac{h_\tau^4}{4} + h_{Nu}^4\right)^{1/2}\right]^{1/2} \tag{11.67}$$

where

$$\frac{h_\tau d}{k_l} = 0.9 \left(\frac{V_v d}{\nu_l}\right)^{1/2}$$

Using h_{hor} from this Eq. (13.63) will give the heat transfer coefficient of a tube bundle with effects of liquid subcooling and vapour shear.

Effect of Non-Condensibles: Non-condensible gases are invariably accompanied along with the vapour. It can be caused by an air leak in the blower or compressor. Often the condensation process occurs at vacuum (as in power plant condensers) and this induces more leakage of air from the atmosphere. The performance of condensers is considerably affected by non-condensibles.

The total pressure of vapour and non-condensible remains constant in a condensation process which is the sum of the partial pressure of the vapour and the non-condensible, $p_{total} = p_0 + p_1$, where p_1 is the partial pressure of the vapour and p_0 is that of the non-condensible. Now, as the vapour condenses at the vapour/gas/liquid

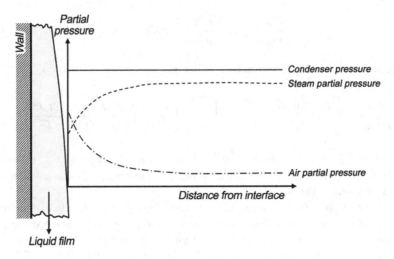

Fig. 11.8 Change in partial pressure of vapour and partial pressure of non-condensible gas near the interface

interface, the concentration of vapour decreases near the interface giving it a gradient of partial pressure from the free stream dropping down at the interface. This makes the partial pressure of non-condensible higher near the interface through which the vapour has to diffuse to get condensed. The situation is somewhat like that of boiling of mixtures where the non-condensible acts similar to that of the less volatile component. This drop in partial pressures near the interface is shown in Fig. 11.8.

It can be concluded that, due to this decrease in vapour pressure near the interface, the saturation temperature also decreases. This impedes heat transfer. For this reason, often the venting system is provided in condensers to extract non-condensible. The heat transfer in case of the presence of non-condensible can be said to be due to condensation as well as sensible cooling of the non-condensible and vapour mixture which flows along the interface. Combining the law of mass diffusion and the famous Lewis relation, one gets the temperature difference between the interface and wall as (Lewis number = 1.0)

$$T_i - T_w = \frac{h_{vo}}{h_l} \left[\frac{h_{fg}}{C_{pv}} \ln \left(\frac{p - p_{li}}{p - p_{vi}} \right) + \zeta \, (T_v - T_i) \right] \qquad (11.68)$$

where correction factor, $\zeta = \frac{h_v}{h_{vo}}$.

Heat transfer coefficient h_v takes care of the heat transferred from vapour–gas mixture to the phase interface and h_{vo} denotes that we are dealing here with heat transfer coefficient of a vapour flow. The correction factor, ζ, allows for the fact that actually vapour does not flow along a solid wall, instead, that part of the flow disappears at the phase interface. Here, p is the total pressure, p_{vi} is the partial pressure of vapour at a rather large distance from the phase interface and p_{li} is that

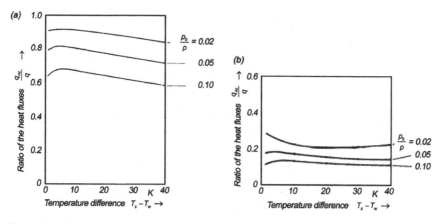

Fig. 11.9 Effect of non-condensible on a **a** Natural convection and **b** Forced Convection condensation (*Courtesy* "Heat Transfer in Condensation and Boiling" by Karl Stephan, with permission from Springer)

at the phase interface. For small inert gas contents, the expression can be simplified to

$$T_i - T_w = \frac{h_{vo} h_{fg}}{h_l C_{pv}} \ln \frac{p - p_{li}}{p - p_{vi}} \tag{11.69}$$

The heat transfer without non-condensible across the film takes place between the vapour and wall temperature while in presence of non- condensible it takes place between interface and wall temperature. Thus, with the help of Eq. (11.69), it can be derived that

$$\frac{q_{nc}}{q} = \frac{T_i - T_w}{T_{sat} - T_w} = \frac{h_{vo} h_{fg}}{\Delta T_{sat} h_l C_{pv}} \ln \frac{p - p_{li}}{p - p_{vi}} \tag{11.70}$$

where q is the heat transfer without non-condensible and q_{nc} with it. The results shown in Fig. 11.9 indicate that the effect of non-condensible is more in natural convection than in forced convection condensation.

11.2.3 Turbulent Film Condensation

Film during condensation may turn turbulent. The Reynolds number of the liquid film is given by

$$Re = \frac{\bar{u}\delta}{\nu_l} = \frac{\bar{u}\delta \rho_l b}{\nu_l \rho_l b} = \frac{\dot{M}}{\mu_l b} \tag{11.71}$$

The heat transfer correlation for Nusselt's theory can be given by

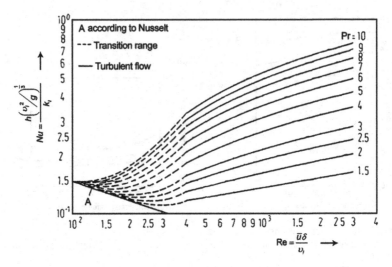

Fig. 11.10 Local Nusselt number for turbulent film condensation on a vertical tube or on a vertical flat wall (*Courtesy* "Heat Transfer in Condensation and Boiling" by Karl Stephan, with permission from Springer)

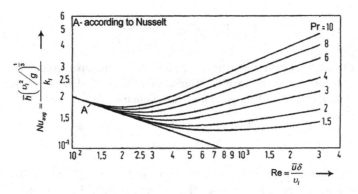

Fig. 11.11 Average Nusselt number for turbulent film condensation on a vertical tube or on a vertical flat wall (*Courtesy* "Heat Transfer in Condensation and Boiling" by Karl Stephan, with permission from Springer)

$$Nu = \frac{h(v_l^2/g)^{1/3}}{k_l} = (3Re)^{-1/3} \qquad (11.72)$$

This correlation for turbulent film condensation is more complex and also Prandtl number plays a role (Grigull 1942). The resulting curves for local and average Nusselt numbers are shown in Figs. 11.10 and 11.11, respectively.

These results can be applied to both vertical plate and horizontal tubes since in turbulent film flow the effect of curvature is less. The curves show a wide transition zone from laminar to turbulence which is dependent upon the Prandtl number. In this zone, the Nusselt number first decreases due to an increase in the film thickness but subsequently increases due to turbulence. However, beyond a Reynolds number of 400, the film is always turbulent. The mean Nusselt number curves show that at smaller values of Pr it is advantageous to remain in the laminar zone. Only at high Pr, turbulence brings an advantage. For practical use, the following equation can be used for turbulent film condensation.

$$Nu = 0.0325 \, Re^{1/4} Pr^{1/2} \tag{11.73}$$

The film Reynolds number is given by

$$Re = \left[89 + 0.024 Pr^{1/2} \left(\frac{Pr}{Pr_w} \right)^{1/4} (z - 2300) \right]^{4/3} \tag{11.74}$$

where

$$z = \frac{C_{p_l}(T_{sat} - T_w)x}{h_{fg} \, Pr \, (v_l^2/g)^{1/3}} \tag{11.75}$$

Here, the properties are to be taken at saturation temperature and Pr at wall temperature.

For the transition zone, one can use

$$h_{trans} = \sqrt[4]{(1.5 h_{laminar})^4 + h_{turbulent}^4}$$

11.2.4 Condensation of Flowing Vapour

Nusselt himself extended his theory to flowing vapour on vertical flat plates by replacing the no shear boundary condition by

$$\mu_l \left. \frac{\partial u}{\partial y} \right|_{y=\delta} = \pm \tau_\delta \tag{11.76}$$

Here, τ_δ is the interfacial shear. A positive sign is used for downward flow and a negative for upward. The value of the shear stress can be calculated from the equality of pressure force and friction force as

$$\tau_\delta \pi d = \frac{\pi d^2}{4} \frac{dp}{dx} \tag{11.77}$$

Introducing friction coefficient

$$\tau_\delta = \zeta \frac{\rho_v V_v^2}{4} \tag{11.78}$$

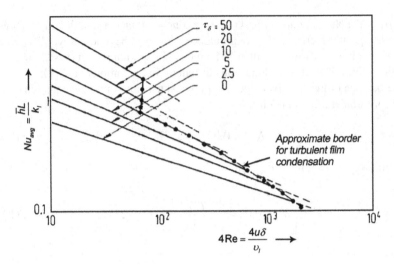

Fig. 11.12 Effect of shear stress on heat transfer during laminar condensation (*Courtesy* "Heat Transfer in Condensation and Boiling" by Karl Stephan, with permission from Springer)

Including the shear effect the film thickness was derived by Nusselt as

$$\delta^4 \pm \frac{4}{3}\frac{\tau_\delta \delta^3}{\rho_l g} = \frac{4k_l \mu_l (T_{sat} - T_w)}{\rho_l^2 g\, h_{fg}}x \qquad (11.79)$$

Using this value of thickness, Jakob et al. (1935) calculated the heat transfer using Nusselt theory which is shown in Fig. 11.12. They improved from Nusselt's calculation by allowing a streamwise decrease in vapour velocity. In this figure, the negative wall superheat corresponds to the non-condensing desuperheating zone.

Rohsenow et al. (1956) introduced the effect of superheating and subcooling along with interfacial shear which can be given by

$$\delta^4 \pm \frac{4}{3}\frac{\tau_\delta \delta^3}{(\rho_l - \rho_v)_g} = \frac{4k_l \mu_l (T_{sat} - T_w)}{\rho_l(\rho_l - \rho_v)g h_{fg}^*}x \qquad (11.80)$$

where
$$h_{fg}^* = C_{p_v}(T_v - T_s) + h_{fg} + 0.68C_{p_l}(T_{sat} - T_w)$$

In Fig. 11.12, the dotted lines show the region when turbulence becomes important. Other correlations include that by Fujii and Uehara (1972)

$$Nu_L Re_L^{1/2} = \left[0.656(1.2 + G^{-1})^{4/3} + 0.79F_L\right]^{1/4} \qquad (11.81)$$

where

$$\mathrm{Re}_L = \frac{u_\infty \rho L}{\mu_l}, \quad F_L = \frac{h_{fg}gL}{k_l \Delta T_{sat} U_\infty^2} \quad \text{and} \quad G = \left(\frac{k_l \Delta T}{\mu_l h_{fg}}\right) \left(\frac{\rho_l \mu_l}{\rho_v \mu_v}\right)^{0.5}$$

For horizontal tube, the equation by Shekriladze and Gomelauri (1966) can be given by

$$Nu \mathrm{Re}_d^{-1/2} = \frac{0.9 + 0.728 F_d^{1/2}}{\left(1 + 3.44 F_d^{1/2} + F_d\right)^{1/4}} \tag{11.82}$$

where

$$Nu = \frac{\bar{h}d}{k_l}, \quad F_d = (\rho_l - \rho_v)\mu_l h_{fg}gd \quad \text{and} \quad \mathrm{Re}_d = \frac{U_\infty \rho_l d}{\mu_l}$$

This equation reduces to the Nusselt equation for $U_\infty = 0$. A general formula suggested by Rose (1988) is in the form

$$Nu_n \mathrm{Re}_L^{-1/2} = 0.436 \left[\frac{1.508}{(1 + J)^{3/2}} + \frac{1}{G}\right]^{1/3} \tag{11.83}$$

where

$$\mathrm{Re}_L = \frac{U_\infty \rho_l n}{\mu_l} \quad \text{and} \quad G = \left(\frac{k_l \Delta T_{sat}}{\mu_l h_{fg}}\right) \left(\frac{\rho_l \mu_l}{\rho_v \mu_v}\right)^{1/2}$$

Examples

Problem 1: A square vertical plate (800 mm + 800 mm) is exposed to saturated steam at atmospheric pressure. The plate temperature is maintained at 84 °C. Check whether the condensate film is laminar or turbulent and calculate the mass of steam condensed per hour.

Solution: Now, we get the average heat transfer coefficient and using that we get the Reynolds number which tells about the film boundary.

$$\mathrm{Re} = \frac{4h_m L(T_v - T_w)}{h_{fg}\mu_l} = \frac{4 \times 7097 \times 0.8 \times (100 - 84)}{(2257 \times 10^3)(306 \times 10^{-6})} = 526$$

As Re < 1800, the film boundary is laminar.
  Mass flow rate of the condensate is found as

$$\dot{m} = \frac{Q}{h_{fg}}$$
$$= \frac{h_m A(T_v - T_w)}{h_{fg}}$$

$$= \frac{7097 \times 0.8 \times 0.8 \times (100 - 84)}{2257 \times 10^3}$$

$$= 115.91 \, \text{kg/hr}$$

Problem 2: Calculate the heat transfer and the condensate mass flow rate per unit width of a wide vertical cooling fin, approximating a flat plate of 0.5 m high, which is exposed to steam at atmospheric pressure. The fin is maintained at 90 °C by cooling water.

$$\text{Re} = \frac{4h_m L(T_v - T_w)}{h_{fg}\mu_l} = \frac{4 \times 9045 \times 0.5 \times (100 - 90)}{(2257 \times 10^3)(3 \times 10^{-4})} = 267$$

As Re < 1800, the laminar assumption is correct.

Heat transfer rate,

$$\frac{q}{A} = h_m \times (T_v - T_w) = 9045 \times 10 = 90450 \, \text{W/m}^2\text{K}$$

Mass flow rate per unit width,

$$m = \frac{\mu_l \text{Re}}{4} = \frac{3 \times 10^{-4} \times 267}{4} = 0.020 \, \text{kg/s.m} = 72.1 \, \text{kg/hr.m}$$

Exercises

1. Calculate the heat transfer coefficient 400 mm from the upper edge of a vertical surface condensing steam is at 105 °C. What is the film thickness at this location?
2. What is the length of vertical tube 8 cm in diameter at 75 °C is required to condense 32 kg/hr of steam at 80 °C?
3. A 60 cm tall vertical plate is exposed to saturated steam at 0.5 atm. The surface temperature of the plate is uniform at 75 °C. Compute the average heat transfer coefficient and rate of condensation per unit width of the plate.
4. Calculate the amount of saturated steam condensing on a horizontal tube of diameter 16 mm and length 1.5 m if the steam pressure is 1.2 MPa and the tube surface temperature is 180 °C.
5. How shall the heat transfer coefficient and amount of dry saturated steam condensing per unit time on the surface of a horizontal tube change if the tube diameter is increased 5 times and steam pressure rises from 0.05 atm to 5 atm and the temperature difference remains unchanged.
6. Derive an expression for the average heat transfer coefficient for laminar film condensation on an isothermal sphere.
7. Assuming the tube diameter is large compared with the condensate thickness, find what l/d ratio will provide the same laminar film condensation controlled heat transfer to a tube in both vertical and horizontal orientations.
8. The cooling water that flows through the tube of a vertical steam-to-water heat exchanger must remove an amount of heat 350 kW. Dry saturated steam con-

denses on the external surface of the tube at a pressure of 1 MPa. Calculate the temperature difference required if the heat exchanger is built up of 30 tubes having a diameter 20 mm and height 1 m.

9. Dry saturated steam undergoes film condensation on a horizontal tube of diameter 20 mm and length 1.5 m at a pressure of 2 MPa. The tube surface temperature is 250 °C. How shall the mean coefficient of heat transfer from the condensing steam to the tube changes if the tube is vertical and all other conditions remain unchanged?

10. Dry saturated steam condenses on the external surface of a vertical tube. The condensate film is in laminar flow along the entire height of the tube. Determine the dependence of the rate of heat flow and heat flux on the height of the tube.

11. A vertical plate 0.8 m square is exposed to steam at 0.065 bar. If the surface temperature of the wall is maintained at 15 °C, calculate the heat transfer and rate of condensation.

12. A condenser is to be designed to condense 80 kg of steam per hour at atmospheric pressure. A square array of 100, 12 mm diameter tubes is available for the design, and the wall temperature of the tube is to be maintained at 98 °C. Estimate the length of the tube required.

13. Calculate the outside diameter of a horizontal pipe which is exposed to dry steam at 100 °C. The pipe surface is maintained at 84 °C by circulating water through it. The condensate rate was found to be 21.94 kg/hr per metre length of the pipe.

14. Find the ratio of heat transfer to a horizontal tube of a larger diameter to that of a vertical tube of the same size for the same temperature assuming laminar film condensation.

15. Calculate local and average heat transfer coefficients for a 30° inclined plate of length 50 cm and diameter 5 cm under saturated steam condensation given that saturation temperature $= 35$ °C and wall temperature $= 25$ °C.

16. Laminar film condensation occurs on a vertical isothermal cone of height 1 m and angle 45°. Find

(a) mass flow rate at a distance x,
(b) mean heat transfer coefficient.

17. There are 200 brass tubes in a shell, each having 20 mm diameter. Cooling water enters the tube at 30 °C with a mass flow rate of 1000 kg kg/s. The heat transfer coefficient for condensation on the outer surface of the tubes is 10000 W/m² K. If the total heat of the condenser is 200 MW when the steam condenses at 60 °C, determine

(a) the overall heat transfer coefficient;
(b) the outlet temperature of the cooling water;
(c) the rate of condensation (take $h_{fg} = 2400$ kJ/kg).

18. A vertical plate of length 1 m is maintained at 60 °C in the presence of saturated steam at atmospheric pressure. Find

(a) the heat transfer per unit width;

   (b) the condensation rate per unit width.

19. A rectangular vertical plate of area $2\,m^2$ is exposed to saturated steam at atmospheric pressure and the plate temperature is maintained at $60\,°C$. Examine the flow, whether it is laminar or turbulent. Also, determine the heat transfer rate and the mass of steam condensed per hour.

20. Dropwise condensation is done by coating a surface of a tube with a non-wetting agent. Find the total rate of heat transfer per unit length if steam at a saturation temperature of $157\,°C$ condenses at the surface of $2\,cm$ diameter vertical tube with a surface temperature of $110\,°C$.

21. Assuming linear temperature distribution within a laminar condensed liquid film on a vertical surface, derive an expression for the thickness of the film in terms of heat transfer coefficient.

22. A condenser having water as a cooling agent enters at $50\,°C$ with a velocity of $1\,m/sec$ acting over $25\,m^2$. Steam is at $125\,°C$. Determine the air exit temperature.

23. Laminar film condensation occurs on a horizontal wall with a constant heat flux q around its periphery that is

$$h = q/(t_w - t_{sat}) = 0.615\,kW/m^2\,K$$

for a horizontal tube with an isothermal wall, derive the differential equation governing the film thickness as a function of angle measured from the top of the tube. Discuss the limiting cases.

24. Ammonia at a saturation temp of $20\,°C$ condenses on the surface of a vertical plate $1\,m$ high and $2\,m$ wide. Surface temp. of the plate is $30\,°C$. Determine the following:

   (a) The total rate of heat transfer;
   (b) Total mass flow rate over the plate surface;
   (c) The above two if the plate is inclined at an angle of $45°$ with the horizontal.

References

Barcozy CJ (1966) A systematic correlation for two phase pressure drop. Chem Eng Prog Symp Ser 62:232–249

Carpenter EF, Colburn AP (1951) The effect of vapor velocity on condensation inside tubes. In: Proceedings of the institution of mechanical engineers, general discussion heat transfer, pp 20–26

Chen MM (1961) An analytical study of laminar film condensation. Part 1, Flat plates. Part 2, Single and multiple horizontal tubes. Trans ASME Ser C J Heat Transf 83:48–60

Colburn AP, Drew TB (1937) The condensation of mixed. Trans Am Inst Chem Eng 33:197

Fieg GP, Roetzel W (1994) Calculation of laminar film condensation in/on inclined elliptical tubes. Int J Heat Mass Transf 37:619–624

Fujji T, Uehara H (1972) Laminar film condensation on a vertical surface. Int J Heat Mass Transf 15:217–233

Grigull U (1942) 'Wärmeübergang bei der Kondensation mit turbulenter Wasserhaut. Forsch. Inge-
    nieurwes, vol 13, pp 49–57, und VDI Z, vol 86, pp 444–445

Jakob M, Erk S, Eck H (1935) Verbesserte Messungen und Berechnungen des Wärmeübergangs
    beim Kondensieren strömenden Dampfes in einem vertikalen Rohr. Phys Z 36:73–84

Kern DQ (1950) Process heat transfer. McGraw-Hill, New York

Koh JCY (1962) Film condensation in a forced-convection-layer flow. Int J Heat Mass Transf
    5:941–954

Nusselt W (1916) Die oberflächenkondensation des wasserdampfes. VDI Z 60:541–546

Roetzel W (1992) Laminare Filmkondensationan einer senkrechtene- Wand mitortlich verander-
    licher Oberflachentemperaturd—und Stoffubertragung 27:173–175

Rohsenow WM (1956) Heat transfer and temperature distribution in laminar film condensation.
    Trans ASME Ser C J Heat Transf 78:1645–1648

Rohsenow WM, Webber JH, Ling AT (1956) Effect of vapor velocity on a laminar and turbulent-film
    condensation. Trans ASME 78:1637–1643

Rose JW (1988) Fundamentals of Condensation Heat Transfer: Laminar Film Condensation. JSME
    Int J Ser II 31(3):357–375

Shekriladze IG, Gomelauri VI (1966) Theoretical study of laminar film condensation of flowing
    vapour. Int J Heat Mass Transf 9:581–591

Sparrow EM, Gregg JL (1959) Laminar condensation heat transfer on a horizontal cylinder. Trans
    ASME 291–296

Chapter 12
In-Tube and Dropwise Condensation

In a number of applications, condensation takes place inside a tube. The 'up flow' and 'down flow' condensers with flow inside tubes are common in industry. In Chap. 15, we will discuss about the usefulness and features of such condensation. Here we try to explore how such condensation process takes place and how we can estimate the heat transfer and rate of condensation in such a process. Condensation can also take place in a horizontal tube. Due to the influence of gravity, asymmetric liquid film can be formed in such cases and as a result a non-uniform peripheral heat transfer rate can be obtained. This chapter discusses all these aspects of in-tube condensation of single component vapour.

Dropwise condensation is observed often in nature and is a fiercely debated topic among researchers. While everyone recognizes its superiority over film condensation, the ways and means of achieving dropwise condensation and more importantly sustaining it over a long period of time is a challenge which the contemporary technology has not been able to solve yet. Result studies on superhydrophobic surfaces or the lotus effect have raised hope that artificially manipulated surface textures, may in future, be able to produce commercially viable technology of dropwise condensation. In this chapter we make an introduction to the available knowledge on dropwise condensation.

12.1 Condensation Inside Vertical Tube

In a vertical tube, the flow direction is aligned with that of the gravity. Obviously, due to the action of gravity the liquid falls as a film around the entire inner periphery of the tube. The film thickness will be almost uniform over the entire periphery but it will increase as it moves downward due to more and more vapour condensation. The scenario is somewhat similar to Nusselt condensation over vertical wall. However, being internal flow, the pressure of the fluids (both vapour and liquid) will contin-

© The Author(s) 2023
S. K. Das and D. Chatterjee, *Vapor Liquid Two Phase Flow and Phase Change*, https://doi.org/10.1007/978-3-031-20924-6_12

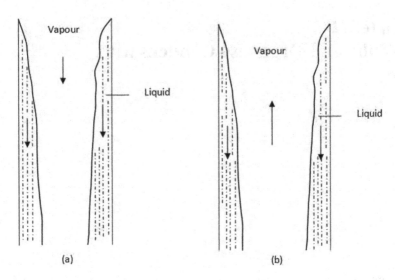

Fig. 12.1 Vertical in-tube condensation. **a** downflow and **b** reflux

uously change along the tube. Two cases occur: (a) the vapour flow is downward where vapour and liquid are flowing in the same direction. This is called the down flow condensation, (b) the vapour flows upwards and the liquid film downward. This is called upflow or reflux condensation. Figure 12.1 shows these two cases.

In case of downflow condensation, the liquid film accelerates under gravity and interfacial shear between liquid and vapour. This causes thinning of the film and as a result increases heat transfer. For reflux condensation it's thickening and heat transfer reduces. In reflux condensation of the vapour, as the velocity is increased beyond a point the vapour entrains the liquid and the film separates from the wall. This causes a mixture of liquid and vapour moving upward and this phenomenon is known as flooding of the condenser. We will discuss about this in the next section. In general three situations may be distinguished:

(a) Case when gravity is dominant over vapour flow and for practical purpose, it can be thought as condensation of stagnant vapour as assumed in the Nusselt theory of condensation.
(b) The most practical case when both vapour flow and gravity play role and hence vapour flow including interfacial shear needs to be included in the analysis.
(c) The case where the fluid dynamics effect dominates over gravity effects as in flooding. Here, the heat transfer becomes independent of the position of the tube and is difficult to predict due to violent interaction of the two phases.

Let us analyse the more common case of (b) where both gravity and interfacial shear play important roles on the condensation process. Let us consider a vertical tube will a liquid film as shown Fig. 12.2. We are considering an annular liquid film and a cylindrical vapour core in the centre over a small control volume as shown in this

Fig. 12.2 Vapour condensation inside a vertical tube (*Courtesy* "Heat Transfer in Condensation and Boiling" by Karl Stephan, with permission from Springer)

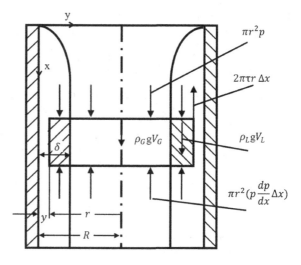

figure. Now, balance of forces over the entire control volume requires viscous force, pressure force and the gravitational forces to be balanced Thus,

$$2\pi r \tau dx + \left(p + \frac{dp}{dx}dx\right)r^2\pi = pr^2\pi + \rho_g V_g g + \rho_l V_l g \qquad (12.1)$$

where r is the radius at a given location, x is the axial position, p is the pressure, and V_g and V_l are the vapour and liquid volume in the control volume. Thus, if R is the tube diameter

$$V_g = \pi(R-y)^2 dx \quad \text{and} \quad V_f = \pi(R-y)^2 dx - V_g \qquad (12.2)$$

Substituting V_g and V_f from Eqs. (12.2) to (12.1) gives

$$\tau = \frac{R-y}{2}\left(\rho_l g - \frac{dp}{dx}\right) - g(\rho_l - \rho_g)\frac{(R-\delta)^2}{2(R-y)} \qquad (12.3)$$

Now, if we apply balance of force only to the vapour volume, we get an expression for vapour liquid interfacial shear, τ_d as

$$\tau_d 2(R-\delta)\pi dx + \left(p + \frac{dp}{dx}dx\right)(R-\delta)^2\pi = p(R-\delta)^2 + \rho_g V_g g$$

giving,

$$\tau_d = \frac{R-\delta}{2}\left(\rho_g g - \frac{dp}{dx}\right) \qquad (12.4)$$

Hence, the shear at any location of the liquid τ can be written from Eqs. (12.3) and (12.4) as

$$\tau = \frac{R - y}{R - \delta}\tau_d + g(\rho_l - \rho_g)\frac{2R(\delta - y) - \delta^2 + y^2}{2(R - y)} \quad (12.5)$$

Now if we consider the liquid film to be much thinner compared to the tube diameter, i.e., $R >> \delta$. Thus in Eqs. (12.4) and (12.5), $R - \delta$ and $R - y$ can be replaced by R, giving

$$\tau_d = \frac{R}{2}\left(\rho_g g - \frac{dp}{dx}\right) \quad (12.6)$$

and

$$\tau = \tau_d + g(\rho_l - \rho_g)(\delta - y) \quad (12.7)$$

The wall shear stress can be obtained by setting $y = 0$

$$\tau_w = \tau_d + g(\rho_l - \rho_g)\delta \quad (12.8)$$

and shear stress in the liquid can be written as

$$\tau = \tau_w - g(\rho_l - \rho_g)y \quad (12.9)$$

Thus, to know interfacial shear, only the knowledge of pressure gradient is enough but to know wall and liquid shear stress, film thickness needs to be known. As we have seen in the Nusselt theory of film condensation, to get the film thickness, we should not only use the viscosity formulation of liquid shear stress but also we must consider mass balance through liquid layer evaporation. At this point we recognize that two cases may occur, the liquid film being in laminar and turbulent regime. Before taking up heat transfer characteristics in these two regimes let us first consider the pressure drop correlation used for it. As stated in Chap. 2, for any two phase flow, the pressure drop can be assumed to comprise three components acceleration pressure drop, frictional pressure drop and geodetic pressure drop,

$$\frac{dp}{dz} = \left(\frac{dp}{dz}\right)_f + \left(\frac{dp}{dz}\right)_a + \left(\frac{dp}{dz}\right)_g \quad (12.10)$$

The acceleration pressure drop can be evaluated as

$$\left(\frac{dp}{dz}\right)_a = -\frac{d}{dz}\left[\frac{G^2 x^{*2}}{\varepsilon\rho_v} + \frac{G^2(1 - x^*)^2}{(1 - \varepsilon)\rho_l}\right] \quad (12.11)$$

Geodetic pressure drop will be same as given in Chap. 2. The frictional pressure drop can be equated to gas or liquid superficial pressure drops through two phase multipliers as

$$\left(\frac{dp}{dz}\right)_f = \left(\frac{dp}{dz}\right)_l \phi_l^2 = \left(\frac{dp}{dz}\right)_g \phi_g^2 \tag{12.12}$$

Many correlations have been suggested for in in-tube condensation, amongst them, the correlation of Fujii et al. (1976) which is used the Fauske (1961) reads

$$\phi_g = 1 + aX_{tt}^{0.2} \tag{12.13}$$

where

$$a = \begin{cases} 1.24\left(\frac{G}{\sqrt{\rho_l \rho_g}}\right)^{0.7}, & \text{for } \frac{G}{\sqrt{\rho_l \rho_g}} < 1.5\,\text{m/s} \\ 1.65, & \text{for } \frac{G}{\sqrt{\rho_l \rho_g}} > 1.5\,\text{m/s} \end{cases}$$

For downflow condensation, the correlation of Soliman et al. (1968) was found to be valid, where

$$\phi_g = 1 + 2.85X_{tt}^{0.2} \tag{12.14}$$

(a) Laminar In-Tube Condensation

The laminar flow film in case of in-tube condensation in vertical tubes is not very common in industrial applications. This is because, usually the tubes are long and the vapour interfacial effect along with pressure drop makes the film turbulent. Also the solution of the Eqs. (12.5)–(12.9) are not as straightforward as the Nusselt theory of condensation because the interfacial shear can not be neglected here. Isachenko et al. (1980) presents the results of laminar in-tube condensation in a vertical tube as given in Fig. 12.3. The data can be correlated as

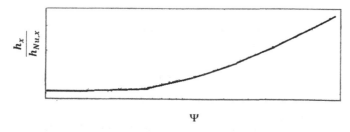

Fig. 12.3 Heat transfer inside a vertical tube for laminar condensate flow

$$\frac{h_x}{h_{N_{u,x}}} = \left[0.0054\psi + (0.005\psi + 1)^{1/2}\right]^{1/2} \tag{12.15}$$

where

h_x = local heat transfer coefficient

$h_{N_{u,x}}$ = local heat transfer coefficient as per Nusselt theory of condensation

and,

$$\psi = \frac{\rho_g}{\rho_l}\left(\frac{\gamma_g}{\gamma_l}\right)^2 \frac{Re_{gd}^2}{Ga_{ld}^{\frac{2}{3}}} Re_{lx}^{-0.28}$$

where

$$Re_{gd} = \frac{\bar{u}_g d}{\gamma_g}$$

$$Ga_{ld} = \frac{gd^3}{\gamma_l^2}$$

$$Re_{ln} = \frac{\bar{u}_l x}{\gamma_l}$$

$\bar{u}_g$ and $\bar{u}_l$ are average vapour and liquid velocity.

(b) Turbulent In-Tube Condensation

The most practical case of condensation is the turbulence film condensation. This is indeed the case in a large number of condensers. The solution for the hydrodynamic aspect of the momentum balance Eq. (12.4) can be obtained by using proper constitutive equation for shear stress. In turbulent flow, in 2D case, it can be shown that the additional contribution due to turbulence to the total stress is given by

$$\tau_{xy,turbulent} = -\overline{\rho u' v'} \tag{12.16}$$

where u' and v' are the velocity fluctuation from mean velocity $\bar{u}$ and $\bar{v}$ in x and y directions respectively. Thus

$$u = \bar{u} + u'$$
$$v = \bar{v} + v'$$

where u and v are instantaneous velocities at a given point in x and y directions. This is shown in Fig. 12.4. Newton's law of viscosity relates laminar viscous stress to shear rate in the form

Fig. 12.4 Fluctuation and mean (i) velocity and (ii) temperature in turbulent flow

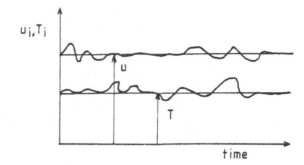

$$\tau_{xy,laminar} = \mu \frac{\partial u}{\partial y} = \rho v \frac{\partial u}{\partial y} \qquad (12.17)$$

where μ is fluid viscosity and v is kinematic viscosity or momentum diusivity.

Since it is difficult to determine $\rho u'v'$ at every point in the flow field, this can be written from analogy as

$$\tau_{xy,turbulent} = \rho \varepsilon \frac{\partial \bar{u}}{\partial y} \qquad (12.18)$$

where ε is the diffusivity of momentum due to turbulence or 'eddy diffusivity' as popularly known.

The total shear stress is the sum of laminar and turbulent stress given by

$$\tau = \tau_{laminar} + \tau_{turbulent} = \rho (v + \varepsilon) \frac{\partial \bar{u}}{\partial y} \qquad (12.19)$$

The total diffusivity can be nondimensionalized as

$$\varepsilon^+ = \frac{v + \varepsilon}{v} = 1 + \frac{\varepsilon}{v} \qquad (12.20)$$

For purely laminar flow $\varepsilon = 0$ giving $\varepsilon^+ = 1$. For purely turbulent flow, $\varepsilon \gg v$ giving $\varepsilon^+ = \varepsilon$. Now dropping the overbar for average velocity, we can write

$$\tau = \rho v \varepsilon^+ \frac{\partial u}{\partial y} \qquad (12.21)$$

Hence, from Eq. (12.9), we can write

$$\rho v \varepsilon^+ \frac{\partial u}{\partial y} = \tau_w - g(\rho_l - \rho_g)y \qquad (12.22)$$

From this, the velocity profile can be obtained as

$$u = \int_o^y \frac{1}{\varepsilon^+} \left[\frac{\tau_w}{\rho_l \nu_l} - \frac{g}{\nu_l} \left(1 - \frac{\rho_g}{\rho_l} \right) y \right] dy \qquad (12.23)$$

It is evident from this equation that to obtain velocity profile we not only require to know the wall shear stress but also the dimensionless eddy diffusivity, ε^+. It should be mentioned here unlike the momentum diffusivity ν, the eddy diffusivity ε is not a fluid property but depends on local flow features of turbulent flow.

In turbulent flow often we use a fictitious velocity known as the friction velocity to non-dimensionalize velocity and coordinate as

$$u_\tau = \sqrt{\frac{\tau_w}{\rho_l}}$$

$$y^+ = \frac{y u_\tau}{\nu_l}$$

$$u^+ = \frac{u}{u_\tau}$$

Using these non-dimensional parameters, Eq. (12.22) can be reduced to

$$u = \int_0^{y^+} \frac{1}{\varepsilon^+} \left[1 - \frac{g \nu_l}{u_\tau^3} \left(1 - \frac{\rho_g}{\rho_l} \right) y^+ \right] dy^+ \qquad (12.24)$$

This describes the velocity profile in the liquid, of course we must know the correlation between ε^+ and y^+ to evaluate it. The next task is to evaluate the film Reynolds number at a given axial location of the tube. To do that we need the liquid flow rate at that section which keeps on changing downward due to more and more condensation. Thus, the local liquid flow can be calculated as:

If we consider an annular area element next to the inner tube,

$$dA = 2\pi R dy$$

The mass flow rate through this area

$$d\dot{M} = \rho_l u \, dA$$

Integrating over the local film thickness we get the local flow rate as

$$\dot{M} = \int_A \rho_l u \, dA$$

$$= \int_0^\delta \rho_l u \, (2\pi R \, dy)$$

$$= 2\pi R \int_0^\delta \rho_l u \, dy \qquad (12.25)$$

The quantity under integration can be readily identified as condensate mass flow rate per unit circumference length, Γ. Thus,

$$\dot{M} = 2\pi R \Gamma$$

$$= 2\pi R \int_o^\delta \rho_l \left[\int_o^y \frac{1}{\varepsilon^+} \left\{ \frac{\tau_w}{\rho_l \nu_l} - \frac{g}{\nu_l} \left(1 - \frac{\rho_g}{\rho_l} \right) y \right\} dy \right] dy \qquad (12.26)$$

If the liquid density is assumed to be constant, we can write

$$\Gamma = \rho_l \int_0^\delta u \, dy = \rho_l \bar{u} \delta \qquad (12.27)$$

where $\bar{u}$ = average film velocity = $\frac{1}{\delta} \int_0^\delta u \, dy$. Hence, film Reynolds number can be written as

$$Re = \frac{\bar{u}\delta}{\nu_l} = \frac{\Gamma}{\rho_l \gamma_l} = \frac{\Gamma}{\mu_l}$$

Please note here that the characteristic length for the film Reynolds number is taken as the film thickness. This makes sense because the film Reynolds number should indicate the transition of the film from laminar to turbulence. From the definition of dimensionless quantities u^+ and y^+, we can see that the Reynolds number can be written as

$$Re = \frac{\Gamma}{\rho_l \nu_l} = \int_0^{\delta^+} u^+ dy^+ \qquad (12.28)$$

where

$$\delta^+ = \frac{\delta_{u_x}}{\nu_l}$$

which can be expanded as

$$Re = \int_0^{\delta^+} \left[\int_0^{y^+} \frac{1}{\varepsilon^+} \left\{ 1 - \frac{1}{Re_\tau^3} \left(1 - \frac{\rho_g}{\rho_l} \right) y^+ \right\} dy^+ \right] dy^+ \qquad (12.29)$$

where Re_τ = friction Reynolds number = $\frac{u_\tau (\nu_l^3/g)^{1/3}}{\nu_l}$.

We can thus write the condensate flow rate, as

$$\Gamma = \rho_l \nu_l \int_0^{\delta^+} \left[\int_0^{y^+} \frac{1}{\varepsilon^+} \left\{ 1 - \frac{1}{Re_\tau^3} \left(1 - \frac{\rho_g}{\rho_l} \right) y^+ \right\} dy^+ \right] dy^+ \qquad (12.30)$$

As stated earlier, to calculate this, we need ε^+ as a function of y^+. In the literature of turbulent flow, this is known as turbulence modelling. While a large number of models do exist in turbulence literature, Rohsenow (1956) employed the well-known universal velocity profile suggested by Nikuradse. In this suggestion the velocity

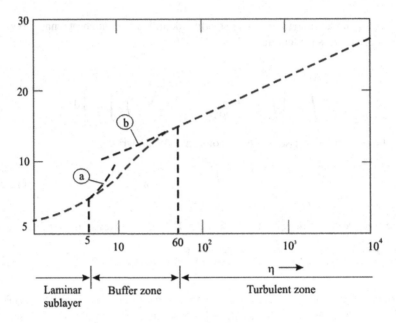

Fig. 12.5 Velocity profile inside a turbulent boundary layer

profile is suggested piecewise for three regions, the laminar sublayer, the buffer layer and the turbulent core as shown in Fig. 12.5.

This is given as

$$u^+ = \begin{cases} y^+ & \text{for } 0 \le y^+ \le 5 \\ -3.05 + 5lny^+ & \text{for } 5 \le y^+ \le 30 \\ 5.5 + 2.5lny^+ & \text{for } 30 \le y^+ \le \delta^+ \end{cases} \qquad (12.31)$$

These correlations can be shown to yield the following expressions for ε^+,

$$\varepsilon^+ = \begin{cases} 1 & \text{for } 0 \le y^+ \le 5 \\ \frac{y^+}{5} & \text{for } 5 \le y^+ \le 30 \\ \frac{y^+}{2.5}\left[1 - \frac{g\gamma_l}{u_\tau^3}\left(1 - \frac{\rho_g}{\rho_l}\right)y^+\right] & \text{for } 30 \le y^+ \le \delta^+ \end{cases} \qquad (12.32)$$

Thus the velocity profile, Reynolds number and the condensate flow rate for turbulent in-tube condensation can be estimated using the above turbulent models.

Estimating velocity profile is the first step to evaluate heat transfer. While evaluating that we can assume, in a manner similar to Nusselt's theory of condensation, that through the liquid film the only mode of heat transfer is heat diffusion (conduction). However, we consider that this heat diffusion is due to both laminar diffusion or pure conduction and diffusion due to turbulence. Thus, the turbulent thermal diffusivity is also included. Thus although only heat diffusion is considered convection effect are not neglected altogether as in the case of laminar condensation analysis. Here, the convection effects are indirectly incorporated through the turbulent diffusivity term. Thus the heat flux through the film can be written as

$$q = \rho c_p (\alpha + \varepsilon_t) \frac{dT}{dy} \tag{12.33}$$

where

$$\alpha = \text{thermal diffusivity of liquid} = \frac{k}{c_p}$$

$$k = \text{thermal conductivity of liquid}$$
$$c_p = \text{specific heat of liquid}$$
$$\varepsilon_t = \text{Eddy diffusivity of heat}$$

We define a turbulent Prandtl number (Pr_t) with turbulent diffusivity just as we do it for laminar case

$$Pr = \frac{\nu}{\alpha}$$
$$Pr_t = \frac{\varepsilon}{\varepsilon_t}$$

Thus,

$$q = -k_l \left[1 + \frac{Pr}{Pr_t} \frac{\varepsilon}{\nu_l} \right] \frac{dT}{dy} \tag{12.34}$$

Noting that $\varepsilon^+ = 1 + \frac{\varepsilon}{\nu_l}$ we can write,

$$q = -k_l \left[1 + \frac{Pr(\varepsilon^+ - 1)}{Pr_t} \frac{\varepsilon}{\nu_l} \right] \frac{dT}{dy} \tag{12.35}$$

If we want to find out an equivalent heat transfer coefficient for heat diffusion, h, we can write

$$q = h(T_{sat} - T_w) \tag{12.36}$$

Equations (12.35) and (12.36) by integrating over film thickness give

$$\frac{k_l}{h} \int_{T_w}^{T_{sat}} \frac{dT}{T_{sat} - T_w} = \int_0^\delta \frac{dy}{1 + \frac{Pr(\varepsilon^+ - 1)}{Pr_t}} \tag{12.37}$$

or

$$\frac{k_l}{h} = \int_0^\delta \frac{dy}{1 + \frac{Pr(\varepsilon^+ - 1)}{Pr_t}} \tag{12.38}$$

The Nusselt number can be defined on the basis of a constant property-based length scale $(\gamma_l^2/g)^{1/3}$ for convenience rather than film thickness which is a variable to be determined,

$$Nu = \frac{h(\gamma_l^2/g)^{1/3}}{k_l} \tag{12.39}$$

If we convert y to y^+ in Eqs. (12.38) and (12.39) can be written as

$$\frac{1}{Nu} = \frac{1}{Re_\tau} \int_0^{\delta^+} \frac{dy^+}{1 + \frac{Pr(\varepsilon^+ - 1)}{Pr_t}} \tag{12.40}$$

It can be really seen that for the solution of this equation, we not only need turbulence model for momentum as $\varepsilon^+(y^+)$ but also a model for the turbulent Prandtl number. Rohsenow (1956) consider a constant value of $Pr_t = 1$ to solve it which is used in many practical cases.

To find out the temperature profile within the film, Eq. (12.35) can be integrated to get

$$T - T_w = -\frac{q\delta}{k_l} \int_0^{y/\delta} \frac{d(y/\delta)}{1 + \frac{Pr(\varepsilon^+ - 1)}{Pr_t}} \tag{12.41}$$

For integrating over the entire film thickness,

$$T_{sat} - T_w = -\frac{q\delta}{k_l} \int_0^1 \frac{d(y/\delta)}{1 + \frac{Pr(\varepsilon^+ - 1)}{Pr_t}} \tag{12.42}$$

Dividing Eqs. (12.41) by (12.42), we get

$$\theta = \frac{T - T_w}{T_{sat} - T_w} = \frac{\int_0^{y/\delta} \frac{d(y/\delta)}{1 + \frac{Pr(\varepsilon^+ - 1)}{Pr_t}}}{\int_0^1 \frac{d(y/\delta)}{1 + \frac{Pr(\varepsilon^+ - 1)}{Pr_t}}} \tag{12.43}$$

This was calculated by Rohsenow (1956) as shown in Fig. 12.6. It can be seen that with the increase of Reynolds number the profile deviates from the Nusselt theory. This analysis gives a fairly good result. Carpenter and Colburn (1951) suggested a simpler equation for average heat transfer for turbulent film condensation inside tubes as

$$\left(\frac{\bar{h}_l \mu_l}{k_l \rho_l^{1/2}}\right) = 0.0065 \left(\frac{C_{pl}\mu_l}{k_l}\right)^{1/2} \tau_l^{1/2} \tag{12.44}$$

where

$$\tau_l = \zeta_l \left[\frac{G_G^2}{2\rho_G}\right]$$

Fig. 12.6 Temperature profile for turbulent condensation in a vertical tube, following Carpenter and Colburn (*Courtesy* "Heat Transfer in Condensation and Boiling" by Karl Stephan, with permission from Springer)

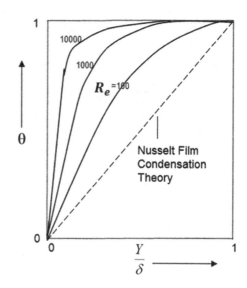

G_G = mass velocity of vapour.

The mass velocity, G_G, is given by

$$G_G = \left[\frac{G_1^2 + G_1 G_2 + G_2^2}{3} \right]^{1/2} \tag{12.45}$$

where G_1 and G_2 are the mass velocities of the vapour at entry and exit. If all the vapour condenses, $G_2 = 0$ and $G_G = 0.58G$. A large number of correlations are available for turbulent film condensation inside tubes. The general trend of condensation in tubes is shown in Fig. 12.7. It can be seen that in the laminar zone the Nusselt number decreases with Reynolds number and then increases in the turbulent zone. The transition point is found to occur at lower Re with an increase in Prandtl number.

Blangetti (1979) suggested exact solution to the turbulent flow condensation problem using the ε^+ profile suggested by van Driest (1956) and Levich (1962). According to his suggestion for laminar flow, Nusselt number is given by

$$Nu_x = 0.707 \left[\frac{(1 - \rho_g/\rho_l)\tau^*}{Re_{f,x}} \right]^{1/2} \tag{12.46}$$

where

$$\tau* = \frac{\tau_0}{g\rho_l(1 - \rho_g/\rho_l)(v^2/g)^{1/3}}$$

τ_0 = interfacial shear

$Re_{f,x}$ = local film Reynolds number

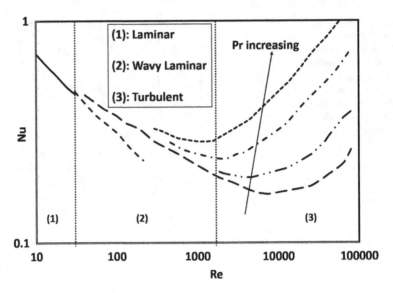

Fig. 12.7 Results of film condensation

For the turbulent part, he suggested a correlation of the form given below and provided table for constants a–f.

$$Nu_{x,t} = aRe_{f,x}^{b}Pr^{c}(1 - e\tau^{*f}) \qquad (12.47)$$

12.2 Condensation Inside Horizontal Tube

The scenario is different inside horizontal tubes. There are several regimes of flow here which are shown in Fig. 12.8. It can be said that as the liquid component increases at low shear rate, the flow changes from stratified to intermittent wavy regime. At higher shear rate annular flow occurs at low liquid fraction which changes to bubbly flow as liquid fraction increases.

A clear picture of the shape and form of the film, as the condensation process progresses, is given in Fig. 12.9. During low liquid fraction low shear condensation the liquid gets stratified as it moves towards the bottom of the tube under gravity. This creates a liquid–vapour cross-section given in Fig. 12.10.

In case of unsymmetrical liquid film, the heat transfer coefficient can be taken as an average between the film trough region as

$$\bar{h}_{total} = \frac{\psi}{\pi}\bar{h}_{\psi} + \left[1 - \frac{\psi}{\pi}\right]\bar{h}_{B} \qquad (12.48)$$

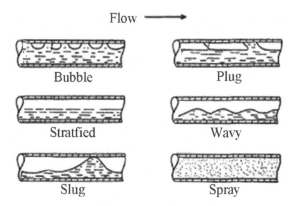

Fig. 12.8 Condensation flow pattern in horizontal tubes (*Courtesy* "Heat Transfer in Condensation and Boiling" by Karl Stephan, with permission from Springer)

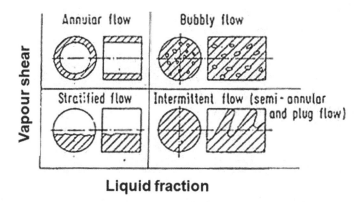

Fig. 12.9 Flow patterns for condensation inside a horizontal tube (*Courtesy* "Heat Transfer in Condensation and Boiling" by Karl Stephan, with permission from Springer)

Fig. 12.10 Stratified flow in condensation inside a tube (*Courtesy* "Heat Transfer in Condensation and Boiling" by Karl Stephan, with permission from Springer)

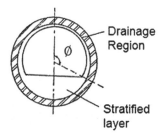

Since due to thick trough, $\bar{h}_\psi \gg \bar{h}_B$

$$\bar{h}_{total} = \frac{\psi}{\pi}\bar{h}_\psi \text{ can be approximated} \tag{12.49}$$

This heat transfer coefficient can be evaluated as

$$\bar{h}_{psi} = F(\psi)\left[\frac{\rho_l^2 g h_{fg} k_l^3}{\mu_l \Delta T_{sat} d}\right]^{1/4} \tag{12.50}$$

F is a function of ψ depending on flow rate as

$$F(\psi) = 0.728\frac{\pi}{\psi}\left[\frac{\psi}{\pi} + \frac{\sin 2(\pi - \psi)}{2\pi}\right]^{3/4} \tag{12.51}$$

which is valid for $\frac{\pi}{2} \leq \psi \leq \pi$.

Inserting Eqs. (12.50) and (12.51) to (12.49) we get

$$\bar{h}_{total} = 0.728\varepsilon\left[\frac{\rho_l^2 g h_{fg} k_l^3}{\mu_l \Delta T_{sat} d}\right]^{1/4} \tag{12.52}$$

where ε is the void fraction given by

$$\varepsilon = \left[1 + \frac{1 - x^*}{x^*}\left(\frac{\rho_g}{\rho_l}\right)^{2/3}\right]^{-1}$$

Annular or Bubbly Flow: In this region heat transfer coefficient can be calculated from

$$h = \left(\frac{\tau_w}{\rho_l}\right)^{1/2}\rho_l C_{pl}\frac{1}{T^+} \tag{12.53}$$

$$T^+ = 8.5Re^{0.1}Pr^{0.57}$$

$$Re = \frac{M}{b\mu_l}$$

where b is tube circumference.

12.3   Flooding in Reflux Condensation

'Flooding' is a phenomenon occurring during condensation when the vapour moves opposite to the direction of flow of the condensate film. The film of liquid moves downward under the action of gravity, and hence flooding takes place when the vapour moves upward. If the vapour velocity is increased beyond a point, the vapour starts entraining the liquid flow and as a result the liquid flow reversal takes place. The liquid thus floods the top of the tube. The sequence of events taking place during flooding are shown in Fig. 12.11.

It is can be seen that as the gas flow increases, ripples are formed in the film. This leads to the formation of large waves which blocks the flow tube partially and a part of the liquid is carried by the vapour to the point of liquid inlet (or film formation point). Thus the liquid–gas mixture moves upward violently causing flooding. This forms an annular film in the upper part of the tube. This is often termed as the hanging film. This reduces the gas flow rate and the second flow reversal takes place when the liquid film again starts moving downwards.

The entire regime and successive phenomena in floods are not completely understood even today. However, the analysis presented below explains some of the features of flooding quite well.

Let us take the liquid film for a counter flow situation as shown in Fig. 12.12 in which we consider the shaded liquid part acted upon by liquid shear and liquid–vapour interfacial shear. Thus making a force balance over the control volume gives

$$(\tau_0 + \tau)dz = \rho_l(\delta - y)gdz$$

or,

$$\tau = \rho_l(\delta - y)g - \tau_0 \tag{12.54}$$

Fig. 12.11 Sequence of events during flooding

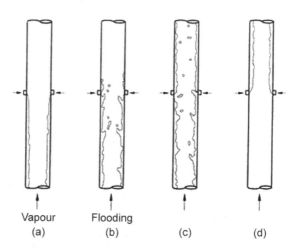

Vapour   Flooding
(a)   (b)   (c)   (d)

Fig. 12.12 Liquid film
under flooding

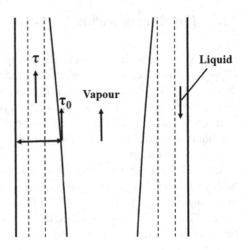

The liquid shear, for laminar flow can be written as

$$\tau = \mu_l \frac{du}{dy}$$

substituting this in Eq. (12.54) gives

$$\frac{du}{dy} = \frac{\rho_l g}{\mu_l}(\delta - y) - \frac{\tau_0}{\mu_l}$$

Integrating this, we get

$$u = \frac{\rho_l g}{\mu_l}\left(\delta y - \frac{y^2}{2}\right) - \frac{\tau_0 y}{\mu_l} + c$$

Due to no-slip condition, the liquid velocity is zero at wall,

$$u = 0 \text{ at } y = 0 \text{ giving } c = 0$$

Thus,

$$u = \frac{\rho_l g}{\mu_l}\left(\delta y - \frac{y^2}{2}\right) - \frac{\tau_0 y}{\mu_l} \tag{12.55}$$

It is seen that for no interfacial shear, the liquid has a parabolic profile downward. As
the shear increases, the velocity at the interface decreases and finally at a particular
value of shear τ_1, the interface becomes stagnant. A further increase in interfacial
shear to τ_2 may take the average liquid velocity in the film to be zero and this can be
taken as the point of flow reversal. The shear τ_1 where the surface velocity becomes
zero can be obtained by putting $y = \delta$ and $u = 0$ in Eq. (12.55), as

$$\tau_1 = \frac{1}{2}\rho_l g \delta \tag{12.56}$$

While to get the flow reversal shear stress, we may set liquid flow equal to zero, as

$$\int_0^\delta u\, dy = 0$$

giving,

$$\tau_2 = \frac{2}{3}\rho_l g \delta \tag{12.57}$$

Although these analytically explain the phenomenon the practical flow reversal points are found to be below this. The instability of the film is needed to be considered to get an accurate analysis.

Flooding also depends on parameters such as length of the tube and condition of vapour entry. A sharp edge of the tube at vapour entry initiates early flooding. This may be due to the turbulence induces across the sharp edge. There are a number of correlations available for flooding vapour velocity. Wallis (1961) suggested,

$$V_l^{*1/2} + V_g^{*1/2} = C \tag{12.58}$$

where

$$V_l* = \frac{v_l \delta_l^{1/2}}{\left[gd(\rho_l - \rho_g)\right]^{1/2}} \quad \text{and} \quad V_g* = \frac{v_g \delta_g^{1/2}}{\left[gd(\rho_l - \rho_g)\right]^{1/2}}$$

where d is the tube diameter V_g, V_l are the vapour and liquid velocities. This correlation is valid for $d < 50$ mm. For larger diameter, Pushkina and Sorokin suggested the criteria based on Kutateladze number, K_g. They suggested for flooding

$$K_g = 3.2 = \frac{v_g \rho_g^{1/2}}{\left[g\sigma(\rho_l - \rho_g)\right]^{1/4}} \tag{12.59}$$

However, McQuillan and Whalley (1985) suggested a single correlation for the entire range of diameter as

$$K_g = 0.286 B_0^{0.26} Fr^{-0.22}\left[1 + \frac{\mu_l}{\mu_{water}}\right]^{-0.18} \tag{12.60}$$

where

$$B_0 = \frac{d^2 g(\rho_l - \rho_g)}{\sigma}$$

$$Fr = V_l \frac{\pi}{4} d \left[\frac{g(\rho_l - \rho_g)^3}{\sigma^3} \right] \qquad (12.61)$$

12.4 Dropwise Condensation

Dropwise condensation is the form of condensation which is most desirable but yet to be practically brought about in an industrial environment at an affordable cost. This essentially means condensation at preferred sites leaving the rest of the cold surface exposed to vapour. This gives a much higher rate of condensation and a heat transfer coefficient of about an order of magnitude higher than that of film condensation under similar condition. This means that the surface wettability should be poor. There are a number of techniques by which this can be promoted. One method is to inject some chemicals known as promoters of dropwise condensation (such as palmitic aid). The other option is to use noble metal coating on the surface such as gold or platinum. This option is ruled out from economic consideration and the option of using promoters is tried but only with partial success. The major problem as Prof. John Rose, an expert on condensation, puts it is "You can get hundreds of hours of sustained drop condensation in the clean laboratory environment, but whenever it is moved to a practical industrial process, it is destroyed and the film is formed" (private discussion with the author). Thus till now no practical application of dropwise condensation could be made because even if we start dropwise condensation, it cannot be sustained and ultimately it transforms into film condensation. For this reason it is not dealt in detail here. However, in recent times a renewed interest in dropwise condensation has been observed due to recent developments in super hydrophobic surfaces. This has given hope that engineered non-wetting surfaces can be built which will promote dropwise condensation. We will discuss this possibility later in this chapter.

Often the heat transfer coefficient in drop condensation is given by comparing it with film condensation as

$$\left(\frac{j_{drop}}{j_{film}} \right) = \left[\frac{\rho_l^2 D^2 g}{24.2 \mu_l j_{film}} \right]^{1/2} \qquad (12.62)$$

where j is the mass flux. For steam at atmospheric pressure, this ratio comes out to be 6.5. Figure 12.13 shows dropwise condensation.

One major factor in dropwise condensation is the surface condition. On one hand, the surface roughness and surface liquid combination decides wettability of the liquid. On the other the adsorption of various molecules, particularly the pollutants affects the condensation process significantly. Hence, the cleanliness is a major factor in

Fig. 12.13 Dropwise
condensation

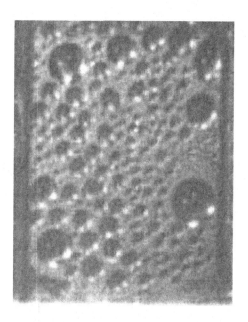

dropwise condensation. It is for this reason, the noble metals, which does not adsorb
pollutants, show excellent dropwise condensation behaviour.

12.4.1   Effects of Various Parameters on Dropwise Condensation

Dropwise condensation has long been investigated by a number of investigators. As
a result, the trends of this type of condensation against various parameters have been
observed. We will describe some of these trends in the following paragraphs.

(a) Effect of superheat: The dropwise condensation curve obtained by a series of
investigators shows considerable amount of scatter mainly due to measurement
inaccuracy. Since in dropwise condensation the heat transfer coefficient is much
higher, the resulting wall subcooling is much smaller. This makes the relative
uncertainty measurement much higher. However, even with the scatter, the data
show a consistent trend as shown in Fig. 12.14.

It is obvious from this figure that at lower subcooling the rate of increase of heat
flux with subcooling is increasing while in the moderate subcooling it attains a
constant value. It is often attributed to the non linear growth of active nucleation
site density particularly at lower wall subcooling.

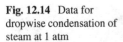

Fig. 12.14 Data for dropwise condensation of steam at 1 atm

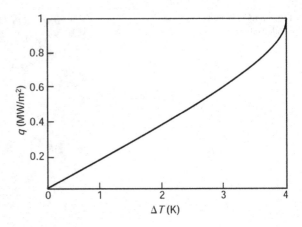

Fig. 12.15 Effect of pressure on dropwise condensation of steam

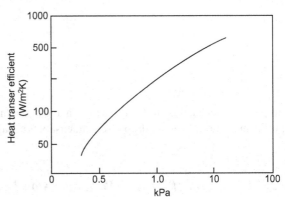

(b) Effect of pressure: The effect of pressure on dropwise condensation can be seen as shown in Fig. 12.15.

It can be seen that with decrease in subcooling the rate of change of heat flux decreases. It is often attributed to the increase of interfacial resistance of the non-condensible at lower pressure. Briggs and Rose (1996) proposed an empirical correlation of the form.

$$h = T_c^{0.8}(5 + 0.3\Delta T) \tag{12.63}$$

Here, heat transfer coefficient h is in $W/m^2 K$, T_c in $°C$ and T in Kelvin.

(c) Effect of departing drop size: The departing drop sizes of liquid differ due to various factors such as gravity (inclination), vapour shear and value of liquid surface tension. A consistent trend of decrease of heat transfer with increase in departing drop size is seen as shown in Fig. 12.16.

(d) Effect of surface material: The surface material particularly, its thermal properties do affect the dropwise condensation process. Mikic (1969) presented a

Fig. 12.16 Effect of departing drop size on heat transfer coefficient in Dropwise Condensation

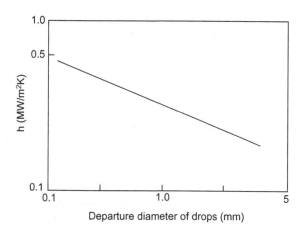

Fig. 12.17 Effect of surface conductivity in Dropwise Condensation

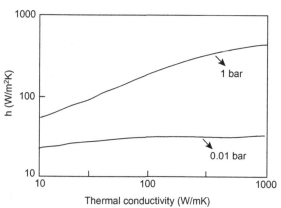

theory based on constriction effect, which proposed that higher surface conductivity promotes dropwise condensation. This is supported by experimental data as shown in Fig. 12.17.

12.4.2 Theories on Dropwise Condensation

A number of theories have been proposed by researchers to explain and analyse dropwise condensation. However, these theories make various assumptions and empiricisms which are open to question. We discuss here a popular theory which appears to be consistent without gross unrealistic assumptions.

Upon contact of saturated vapour with the base wall, the vapour gets adsorbed as a poly-molecular thin liquid film on the surface. This film is extremely thin, of the order of tens to hundreds of nanometer. This is also a more ordered arrangement of molecules which results in different thermophysical properties of the film than usual liquid. This film is acted upon by a pressure known as the structural disjoining pressure, P_{sd}. This pressure varies with the film thickness of this thin layer as

$$P_{sd} \sim \frac{1}{\delta^3}$$

Now, since the thickness of this film varies on the surface due to imperfections or undulations on the surface, the structural disjoining pressure on the surface also varies and the film attracts more at some locations than other. This creates preferred sites of droplet nucleation.

As the droplet grows on the surface, at the top of the droplet an effective upward pressure must act due to curvature. This is due to the surface tension between the liquid and vapour. Thus, in the presence of structural disjoining pressure, we get a modified Young–Laplace equation from the one we have seen for bubbles in Chap. 3. Here,

$$\Delta p = p_l + p_g = \frac{2\sigma}{R} - P_{sd} \tag{12.64}$$

Using Clausius–Clapeyron equation, we get the critical temperature difference sub-cooling corresponding to the critical radius below which condensation cannot take place (R_c), is given by

$$\Delta T_{cr} = \frac{\Delta p T_{sat}}{\rho_l h_{fg}} \tag{12.65}$$

Using Eq. (12.63) we get,

$$\Delta T_{cr} = \frac{2\sigma T_{sat}}{\rho_l h_{fg} R_c} - \frac{P_{sd} T_{sat}}{\rho_l h_{fg}} \tag{12.66}$$

It can be said that the first term on the right-hand side is due to the primary capillary effect, while the second term is the secondary structural disjoining effect. In absence of structural disjoining pressure, this equation reduces to the subcooling equation similar to liquid superheat equation in boiling.

$$\Delta T_{cr} = \frac{2\sigma T_{sat}}{\rho_l h_{fg} R_c} \tag{12.67}$$

Now, to calculate the heat transfer in dropwise condensation we need to utilize the data on the above description. Let us assume that at steady state, at any moment large number of droplets are growing from the critical size R_c to the drop departure size R_0 (under gravity or shear). The instantaneous size distribution of droplets can be given by a function.

$$\psi(R) = \frac{dn}{dR}$$

where n is the number of droplets between the size of R and $(R + dR)$ per unit surface area. At steady state, the droplet formation and removal may be assumed to result a quasi-stationary situation where it is constant.

Thus, the increase in volume of liquid is

$$\Delta V = A_s \frac{dR}{dt} = A_s w(R) \tag{12.68}$$

where

$$A_s = \text{droplet surface area}$$
$$w(R) = \text{growth rate of droplet} = \frac{dR}{dt}$$

If we consider the droplet as hemisphere, then for growth from R to $R + dR$, the heat transfer is given by

$$dq_R = \rho_l h_{fg} (2\pi R^2) \frac{dR}{dt} dx$$
$$= 2\pi \rho_l h_{fg} R^2 w(R) \psi(R) dR$$

Integrating between critical and departure radius

$$q = 2\pi \int_{R_e}^{R_0} \rho_l h_{fg} R^2 w(R) \psi(R) dR \tag{12.69}$$

We can determine the growth rate by considering that the droplet is stagnant, as

$$\frac{dR}{dt} = 2K_l \left(\frac{T_{sat} - T_w}{\rho_l h_{fg} R} \right) \tag{12.70}$$

Knowing the size distribution, $\psi(R)$, one can calculate the rate of heat transfer. However, a large number of experimental correlations are available for dropwise condensation. A plot of such correlation is shown in Fig. 12.18.

There are also more complex theories developed by le Fevre and Rose (1956) and Tanaka (1975) which considers various factors such as interfacial heat transfer resistance. However a complete theory on dropwise condensation is yet to emerge.

12.4.3 Methods to Promote Dropwise Condensation

Because of its attractive heat transfer characteristics and large amount of condensation, researchers have been trying for decades to evolve methods which will sustain dropwise condensation. Some of the options tried are as follows:

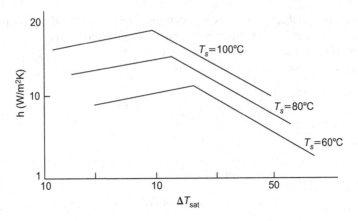

Fig. 12.18 Heat transfer in dropwise condensation of steam

1. Surface coatings: One approach is to coat the solid condensing surface with materials which imparts non-wetting features to the surface. This has been tried in a number of ways like using organic polymers like polytetrauoroethylene (PTFE, Teylon), using inorganic compounds like metal Sulfhides or noble metals like gold or platinum. While some of these (like Teflon) adds to thermal resistance due to poor thermal conductivity, the option of noble metal is prohibitive from cost consideration.

2. Injection of promoters: Another method tried is the injection of promoting agents such long chain hydrocarbon molecules. Under laboratory conditions they give satisfactory dropwise condensation. However in industrial ambience this is found to be non-sustainable.

3. Engineered surface: In recent times various types of engineered surfaces have been manufactured to import hydrophobicity. The particularly important family of surfaces are the superhydrophobic surfaces mimicking the well known Lotus Effect.

What makes a surface superhydrophobic are the micron-sized protrusion on the surface as shown in Fig. 12.19. Figure 12.19a shows the case when the droplet remains in non-wetting state known as the Cassie state while Fig. 12.19b shows the wetting state known as the Wenzel state. Although studies on such surfaces with respect to dropwise condensation has been initiated (Varanasi et al. 2012), a conclusive view and evidence appears to be still far away.

Fig. 12.19 Droplet on a
superhydrophobic surface. **a**
Cassie state and **b** Wenzel
state

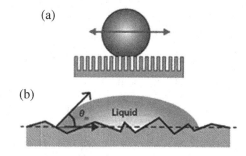

Examples

Problem 1: Propane is condensing inside a horizontal tube whose internal diameter is
15 mm. The refrigerant enter at its saturation temperature of 2 °C as a saturated vapour
and leaves as a saturated liquid. The flow rate of vapour entering is 0.03534 kg/s and
the tube wall has a mean internal temperature of −10 °C. Determine the local condensing heat transfer coefficient at a vapour quality of 0.5 using (a) Shah and (b)
Dobson-Chato methods.

Solution: The physical properties of propane at 2 °C.

$\rho_L = 528 \, \text{kg/m}^3$;
$\rho_G = 11 \, \text{kg/m}^3$;
$\mu_L = 0.0001345 \, \text{Pa-s}$;
$\mu_G = 0.0000075 \, \text{Pa-s}$;
$h_{fg} = 373100 \, \text{J/kg}$;
$k_L = 0.108 \, \text{W/m K}$;
$C_{p_L} = 2470 \, \text{J/kg K}$;
$p_{crit} = 4264 \, \text{kPa}$;
$k_G = 0.0159 \, \text{W/m K}$;
$C_{p_G} = 2470 \, \text{J/kg K}$;

The mass velocity of the total flow of liquid plus vapour is

$$\dot{m} = \frac{\dot{M}}{\frac{\pi}{4}D^2}$$

According to Shah:

$$Re_L = \frac{\dot{m}D}{\mu_L} = 22305$$

Reduced pressure, $p_r = \frac{p_{sat}}{p_{crit}} = 0.119$

$$\frac{h_x d}{k_L} = 0.023 Re_L{}^{0.8} Pr_L{}^{0.4} \left[(1-x)^{0.8} + \frac{3.8^{0.76}(1-x)^{0.04}}{p_r{}^{0.38}} \right]$$

Upon simplifying, we get, $h_x = 4280$ W/m$^2 K$.

Dobson and Chato method:

$$Re_L = \frac{\dot{m}D(1-x)}{\mu_L} = 11152$$

$$X_{tt} = \left(\frac{(1-x)}{x} \right) \left[\frac{\rho_G}{rho_L} \right]^{0.5} \left[\frac{\mu_L}{\mu_G} \right]^{0.1}$$

$$= 0.1926$$

$$Ga_L = \frac{g\rho_L D^3 (\rho_L - \rho_G)}{\mu_L{}^2} = 4.996 \times 10^8$$

When $Fr_{SO} > 20$, the annular flow correlation is used:

$$Nu_x = \frac{h_x D}{k_L} = 0.023 Re_L{}^{0.8} Pr_L{}^{0.4} \left[1 + \frac{2.22}{X_{tt}{}^0 .89} \right]$$

Upon simplifying, we get, $h_x = 4766 \ W/m^2 K$.

Problem 2: The steam with mass flux of 200 kg/m^{2}s condenses in horizontal tube with inner diameter of 100 mm. The steam is saturated at the inlet and is at pressure of 8 bar. The pressure is uniform and the wall temperature is at 110 °C. When the quality is 0.5, specify the flow regime.

Given: The properties of steam at 8 bar:

$T_{sat} = 170°$, $h_{fg} = 2046.5$ kJ/kg,
$\rho_L = 896$ kg/m^3, $\rho_G = 4.16$ kg/m^3,
$\mu_L = 190 \times 10^{-6}$ Pa-s, $\mu_G = 15 \times 10^{-6}$ Pa-s.

Solution: Dimensionless vapour velocity introduced by Wallis

$$j_G{}^* = \frac{x^* G}{[g D \rho_G (\rho_L - \rho_G)]^{0.5}} = 0.1129$$

The vapour volumetric content (void fraction) is

$$\varepsilon = \left[1 + \left(\frac{(1-x)}{x} \right) \left(\frac{\rho_G}{\rho_L} \right)^{0.795} \left(\frac{GD}{\mu_L} \right)^{-0.016} \right]^{-1} = 0.9893$$

Since, $\left(\frac{(1-\varepsilon)}{\varepsilon} \right) = 0.0108 << 0.5$ and $j_G{}^* << 1$, the flow regime is stratified (Fig. 12.20).

Fig. 12.20 Condensation in the fully stratified flow

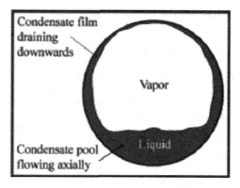

Fig. 12.21 Problem 1

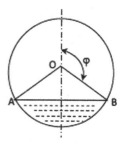

Exercises

1. The flow configuration for a particular tube is given in Fig. 12.21. Derive the expression for void fraction in terms of circumferential angle ϕ assuming the flow to be stratified throughout the tube.
2. Refrigerator manufacturing units use Iso-butane (R600a) as an eco-friendly refrigerant for domestic Refrigerators. A design engineer interested in calculating the heat transfer coefficient for the condenser system. Before that he wants to check the flow regimes for this particular configuration so that appropriate correlations can be used. The refrigerant enters the pressure (4.08 bar), and it is passed through a horizontal tube of 4 mm diameter at 2 g/s. Assume that saturated Vapour entering the condenser and saturated liquid is leaving the condenser.
3. The refrigerant R22 flowing through a 5 mm diameter tube with a pressure of 15 bar and mass flux of 150 kg/m^2 s. The quality of the refrigerant is 0.7. Calculate the void fraction using homogeneous model, and Lockhart–Martinelli correlations.
4. Dry saturated steam at 10 bar condenses on a vertical tube of length 5 m. The surface temperature of tube is 165 °C. Determine the film thickness and local heat transfer coefficient for various vertical distances from the upper end of the tube. Comment on the results.

References

Anand S, Paxson AT, Dhiman R, Smith JD, Varanasi KK (2012) Enhanced condensation on lubricant-impregnated nanotextured surfaces. ACS nano 6(11):10122–10129

Blangetti F (1979) Local heat transfer during condensation with forced convection in a vertical tube

Briggs A, Rose JW (1996) Condensation on low-fin tubes: effects of non-uniform wall temperature and interphase matter transfer. In Process, enhancement, and multiphase heat transfer. Begell House, New York, pp 455–460

Carpenter EF, Colburn AP (1951) In: Proceedings of General Discussion on Heat Transfer, Institution of Mechanical Engineers, London and ASME, New York, p 20

Fauske H (1961) Critical two-phase, steam-water flows. In: Proceedings of the 1961 Heat Transfer and Fluid Mechanics Institute. Stanford University Press, Stanford, CA, 79–89

Fujii T, Honda H, Nagata S, Fujii M, Nozu S (1976) Condensation of RII in horizontal tube (1st report: flow pattern and pressure drop). Trans JSME (B) 42(363):3541–3550 (in Japanese)

Isachenko VP, Salomzoda F, Shalakhov AA (1980) Investigation of heat transfer with dropwise condensation of steam in a vertical. Tube Therm Eng (Engl Transl); (United Kingdom) 27(4)

Kandlikar SG, Boiling condensation and gas-liquid flow, Taylor & Francis Press

le Fevre EJ, Rose JW (1965) An experimental study of heat transfer by dropwise condensation. Int J Heat Mass Transf 8(8):1117–1133

Levich VG (1962) Physicochemical hydrodynamics

McQuillan KW, Whalley PB (1985) Flow patterns in vertical two-phase flow. Int J Multiph Flow 11(2):161–175

Mikic BB (1969) On mechanism of dropwise condensation. Int J Heat Mass Transf 12(10):1311–1323

Rohsenow W (1956) Effect of vapor velocity on laminar and turbulent film condensation. Transactions of ASME, 1645–1648

Soliman M, Schuster JR, Berenson PJA (1968) General heat transfer correlation for annular flow condensation

Tanaka HA (1975) Theoretical study of dropwise condensation

van Driest ER (1956) On turbulent flow near a wall. J Aeronaut Sci 23(11):1007–1011

Wallis GB (1961) Flooding velocities for air and water in vertical tubes. United Kingdom Atomic Energy Authority

Chapter 13
Boiling and Condensation of Mixtures

Phase change can also occur in a mixture of fluids which can be miscible or immiscible. In this chapter, we will deal with the miscible binary mixture of fluids. The features of phase change in these mixtures are quite different from that of single component fluids. The change of phase here takes place over a range of temperature (known as the 'boiling range') rather than at a fixed temperature. This is because, in general, one of the components is likely to be more volatile than the other. The process of phase change here is dominated by mass transfer between the two components. Binary mixtures find their applications in chemical process industry, petroleum refining and food processing where even more than two components(ternary or multi-component mixtures) may be involved.

13.1   Boiling and Condensation of Mixtures

Two miscible liquids form binary mixture and has got distinctive features with respect to boiling and condensation. According to the Gibbs phase rule described in Chap. 3, if we have two components ($m = 2$), we get that the maximum numbers of independent intensive properties are three. These are usually pressure, temperature and mole fraction of one component for a binary mixture. Thus, we can construct 3D surfaces for the mixture. However, if we consider boiling or condensation at a given pressure, a 2D plot of temperature versus composition (mole fraction) is sufficient.

13.1.1   Definitions and Equilibrium of Phases

To proceed with the boiling of mixtures, some definitions are important. The molar concentration of component i is defined as

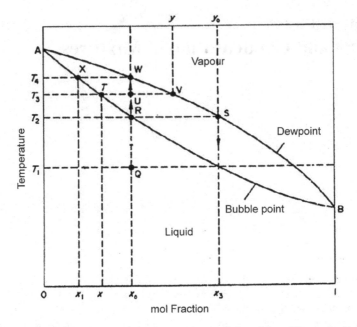

Fig. 13.1 Phase equilibrium diagram of a binary mixture (Adapted from "Convective Boiling and Condensation" by John G. Collier and John R. Thome, with permission from Oxford University Press)

$$c_i = \frac{\rho_i}{M} \tag{13.1}$$

where M is the molecular weight and ρ in its density. Usually, mole fraction in vapour is indicated by 'y' and that in liquid is indicated by 'x'.

$$x_i \quad \text{or} \quad y_i = \frac{c_i}{\Sigma c_i} \tag{13.2}$$

The phrase equilibrium of a binary mixture at a given pressure is indicated by the binary phase diagram given by Fig. 13.1. For a mixture, unlike pure liquid, boiling does not take place at a fixed saturated temperature because the composition of the liquid goes on changing due to the fact that the more volatile component evaporates faster than the less volatile component. For example, if we take a binary mixture of liquids at a temperature T_1 with a mole fraction of x_0 (point Q) and it is heated, then at point R, it starts boiling first. This point is called *bubble point..* Here the liquid composition remains the same but the first bubble of vapour formed will have a mole fraction of y_0 under equilibrium condition which is much richer in more volatile component than the liquid. Now, if it is heated further, more vapour, rich in more volatile component, will leave, making the liquid rich in less volatile component, and as a result, the temperature will also increase. For example, at point U, the liquid

Fig. 13.2 Binary equilibrium curve for azeotropes (*Courtesy* "Heat Transfer in Condensation and Boiling" by Karl Stephan, with permission from Springer)

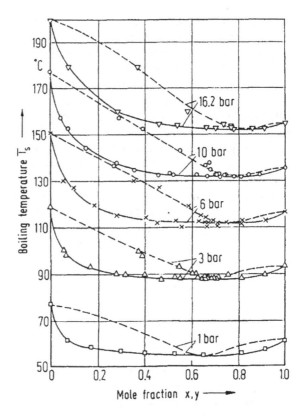

will have a mole fraction of x and the vapour y of the more volatile component. Finally, at W, the entire liquid evaporates and the vapour will have the same mole fraction x_0 as the initial liquid (x_0) and the last drop of liquid evaporated will have a composition x_1. This point W at which the last drop of liquid evaporation will be the same point at which condensation will start or first drop of liquid will appear when the vapour mixture is cooled at constant pressure. This point is called the *dew point*.

However, there are some mixtures that have the temperature–composition curves as shown in Fig. 13.2. The interesting feature of these mixtures is a particular composition at which the liquid and vapour composition are the same and the mixture behaves like a pure liquid with a unique saturation temperature T_0. This composition is called an *azeotrope* , e.g., mixture of methanol and benzene.

13.1.2 Boiling of Mixtures

Industrial processes often involve boiling of mixture of liquids. Petrochemicals are the most important example. Other processes like concentrating of solutions, thickening of dyes, separation by distillation, recovery of solvents and desalination of seawater involve boiling of multi-component mixtures. Another area in which mixtures are becoming important is refrigeration. Almost all the alternative refrigerants suggested with lower ozone depleting potential (ODP) and global warming potential (GWP) are mixtures of a number of liquids. Apart from this, traces of lubricating oil are always present in refrigeration process. A mixture of two components is called *binary mixture*, three components is called *ternary mixture* and higher number of components is simply called *multi-component mixture*. Very few studies are available on boiling with more than two components. Here we will focus on binary mixtures to understand the basic phenomenon of boiling of mixtures.

Usually, heat transfer coefficient for boiling of mixtures is remarkably smaller compared to that for boiling of pure components. However, if one component is surface active, boiling heat transfer can be enhanced due to reduction of surface tension and formation of small large number of bubbles. In boiling of mixtures, heat and mass transfer processes are closely related. Mass transfer impedes the vapour formation rate to a large extent. In the following section, we will give a simple overview of boiling of mixtures.

13.1.2.1 Pool Boiling of Mixtures

The bubble growth characteristics of binary mixture is quite complex. As a result, complicated bubble growth equations can be developed which are beyond the scope of this chapter. Instead, we will try to look at some physical features which will be helpful in understanding the process of heat transfer. Firstly, one must realize that during the process of bubble growth equilibrium does not exist even inside the liquid phase. Since the more volatile component vaporizes more, the liquid surrounding the bubble become more concentrated with less volatile component. This part is shown as a liquid layer in Fig. 13.3. Across this layer, the temperature rises and reaches the temperature of the bulk liquid composition. Now, for the more volatile component to vaporize at the liquid–vapour interface, it has to diffuse through the layer of liquid with high concentration of less volatile component. This is difficult because of mass diffusivity and hence the heat transfer is also impeded. The specific nature of micro-layer evaporation also contributes to the deterioration of boiling heat transfer.

As a result of the nature of equilibrium diagram described above, the pool boiling curve of a binary mixture gets substantially altered as shown in Fig. 13.4. Firstly, the inception of nucleate boiling is delayed due to the temperature gradient caused by the concentration gradient near the wall. Then the nucleate boiling curve gets shifted towards right indicating deterioration of heat transfer because the same heat flux will now require higher wall superheat giving lower heat transfer coefficient. The critical heat flux is usually increased and the heat flux at *Leidenfrost point* is also increased.

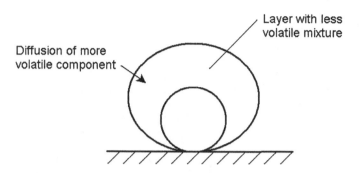

Fig. 13.3 Vapour bubble in a binary system

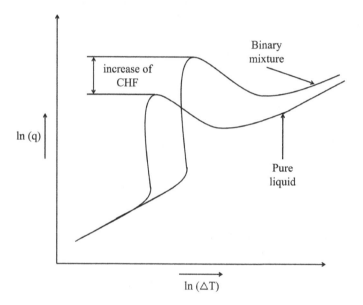

Fig. 13.4 Changes in pool boiling characteristics of binary mixture

The film boiling heat transfer is marginally higher for binary mixtures compared to single component fluid.

Pool boiling of binary mixture has been studied extensively. By observing these results, Stephan and Korner (1969) proposed a simple method for the calculation of heat transfer coefficient for boiling of binary mixtures. First, we recognize that the heat transfer coefficient for a binary mixture is much lower than what is expected by linear interpolation of the heat transfer coefficient of the two mixtures. As a result, the wall superheat is higher than what is predicted by interpolation as shown in Fig. 13.5.

Fig. 13.5 Actual and
interpolated values of heat
transfer (Adapted from
"Convective Boiling and
Condensation" by John G.
Collier and John R. Thome,
with permission from Oxford
University Press)

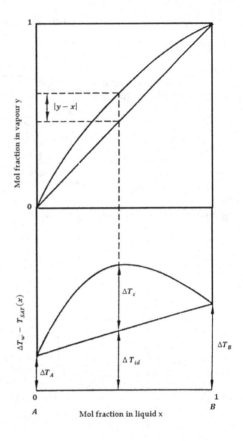

Fig. 13.5 Actual and interpolated values of heat transfer (Adapted from "Convective Boiling and Condensation" by John G. Collier and John R. Thome, with permission from Oxford University Press)

Now the ideal wall superheat for a binary mixture is the interpolated value of wall superheat of the two pure liquids at the same heat flux.

$$\Delta T_{id} = x_1 \Delta T_1 + x_2 \Delta T_2 \tag{13.3}$$

where

$$\Delta T_1 = \frac{\dot{q}}{h_1}$$

$$\Delta T_2 = \frac{\dot{q}}{h_2}$$

The heat transfer coefficient and wall superheat for a binary mixture is defined as

$$\dot{q} = h(T_w - T_{sat}) = h \Delta T$$

However, the actual driving of temperature difference is much higher than ΔT_{id}. Thus

$$\Delta T = \Delta T_{id} + \Delta T_E \tag{13.4}$$

$$\Delta T = \Delta T_{id}(1 + \theta)$$

where

$$\theta = \frac{\Delta T_E}{\Delta T_{id}}$$

They proposed a correlation for θ in the form

$$\theta = K_{12}|y - x| \tag{13.5}$$

The constant K_{12}, depending on binary combination, is given by

$$K_{12} = K_{12}^0 \left(0.88 + 12\frac{P}{P_0}\right) \tag{13.6}$$

for 1 bar $< P <$ 10 bar and $P_0 = 1$ bar, the value of K_{12}^0 is different for every mixture as given in Table 13.1. In absence of data $K_{12}^0 = 1.4$ can be used. Another empirical correlation which fits well to experimental data was given by Schlunder (1986) in the form

$$\theta = 1 + \frac{1}{\Delta T_{id}}(T_{s2} - T_{s1})(y - x)\left[1 - \exp\left(\frac{-q}{\rho_L \beta_L h_{fg}}\right)\right] \tag{13.7}$$

where β_L = liquid mass transfer coefficient $\simeq 2 \times 10^{-4} m/s$.

T_{s1}, T_{s2} = saturation temperature of pure components

For pool boiling of mixture with more than two (N) components, the Stephan's method mentioned above can be extended as

$$\Delta T_{id} = \Sigma x_i \Delta T_i \tag{13.8}$$

and

$$\theta = |\Sigma K_{iN}(y_i - x_i)| \tag{13.9}$$

This method that predicts the measured values of ternary and other mixtures of *azeotropes* is special since the values can be well approximated by the interpolated values.

13.1.2.2 Flow Boiling of Mixtures

Similar to single component liquid, during flow boiling, the pressure decreases in the direction of flow resulting in a decrease in saturation temperature. However, if the tube pressure drop is not too much (not for very small diameter or roughened tube), this pressure drop effect can be compensated in a binary mixture due to rise in temperature due to increasing concentration of the less volatile component in the

Table 13.1 Values of K_{12}^0

Mixture	K_{12}^0
Acetone/Ethanol	0.75
Acetone/Butanol	1.18
Acetone/Methanol	1.19
Acetone/Water (Copper heating surface)	1.40
Acetone/Water (Nickel heating surface)	0.81
Acetone/Benzene	0.42
Ethanol/Cyclohexane	1.31
Ethanol/Water (Copper heating surface)	1.21
Ethanol/Water (Nickel heating surface)	0.71
Benzene/Toluene	1.44
n-Heptane/Methylcyclohexane	1.95
i-Propanol/Water	2.04
Methanol/Ethanol	1.39
Methanol/Benzene	1.08
Methanol/Amylalcohol	0.80
Methanol/Water	0.56
Methylethylketone/Water	1.21
Propanol/Water	3.29
Water/Glycol	1.47
Water/Glycerol	1.50
Water/Pyridine	3.56

liquid as vaporization progresses down the tube. It is possible that the temperature increases downstream.

In general, the heat transfer decreases in binary mixture. Hence, a correction of the method suggested by Chen (1963) for convective boiling of pure liquid in vertical tube is needed. This can be done by a modification of suppression factor S and enhancement factor F, as

$$H_{binary} = S^* h_{FZ} + F^* h_{con} \qquad (13.10)$$

The changed value of suppression factor S given by S^* appears as

$$S^* = S\frac{\Delta T^*}{\Delta T} = \frac{1}{1 + 2.53 \times 10^{-6} Re_{2\phi}^{1.17}} \frac{1}{1 - (\xi_G - \xi_L)\dfrac{C_{pl}}{h_{fg}}\left(\dfrac{\partial T}{\partial T}\right)_P} \qquad (13.11)$$

where ΔT^* is effective superheat with mass diffusion effect and ΔT is hypothetical superheat neglecting mass diffusion

$$Re_{2\phi} = Re_L F^{1.25}$$

The modification of convective heat transfer coefficient is given by

$$h^*_{con} = h_{con} \frac{T_w - T_i}{T_w - T_s} \tag{13.12}$$

where T_i is the interfacial temperature and T_B is the bulk liquid temperature. This additional factor is given by

$$\frac{T_w - T_i}{T_w - T_s} = 1 - \frac{q}{h_{fg}\rho_L h_{mL}} \left(\frac{\partial T}{\partial \xi_L}\right)_P \frac{\xi_{G,i}}{T_w - T_i} \tag{13.13}$$

The mass transfer coefficient h_{mL} can be found out as

$$Sh = \frac{h_{mL}d}{D} = .023 Re_{2\phi}^{0.8} Sc^{0.4} \tag{13.14}$$

where

Sh = Sherwood number
D = mass diffusion coefficient
Sc = Schmidt number
ξ = mass fraction (G of vapour, L liquid) of more volatile component.

13.1.3 Condensation of Binary Mixtures

Like boiling, in condensation also there is a wide range of applications for binary and multi-component mixtures. These applications include separation, distillation of simple impurities during condensation as in the case of presence of lubricating oil in refrigerants. The analysis for condensation of binary mixtures is quite involved in which mass transfer plays a major role. Usually, the effect of mass transfer can be incorporated by multiplying a correction factor to the heat transfer coefficient without mass transfer effect.

$$F = \frac{\frac{\dot{m}C_p}{h_0}}{\exp(\frac{\dot{m}C_p}{h_0}) - 1} \tag{13.15}$$

where $\dot{m}$ is the mass flow rate of the vapour and h_0 is the heat transfer coefficient without mass transfer.

Another assumption often used for low mass transfer rate is a unity Lewis number. Here, Lewis number (Le) is given as

$$Le = \frac{\alpha}{D}$$

where α = thermal diffusivity and D = mass diffusion coefficient.

One more major problem in dealing with the condensation of mixture is the determination of interface temperature which is different from the bulk fluid temperature. This is given by

$$\frac{h_L}{\frac{h_L}{h_C}+1}(T_i - T_c) = h_G(T_G - T_i) + \rho_G \beta_G \frac{\xi_{1G} - \xi_\infty G}{\xi_{1G} - \xi_{1L}} h_{fg} \qquad (13.16)$$

where

h_C = coolant side (other side of the wall) heat transfer coefficient
T_c = coolant temperature
ξ = mass fraction
ξ_∞ = mass fraction distant from wall
i stands for at the interface

This equation is arrived at by considering the liquid mass flux at the wall for the volatile component equal to the mass flux of the same component in vapour, plus the one received by diffusion from far off

$$\dot{M}\xi_{1L} = \dot{M}\xi_{1G} + j_G A \qquad (13.17)$$

where

$$j_G = \rho_G \beta_G (\xi_{1G} - \xi_{\infty G})$$

From this expression, Colburn and Drew (1937) developed the following method for the calculation of heat flux during condensation of binary mixtures:

1. Assume mole fraction of component 1(z) in the condensing flux.
2. Approximate the inter facial concentration of 1 (n_{io}) as the completely mixed concentration of the liquid phase.
3. With the value of total pressure, calculate the interfacial concentration of the other component y_{io}.
4. Calculate inter facial temperature.
5. Calculate the molar flux from

$$\dot{n}_T = \beta_{CT} \ln[\frac{Z - y_{io}}{Z - y_{ib}}]$$

where $\beta_{CT} = p/RT$.
6. Calculate the mass transfer correction factor, and heat flux as

$$q = \dot{n}_T h_{GLn} + h_0(T_i - T_l) \qquad (13.18)$$

$$\phi = \frac{\dot{n}_1 C_{PG2} + \dot{n}_2 C_{PG2}}{h_0}$$

$$F = \frac{\phi}{\exp(\phi) - 1}$$

and

$$h^*_{G_{Ln}} = Z h_{G_{l1}} + (1 - Z) h_{G_{l2}}$$

7. Calculate

$$q^* = -U(T_i - T_2) \tag{13.19}$$

where U is the overall heat transfer coefficient.
8. If q^* is close to q, it is the heat flux, otherwise iterate.

References

Chen JC (1963) A correlation for boiling heat transfer to saturated fluids in convective flow. ASME 744 63-HT-64. In Presented at 6th National heat transfer conference, Boston
Colburn AP, Drew TB (1937) The condensation of mixed vapors. Trans AIChE 33:197–215
Stephan K, Korner M (1969) Calculation of heat transfer in evaporating binary liquid mixtures. Chem-Inge-Tech 41(7):409–417

Chapter 14
Numerical Modelling of Boiling

In this chapter, an overview of the numerical simulation of boiling phenomena is presented. For the understanding of flow and heat transfer characteristics of boiling flows, the interface dynamics and the associated heat and mass transfer have to be accurately modelled. Many interface tracking techniques have been developed to obtain the exact movement of interfaces in two-phase flows. After a brief overview of these methods, a class of interface tracking methods is discussed in detail. The modifications required in the governing equations to account the interfacial surface tension, heat and mass transfer for the numerical simulation of boiling flows are discussed. At the end of the chapter, the current state of the modelling of various regimes of boiling flows with accurate interface tracking methods are presented.

14.1 Numerical Modelling of Interfacial Flows

Many methods have been developed for the modelling of interfacial flows to capture the moving interface accurately. These methods are classified into: (a) moving mesh methods (b) front tracking methods and (c) interface capturing methods (Fig. 14.1). Understanding the advantages and the limitations of these methods helps us in selecting a suitable method for modelling boiling flows.

14.1.1 Moving Mesh Methods

In a moving mesh method, the mesh is moved as the interface evolves with time. Moving mesh methods require continuous deformation of the mesh to attach the interface with faces of the computational cells as shown in Fig. 14.1a. Even though the interface is tracked accurately with time, modelling of the merging and break up

© The Author(s) 2023
S. K. Das and D. Chatterjee, *Vapor Liquid Two Phase Flow
and Phase Change*, https://doi.org/10.1007/978-3-031-20924-6_14

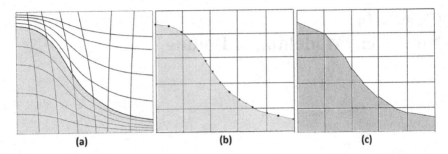

Fig. 14.1 Modelling of interfacial flows **a** moving mesh methods **b** front tracking methods and **c** interface capturing methods

of bubbles or droplets are not possible with these methods. Hence, these methods are only used for modelling flows in which the interface break up and merger are not expected. These methods have only very limited usage in modelling boiling flows.

14.1.2   Front Tracking Methods

In the front tracking methods, a fixed mesh is used to obtain the velocity field and connected marker points are used to represent the moving interface. The marker points are connected by line elements and forms a moving front as shown in Fig. 14.1b for 2D problems. In the case of 3D problems, the moving front is formed by generating a surface mesh with unstructured triangular elements using the marker points as shown in Fig. 14.2. Based on the velocity field obtained with the fixed grid, the marker points are advected for each time step to obtain their new locations. After each time step, the new information of the front is passed to the fixed cells. With this information, the properties of fluid in each cell is updated. After each time step, the distance between the connected marker points may decrease or increase. Hence, some of the marker points will be deleted or added to ensure that the front with the interface mesh formed by the connected marker points represents the interface shape accurately. Whenever the marker points are added or deleted, the elements which represent the interface should be renumbered. Even though the front tracking methods provide the interface accurately, the implementation of these methods for 3D problems is challenging, in particular, the cases in which bubble merger or break up is common.

Juric and Tryggvason (1998) developed a front tracking method for boiling flows. The details of the implementation of their front tracking method and its applications for two-phase flows including boiling and solidification are presented by Tryggvason et al. (2001).

Fig. 14.2 The interface
representation of a rising gas
bubble in the front tracking
method

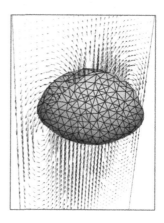

14.1.3 Interface Capturing Methods

In these methods, a colour function, C is used to specify two different fluids in the
interfacial flows as shown in Fig. 14.1c. The widely used colour functions are the
volume fraction, F of a fluid in each cell of a computational domain and the level
set function, ϕ which is the signed shortest distance from the interface. In these
methods, the colour function is advected passively with the known velocity field.
The advection of a colour function is implemented using the advection equation

$$\frac{\partial C}{\partial t} + u \bullet \nabla C = 0 \tag{14.1}$$

As the velocity field is usually obtained using fixed grid methods and the advection equation is solved based on the known velocity field, an interface capturing
method can capture the break up and merger of bubbles or droplets without any
mesh movement. Hence, the interface capturing methods are popular among the
methods available for modelling of interfacial flows.

14.2 Interface Capturing Methods

14.2.1 Volume of Fluid (VOF) Method

The VOF method is one of the widely used interface capturing methods for modelling
two-phase flow problems. In this method, the volume fraction of a fluid, F is used as
a colour function. Consider two fluids, fluid 1 (a liquid) and fluid 2 (air or a gas) in a
two-phase flow problem and the computational domain is discretized with a number
of cells. In the VOF method, the volume fraction, F, is calculated as the ratio of the

volume of fluid 1 in a cell to the total volume of the cell. If fluid 1 is fully filled in a computational cell, the volume fraction, F is specified as one. If fluid 1 is not present in a cell, then the volume fraction, F, of the cell is specified as zero. The partially filled cells are interfacial cells. If fluid 1 is partially filled in a cell, then the volume fraction, F, is between zero and one. In the volume of fluid method, the volume fraction, F, is advected using the following equation passively:

$$\frac{\partial F}{\partial t} + \mathbf{u} \bullet \nabla F = 0 \tag{14.2}$$

For incompressible flows, the above equation can be written as

$$\frac{\partial F}{\partial t} + \nabla.(\mathbf{u}F) = 0 \tag{14.3}$$

To discretize the advection terms of Eq. 14.3, either a high-resolution/compressive scheme or a geometric-based scheme is used.

Many algebraic methods are used for the discretization of the advection terms of volume fraction equation. The algebraic methods are easy to implement compared to the geometric-based advection methods. However, the algebraic methods produce diffuse interfaces. Some of the algebraic methods are listed below:

1. *"High-Resolution Interface Capturing (HRIC)"* proposed by Muzaferija et al. (1999)
2. *"Compressive Interface Capturing Scheme for Arbitrary Meshes (CICSAM)"* proposed by Ubbink and Issa (1999)
3. *"Switching Technique for Advection and Capturing of Surfaces (STACS)"* proposed by Darwish and Moukalled (2006)
4. *"Bounded Gradient Maximization (BGM)"* proposed by Walters and Wolgemuth (2009)

To avoid diffusive interfaces, the advection of fluid fraction, F and the interface construction can be carried out geometrically. To specify the interface orientation in each cell, the interface normal vector is required. The interface normal can be calculated from the distribution of volume fraction, F surrounding to a cell. In the method of *"Simple Line Interface Calculation (SLIC)"* proposed by Noh and Woodward (1976), the true interface (Fig. 14.3a) is approximated by piecewise linear interfaces positioned vertically or horizontally and in parallel with any one of the cell faces as shown in Fig. 14.3b. To improve the interface reconstruction further, Youngs (1982) developed the *"Piecewise Linear Interface Calculation (PLIC)"* method in which the interface in a cell is positioned according to the normal vector calculated from the volume fraction, F of the adjacent cells as shown in Fig. 14.3c. Although the interfaces obtained in the interfacial cells are not continuous, the PLIC method represents the interface better than the SLIC method.

Interface Reconstruction Using PLIC Method: The unit normal to the interface is obtained using the volume fraction (F) of surrounding cells to find out the orientation of the interface. The normal vector is calculated as

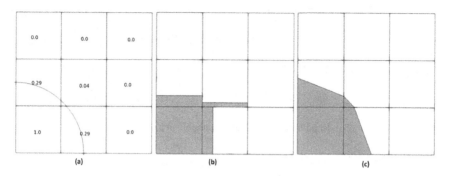

Fig. 14.3 Interface reconstruction in the VOF method **a** actual interface **b** SLIC method **c** PLIC method (Tryggvason et al. 2005)

Fig. 14.4 Calculation of unit normal using a stencil of 3 × 3 cells

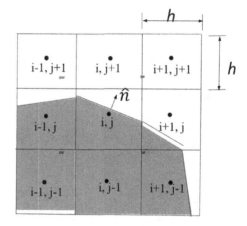

$$\mathbf{n} = -\nabla F \qquad (14.4)$$

From the normal vector, $\mathbf{n}$, the unit normal vector, $\widehat{\mathbf{n}}$ can be calculated as $\frac{\mathbf{n}}{|\mathbf{n}|}$. The normal vector calculation, based on the work of Youngs (1982) using a 3 × 3 cell stencil (Fig. 14.4), is presented here for a uniform gird with the square cells of size $h \times h$.

For the interface located in the cell (i, j), the interface normal is calculated using the 3 × 3 cell stencil as

$$n_x = \frac{\partial F}{\partial x} = \frac{1}{2h} \left[(F_{ne} + F_{se}) - (F_{nw} + F_{sw}) \right] \qquad (14.5)$$

Here

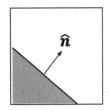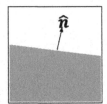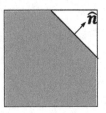

Fig. 14.5 Four out of sixteen interface orientations possible in a square computational cell

$$F_{ne} = \frac{1}{4}\left(F_{i,j+1} + F_{i+1,j+1} + F_{i,j} + F_{i+1,j}\right),$$

$$F_{nw} = \frac{1}{4}\left(F_{i-1,j} + F_{i,j} + F_{i,j+1} + F_{i-1,j+1}\right),$$

$$F_{se} = \frac{1}{4}\left(F_{i,j} + F_{i+1,j} + F_{i,j-1} + F_{i+1,j-1}\right), \quad \text{and}$$

$$F_{sw} = \frac{1}{4}\left(F_{i-1,j-1} + F_{i,j-1} + F_{i,j} + F_{i-1,j}\right)$$

Equation 14.5 can be written as

$$n_x = \frac{\partial F}{\partial x} = \frac{1}{8h}\left(F_{i+1,j+1} + 2F_{i+1,j} + F_{i+1,j-1} - F_{i-1,j+1} - 2F_{i-1,j} - F_{i-1,j-1}\right)$$
(14.6)

Similarly

$$n_y = \frac{\partial F}{\partial y} = \frac{1}{8h}\left(F_{i+1,j+1} + 2F_{i,j+1} + F_{i-1,j+1} - F_{i+1,j-1} - 2F_{i,j-1} - F_{i-1,j-1}\right)$$
(14.7)

After the n_x and n_y are obtained, the intersection points of the linear interface with the faces in a cell for which the value of F is known can be calculated. There are 16 possible ways of interface orientation in a two-dimensional square cell. In Fig. 14.5, four possible orientations of the interface are shown for a square cell. For calculating the interface position in a general rectangular cells, one may refer to the analytical method proposed by Scardovelli and Zaleski (2000).

The above procedure results in discontinuous piecewise linear interfaces as presented in Fig. 14.4. This will lead to inaccuracies in the calculations of the normal and the curvature of the interface. Many methods were proposed to improve the accuracy of interface reconstruction and a few of them are listed below:

1. *"Least squares VOF interface reconstruction algorithm* (LVIRA)*"* proposed by Puckett et al. (1997)
2. *"Parabolic Reconstruction Of Surface Tension* (PROST)*"* proposed by Renardy and Renardy (2002)
3. *"Spline-based interface reconstruction* (SIR)*"* proposed by López et al. (2004)

4. *"Quadratic Spline based Interface (QUASI) reconstruction"* proposed by Diwakar et al. (2009).

Advection of Volume Fraction

The volume fraction equation of kth fluid, F_k, can be written as

$$\frac{\partial F_k}{\partial t} + \nabla.(\mathbf{u} F_k) = 0 \tag{14.8}$$

Integrating this equation over a control volume, V, gives

$$\int_V \frac{\partial F_k}{\partial t} dV = - \int_V \nabla \bullet (\mathbf{u} F_k)\, \mathrm{dV} \tag{14.9}$$

The above equation can be written as follows after using the divergence theorem on the right-side term:

$$\int_V \frac{\partial F_k}{\partial t} dV = - \int_S (\mathbf{u} F_k)_f \bullet \hat{\mathbf{n}} \mathrm{dS} \tag{14.10}$$

Here, f represents a cell face. Using the volume of kth fluid,

$$V_k = \int F_k dV = F_k V,$$

and the total volume of the cell, $V = \sum_k V_k$, the above equation is written as

$$V_k^{n+1} = V_k^n - \delta t \int_S (\mathbf{u} F_k)_f \bullet \hat{\mathbf{n}} \mathrm{dS} \tag{14.11}$$

By dividing the above equation by V, the expression for F_k is written as

$$F_k^{n+1} = F_k^n - \frac{\delta t}{V} \int_S (\mathbf{u} F_k)_f \bullet \hat{\mathbf{n}} \mathrm{dS} \tag{14.12}$$

The second term in the right side of the above equation is calculated geometrically. Many geometric advection methods are employed to solve Eq. 14.12. Figure 14.6 shows a simple unsplit advection method. The volume fraction, F obtained in a cell with this scheme may be greater than one (overshoot) or less than zero (undershoot) as the fluxing of a small volume occurs twice from a cell due to the overlap of two adjacent flux polygon. This leads to inaccuracy in the interface capturing.

To avoid fluxing the same volume twice, the following two methods are used: (a) Operator split advection (b) Unsplit advection (with non-overlapping flux polygon). These two methods are described here.

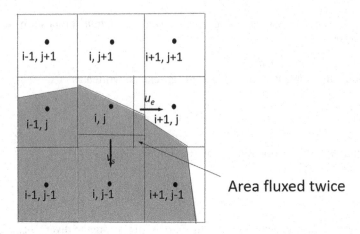

Fig. 14.6 Simple unsplit advection with the problem of double fluxing

(a) Operator Split Algorithm

Consider the following equation for the VOF advection instead of the form given in Eq. 14.8:

$$\frac{\partial F}{\partial t} + \mathbf{u} \bullet \nabla F = F \nabla . \mathbf{u} \tag{14.13}$$

In the operator split method, the above equation is decomposed into the following equations:

$$\frac{F^* - F^n}{\delta t} + \frac{\partial(u F^n)}{\partial x} = F^* \frac{\partial u}{\partial x} \tag{14.14}$$

$$\frac{F^{n+1} - F^*}{\delta t} + \frac{\partial(v F^n)}{\partial y} = F^* \frac{\partial v}{\partial y} \tag{14.15}$$

Integrating Eqs. 14.14 and 14.15 yield

$$F_{i,j}^* dx dy = F_{i,j}^n dx dy + \left(u \delta t F^n dy\right)_w - \left(u \delta t F^n dy\right)_e$$
$$- F_{i,j}^* \left[(u \delta t dy)_w - (u \delta t dy)_e\right] \tag{14.16}$$

$$F_{i,j}^{n+1} dx dy = F_{i,j}^* dx dy + \left(v \delta t F^* dx\right)_s - \left(v \delta t F^* dx\right)_n$$
$$- F_{i,j}^* \left[(v \delta t dx)_x - (v \delta t dx)_n\right] \tag{14.17}$$

In the first step, Eq. 14.16 is solved to get the intermediate value of volume fraction, $F_{i,j}^*$ of all the cells in the computational domain by an x-direction sweep. Using $F_{i,j}^*$, in the second step, the interface is reconstructed and $F_{i,j}^{n+1}$ is obtained throughout the

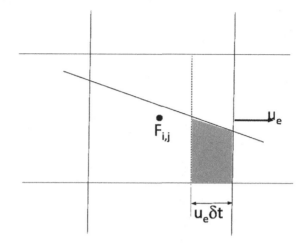

Fig. 14.7 Advection of volume fraction at a face of an interfacial cell

computational domain by sweeping in the y-direction. With this step, the interface reconstruction and the advection step are completed for a particular time step.

In Eqs. 14.16 and 14.17, the face values of F are unknown. These values are calculated geometrically. Let $\left(\delta V_{F,\,e}\right)$ and (δV_e) be the liquid volume and the total volume advected through the east face of the cell. The liquid volume, $\delta V_{F,\,e}$, is the area of the shaded region in Fig. 14.7.

The total volume, δV_e advected at face e is calculated as

$$\delta V_e = |u_e| \delta t dy \tag{14.18}$$

The area of the liquid region, $\delta V_{F,\,e}$ and the total volume, δV_e are calculated geometrically. Based on these values, the face value F_e is calculated as

$$F_e = \frac{\left(\delta V_{F,\,e}\right)}{(\delta V_e)} \tag{14.19}$$

Similarly, the values of F_w, F_n and F_s are also calculated geometrically.

In this method, the interface reconstruction and advection steps are performed twice for 2D computational domain and thrice for 3D computational domain. To avoid repetitive interface reconstruction and advection steps carried out in operator split method, various unsplit advection methods have been developed.

(b) Unsplit Advection Algorithm

The operator split advection method requires reconstruction and advection twice for 2D cases and thrice for 3D cases due to which it is computationally expensive. On the other hand, in the simple unsplit advection method (see Fig. 14.6), some volume of fluid is fluxed twice, which leads to inaccurate results. Hence, an improved unsplit advection algorithm was introduced by Rider and Kothe (1998). In this algorithm, a trapezoidal flux polygon at each edge of the cell is constructed with the width of

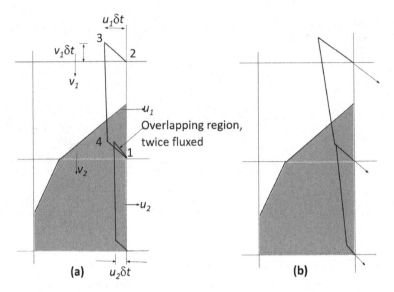

Fig. 14.8 Unsplit advection algorithms **a** method of Rider and Kothe (1998) and **b** method of López et al. (2004)

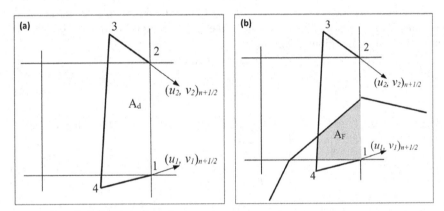

Fig. 14.9 The flux polygon proposed by López et al. (2004)

$u\delta t$ and the angles of the trapezoid are calculated using the cell face velocities as depicted in Fig. 14.8a. However, this method does not completely eliminate twice fluxing of the same fluid.

To completely eliminate fluxing the same volume of fluid twice, the *"Edge Matched Flux Polygon Advection (EMFPA)"* algorithm (Fig. 14.8) was proposed for 2D problems by López et al. (2004). The same approach was extended to 3D problems as *"Face Matched Flux Polyhedron Advection (FMFPA)"* by Hernández et al. (2008).

Consider the face with the vertices 1 and 2 of a 2D cell presented in Fig. 14.9a. The flux polygon is constructed such that its volume is given as

$$A_d = \frac{1}{2} \left(u_e^n + u_e^{n+1} \right) \Delta y \delta t,$$

where u_e is the face velocity of edge 1–2. The slopes of the edges other than the cell face 1–2 are obtained using the velocities at the cell vertices. The velocities at the cell vertices are calculated using the face velocities of shared faces. The velocities at the vertices of the cell at time, $t^{n+\frac{1}{2}}$ are obtained through the linear interpolation of the velocities obtained at t^n and t^{n+1}. Using these velocity values, the slopes of a flux polygon are calculated.

$$\left. \frac{dx}{dy} \right|_{2-3} = \frac{u_2^{n+1/2}}{v_2^{n+1/2}} \tag{14.20}$$

$$\left. \frac{dx}{dy} \right|_{3-4} \frac{(x_1 - x_2) - \left(u_1^{n+1/2} - u_2^{n+1/2} \right) \delta t}{(y_1 - y_2) - \left(v_1^{n+1/2} - v_2^{n+1/2} \right) \delta t} \tag{14.21}$$

$$\left. \frac{dx}{dy} \right|_{4-1} = \frac{u_1^{n+1/2}}{v_1^{n+1/2}} \tag{14.22}$$

The edges 2–3 and 4–1 match with the adjacent face flux polygon as shown in Fig. 14.8b. Hence, the problem of fluxing of a fluid twice is avoided. For the advection of volume fraction, the shaded area A_F, shown in Fig. 14.9b, has to be advected through the face 1–2. It is determined from the intersection of the flux polygon. Similarly, advection through the other faces of a cell can be calculated. For a quadrilateral cell with area A_Ω, the F value at t^{n+1} is then calculated as

$$F^{n+1} = F^n - \frac{1}{A_\Omega} \left(A_{F_T} \right) \tag{14.23}$$

where,

$$A_{F_T} = \sum_{i=1}^{4} A_{F_i}$$

Instead of using the face velocity for the construct a flux polygon, Jofre et al. (2014) used the velocities of the vertices of a face to construct a flux polygon in their unsplit advection algorithm. When the velocities at the vertices of a face is used, the flux polygon may be convex or concave. For calculating the volume of flux polygon, they subdivided the flux polygon into triangle or tetrahedral based on whether the flux polygon is 2D or 3D. This would increase the complexity in the calculation of the volume of truncated flux polygon. However, the method avoids any iterative step involved in the construction of flux polygon.

14.2.2 A Quadratic Spline-Based Interface (QUASI) Reconstruction Algorithm

Out of the various multiphase flow modelling techniques, the VOF method has been widely adopted owing to its excellent volume conservation characteristics within the purview of fixed grid techniques. Here, a discrete volume fraction distribution, F, is utilized to represent the different phases in the system. In a two-phase scenario, F is given as the ratio between the primary fluid volume and the total volume of the computational cell. The F-value, thus, typically varies between 0 and 1. Since F is discontinuous in nature, the fluid volumes are often tracked through geometric procedures involving interface reconstruction and fluid advection. In the interface reconstruction procedure, the interface's approximate position and orientation are obtained within the cell subjected to the local volume constraint. Subsequently, in the advection process, one geometrically estimates the volumetric flux of the primary fluid through all the cell faces.

The primary issue in VOF method arises from the non-uniqueness of the interface reconstruction procedure. In order to overcome this, a Quadratic Spline-based Interface (QUASI) reconstruction algorithm was developed by Diwakar et al. (2009), wherein piecewise quadratic curves represent the fluid interface. A sequence of correction operations is performed to convert the piecewise curves into a continuous quadratic spline (Fig. 14.10) across the domain. The procedure begins with the PLIC representation of the interface. The endpoints of these discontinuous PLIC interfaces are then arbitrarily connected to enforce the function (C^0) continuity at the cell boundaries. Such a step is feasible here on account of the three-point quadratic

Fig. 14.10 **a** Original shape of the sector of a circle; **b** Initial interface approximation using PLIC reconstruction; **c** After implementing C^0 continuity (d) After Implementing C^1 correction

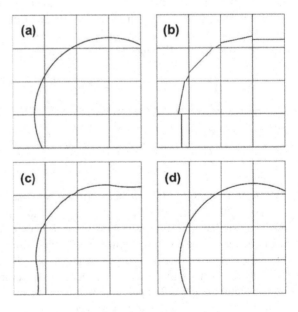

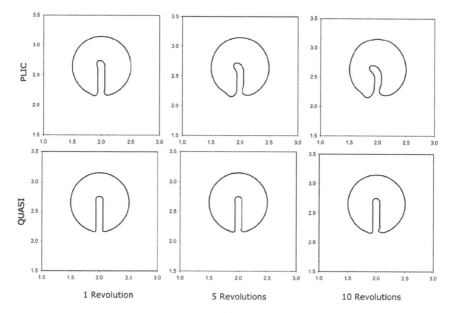

Fig. 14.11 Rotation of Zalesak slotted circle

representation of the interface in a cell that allows for the adjustment of the central point so that the volume constraint is satisfied. In the next step, the interfaces in the neighbouring cells are moved together on their common boundary to enforce the continuity of the derivatives (C^1) throughout. As evident from Fig. 14.10, enforcing C^0 and C^1 continuities yields a unique and accurate interfacial representation and also provides a direct measure of local interfacial curvature. Note that the seemingly complicated tasks of enforcing continuities have been simplified by utilizing simple analytical expressions that greatly minimize the associated computational effort. Here, the fluid advection process was effectuated using the EMFPA algorithm of López et al. (2004).

The efficacy of the QUASI reconstruction, along with the EMFPA algorithm, is now demonstrated using the slotted circle benchmark problem of Zalesak (1979). Here, a slotted circle of the primary fluid revolves in a uniform vorticity field. A comparison of errors obtained for various reconstruction and advection techniques (Diwakar et al. 2009) reveals that the present combination of QUASI reconstruction and EMFPA advection procedures has the lowest magnitude of error. Figure 14.11 shows the interface profile after 1/5/10 complete revolutions of the slotted circle for Youngs' (Youngs 1982) PLIC-EMFPA and the present QUASI-EMFPA combination. It is evident that the slotted disc configuration obtained using the PLIC procedure progressively undergoes a severe shape distortion. In contrast, the QUASI reconstruction procedure yields a fluid configuration identical to the original slotted disc shape, even after ten revolutions. Here, the sharp corners of the slot are smoothed as the quadratic curve cannot sufficiently represent the corner singularity. Neverthe-

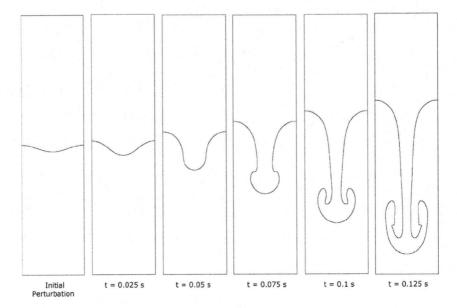

Initial   t = 0.025 s   t = 0.05 s   t = 0.075 s   t = 0.1 s   t = 0.125 s
Perturbation

Fig. 14.12 Evolution of Rayleigh–Taylor instability

less, one may observe the absence of shape distortion with the increasing number of revolutions after the slot gets transiently smoothed in the first few revolutions.

It may be noted that a proper discretization of pressure gradients and surface tension force in the computational domain is essential for harnessing the full benefits of the QUASI procedure in realistic flow problems. In this regard, the strategy of Renardy (2002) has been extended presently to a semi-staggered grid-based finite volume scheme. This extension uses two pressure components (P_1 and P_2) to obtain the overall conservation. First, the capillary pressure component, P_1, is exactly balanced with the singular surface tension force, FST, by forcing ($F_{ST} - \nabla P_1$) to be divergence free in the domain. Following the conventional flow solver algorithms, the other pressure component, P_2, accounts for the mass conservation in the domain and is evolved using the typical pressure Poisson equation.

The above multiphase solver's ability to accurately track interfacial instabilities' evolution is evident in Fig. 14.12, which shows the growth of perturbation in unstably stratified fluid layers. The fluid properties and the other computational parameters used here are the same as those used by Puckett et al. (1997). The time-wise interface pattern shown in Fig. 14.12 represents the actual continuous interface obtained using the QUASI reconstruction, and they agree well with the published results of Puckett et al. (1997).

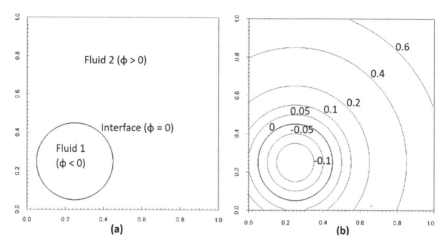

Fig. 14.13 Level set initialization **a** a two-phase region and **b** contours of ϕ

14.2.3 Level Set Method

Another popular interface capturing method is the level set method which was introduced by Osher and Sethian (1988). This method is used in a wide range of applications in engineering. In this method, the level set function, ϕ varies in the domain of interest as a signed distance function from an interface. At the interface, the level set is assigned as zero. The level set function is negative in fluid 1 and positive in fluid 2 as shown in Fig. 14.13. To capture the interface movement, the following advection equation is used:

$$\frac{\partial \phi}{\partial t} + \mathbf{u} \bullet \nabla \phi = 0 \tag{14.24}$$

In this method, ϕ is advected passively using the known velocity field. In the computational domain, ϕ varies smoothly and the interface ($\phi = 0$) is also continuous and smooth. Hence, the interface normal and curvature calculated from the LS method are accurate compared to those of the VOF method. The level set method was extended for the two-phase flows by Sussman et al. (1994). In their implementation, the 2nd order accurate ENO scheme was used for advection terms. The transient term was discretized using Adams–Bashforth method which is of 2nd order accurate. After few times steps, as the LS function, ϕ did not represent as a distance function, they reinitialised ϕ using the following equation:

$$\frac{\partial \phi}{\partial t} = S(\phi_o)\left(1 - |\nabla \phi|\right) \tag{14.25}$$

Here, $S(\phi_o)$ represents a smoothed sign function and ϕ_o is obtained from Eq. 14.24. $S(\phi_o)$ is calculated as $\frac{\phi_o}{\sqrt{\varphi_o^2 + \epsilon^2}}$ where ϵ is usually taken as a grid spacing. The above

Fig. 14.14 The contours φ
for the case of a rising
bubble **a** without
reinitialisation of φ and **b**
with reinitialisation of φ

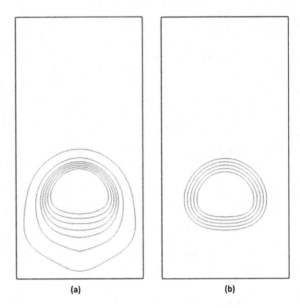

(a)   (b)

equation is solved till the steady state condition is achieved. Figure 14.14 presents
the contours of φ obtained without and with redistancing for the case of falling of a
large water drop (Reynolds number, $Re = 10$) by Sussman et al. (1994).

Even though the redistancing of level set function done through Eq. 14.25 ensures
that φ represent as the distance from the interface ($\varphi = 0$), the interface position
may change during redistancing. Hence, the conservation of mass is not ensured.
To ensure mass conservation, a global mass correction method was proposed by
Son (2001). Apart from the reinitialisation with Eq. 14.25, the following volume
correction equation was also solved:

$$\frac{\partial \varphi}{\partial t} = (V - V_0)\,|\nabla \phi| \qquad (14.26)$$

Here, V represents the bubble volume which is calculated using the level set function,
ϕ and V_0 is the bubble volume which satisfies the conservation of mass.

Mesh adaptation near the interface may reduce mass loss significantly when the
interfacial flows are modelled with the LS method. The other approach to reduce
the mass loss problem of the LS method is to use a higher order schemes for the
spatial and temporal discretisation. Nourgaliev et al. (2005) investigated the mass
loss issue by using the higher order linear WENO schemes of Balsara and Shu (2000)
in combination with mesh adaptation. Their study reveals that even the combination
of adaptive mesh refinement and the use of higher order schemes does not provide
satisfactory results for the time-reversed 2D and 3D deformation test cases. As the
discretization of convective terms with the 7th or 11th order WENO scheme and
the use of the 2nd order Runge–Kutta method for time marching did not completely

eliminate the mass loss issue, in the reinitialisation equation, Kurioka and Dowling (2009) added a volume correction term.

Instead of using local mesh refinement around the interface region using adaptive grid method, which requires complex data structure, Gada and Sharma (2011) introduced a dual-grid level set method in which the momentum equations are solved using a coarse grid and the level set equation is solved in a finer grid to improve the conservation of mass. This approach does not require complex data structure and is easy to implement compared to the adaptive mesh refinement technique.

14.2.4 Coupled Level Set and Volume of Fluid (CLSVOF) Method

In the case of modelling of interfacial flows with the VOF method, the mass conservation is implicitly satisfied to a very good accuracy. However, the interface curvature obtained with this method is not accurate due to the discrete distribution of F. In the LS method, the interface curvature is accurately calculated because of the smooth variation of ϕ. However, the mass conservation error is significant if adaptive mesh coupled with higher order schemes are not used. Taking the positive aspects of both the methods, Bourlioux (1995) introduced a CLSVOF method. Later, another version of the CLSVOF was proposed by Sussman and Puckett (2000). This approach is widely adapted nowadays. The advantages of the VOF and LS methods are exploited for the development of the CLSVOF method to get an accurate curvature estimation along with good mass conservation.

In the CLSVOF method, the ϕ and F are advected. The geometric approach is widely used for the advection of the F in the CLSVOF method, although the algebraic method was also used by Albadawi et al. (2013) and Ferrari et al. (2017). Here, we will consider the widely used geometric approach for the advection of F. As discussed earlier, both the operator split and the unsplit advection methods are used for the advection of F. Tomar et al. (2005) used the operator split advection method for advecting F in their CLSVOF method for the numerical modelling of horizontal film boiling on structured grids. For the same film boiling case, Ningegowda and Premachandran (2014) used the method of unsplit advection of F for their CLSVOF method with non-uniform grids. As the CLSVOF method with unstructured grid supports simulations of two-phase flows in complex geometries, efforts have also been directed towards the development of the CLSVOF approach with unstructured grids (Balcázar et al. (2016) and Singh and Premachandran (2018)).

At time $t = 0$, the ϕ and F are initalised throughout the computational domain. After one time step, the new ϕ is obtained. Using the new ϕ field, interface normal and its curvature are calculated for each cell. The interface normal obtained from the level set function is used for the reconstruction of the interface for the new F obtained in $n+1$ time step in the VOF method. As discussed in the previous subsection, ϕ do not represent as the signed normal distance from the interface after the advection

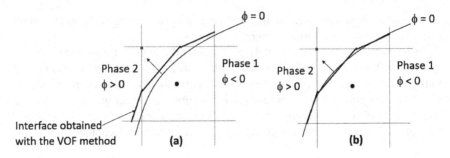

Fig. 14.15 Level set redistancing **a** before redistancing **b** after redistancing

Fig. 14.16 Calculation of shortest distance

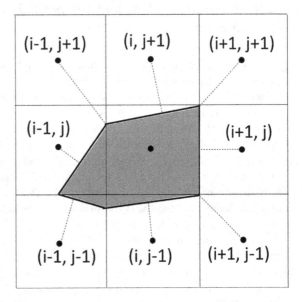

step. Due to this reason, the zero level set contour does not align with the interface obtained using the VOF function as shown in Fig. 14.15a. Using the position of the piecewise linear interface obtained with the VOF method, ϕ is reset to represent the shortest normal distance from the interface. After this geometric redistancing, the contour of zero level set aligns with the interface obtained by the VOF method as shown in Fig. 14.15b.

Figure 14.16 shows the shortest distance from various interface segments to the surrounding cell centres with a 3×3 cell stencil. If the interface and a cell face are aligned, for example, interface at $(i+1/2, j)$ as shown in Fig. 14.16, the distance between the centre of the cell face $(i+1/2, j+1/2)$ to the centre of the cell $(i+1, j)$ is considered as the shortest distance. For more details on obtaining the shortest distance between the interface and the cell centres for uniform structured grids, one may refer to Son and Hur (2002).

14.3 Modelling of Interfacial Flows with Surface Tension

For two-phase flow modelling with moving interfaces, surface tension force has to be incorporated in the momentum equations for the interfacial cells. In this section, the calculation of thermophysical properties for the interfacial cells and the implementation of widely used surface tension models for modelling two-phase flows are discussed.

14.3.1 Calculation of Thermophysical Properties for the Interface Cells

In the VOF method, any thermophysical property, ϕ, for an interface cell is obtained as follows:

$$\phi = F\phi_g + (1 - F)\phi_l \tag{14.27}$$

Here, g denotes gas or vapour and l denotes liquid. As shown in Figs. 14.13 and 14.14, when the LS method is used, the variation of ϕ in the flow field is smooth in two-phase flow modelling. For the numerical stability, to calculate thermophysical properties in the two-phase flow modelling with the LS method, a smoothed Heaviside function similar to the one given below is employed instead of the sharp Heaviside function.

$$H = \begin{cases} 0, & \text{if } \phi < -\epsilon \\ \frac{\phi+\epsilon}{2\epsilon} + \frac{1}{2\pi}\sin\left(\frac{\pi\phi}{\epsilon}\right), & \text{if } |\phi| \le \epsilon \\ 1, & \text{if } |\phi| \le \epsilon \end{cases} \tag{14.28}$$

In the above expression, 2ϵ denotes the width of the interface region and in general this is taken to be $3\Delta x$. In the interfacial region of 2ϵ width, the density (ρ) and viscosity (μ) vary smoothly and these properties are calculated as follows:

$$\rho = \rho_l H(\phi) + (1 - H(\phi))\rho_g, \tag{14.29}$$

and,

$$\mu = \mu_l H(\phi) + (1 - H(\phi))\mu_g \tag{14.30}$$

14.3.2 Modelling of Surface Tension Force

In the VOF, LS and CLSVOF methods, the surface tension force, $\mathbf{f}_{sa}$ is transformed as a body force in the momentum equations. Among many models proposed for the surface tension calculation, the model proposed by Brackbill et al. (1992) is widely used for two-phase flow simulations. In their *"Continuum Surface Force*

(CSF) model", $\mathbf{f}_{sa}$ is written as the interfacial volume force, $\mathbf{f}_{sv}$ using a delta function, δ as follows:

$$\mathbf{f}_{sv} = \mathbf{f}_{sa}\delta = (\sigma\kappa\widehat{\mathbf{n}} + \nabla_s\sigma)\,\delta \qquad (14.31)$$

where, κ is the curvature of the interface, σ is the coefficient of surface tension and ∇_s is the surface gradient.

The coefficient of surface tension may vary with temperature or concentration. If the coefficient of surface tension is considered to be a constant, for simplicity, the second term of the above equation is zero. Hence, the surface tension force, Eq. 14.31 is simplified as

$$\mathbf{f}_{sv} = \mathbf{f}_{sa}\delta = \sigma\kappa\widehat{\mathbf{n}}\delta \qquad (14.32)$$

In the above expression, the only unknown is the interface curvature, κ.

In the framework of VOF methods, large inaccuracies in the calculation of curvature, κ are observed if the volume fraction, F is directly used with the standard finite difference method. Hence, Brackbill et al. (1992) suggested that the VOF function, F is smoothed in a finite thick region around the interface and the mollified, $\widetilde{F}$ function is used for the calculation of curvature. The interface normal is calculated with the mollified volume fraction, $\widetilde{F}$ using the central difference method as $\mathbf{n} = \nabla\widetilde{F}$. The curvature is then calculated as

$$\kappa = -\nabla.\widehat{\mathbf{n}} \qquad (14.33)$$

Here, the unit vector, $\widehat{\mathbf{n}}$ is calculated as $\frac{\mathbf{n}}{|\mathbf{n}|}$.

The finite difference method with the smoothened F value, $\widetilde{F}$ results in high spurious currents in the interface region. These spurious currents affect the interface adversely, which can be severe in certain cases, leading to a reduction in the solution accuracy. The calculation of curvature using the height function is more accurate and the spurious currents generation is less. Hence, only the height function method is provided here in detail for the curvature calculation for the VOF based two-phase flow solver. For the LS and the CLSVOF methods, the interface curvature, κ is calculated using the level set function.

Curvature Calculation Using Height Function for VOF Method
In the two-phase flow modelling, the interface can be specified with a height function of one (in 2D) or two coordinates (in 3D). Figure 14.17 shows the representation of curvature with the height function, $h(x)$ at time t.

Here, for the sake of simplicity, only 2D geometry has been considered. When a uniform grid is used, the height function is usually calculated with a stencil of 7×3 or 3×7 cells around a cell having interface. If $n_x > n_y$, the stencil is selected as shown in Fig. 14.18.

For this case, the height function is calculated using the volume fraction values as follows:

$$h_i = \sum_{k=j-3}^{k=j+3} F_{i,k}\Delta y_k \qquad (14.34)$$

Fig. 14.17 Representation of an interface with height function in 2D

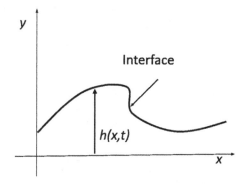

Fig. 14.18 Height function calculation using a 3 × 7 stencil

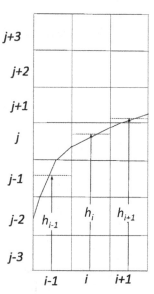

Using height function, the interface curvature, κ is calculated in 2D as

$$\kappa = \frac{h_{xx}}{\left(1 + h_x^2\right)^{3/2}} \tag{14.35}$$

For the uniform grid, the first order derivative, A_x and the second order derivative, h_{xx}, of the height function are calculated using the central differencing scheme as

$$h_x = \frac{h_{i+1} - h_{i-1}}{2\Delta x} \quad h_{xx} = \frac{h_{i+1} - 2h_i + h_{i-1}}{(\Delta x)^2} \tag{14.36}$$

The selection of a 3 × 7 or 7 × 3 stencil does not always give accurate results. To obtain accurate results, the volume of fluid fraction values should monotonically

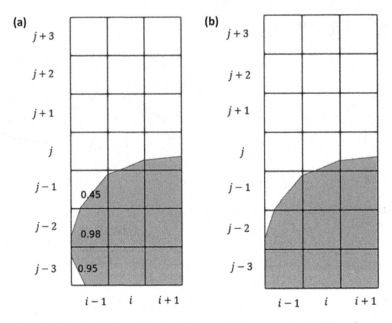

Fig. 14.19 Modification of F_{ij} or the size of the stencil for accurate calculation of curvature **a** initial F distribution **b** modified F distribution

vary along the direction of height function calculation in all 3 rows or 3 columns depending on the stencil. The case shown in Fig. 14.19a does not satisfy this condition for the calculation of κ at cell (i, j) due to a non-monotonic variation of value in j direction at $i + 1$. In order to obtain accurate results, Malik et al. (2007) suggested a modification of F value as shown in Fig. 14.19b.

Popinet (2009) proposed various stencils to improve the curvature calculation using height function. If the curvature and a cell size are of the same order, the height function method cannot be used for the calculation of curvature, and other methods such as parabolic fitting for 2D and paraboloid for 3D may be used as suggested by Popinet (2009).

Curvature Calculation for the LS and CLSVOF Methods
As the level set function smoothly varies without a jump in the values, for both the LS and CLSVOF methods, the curvature, κ is calculated for 2D problems as

$$\kappa = -\nabla \bullet \widehat{\mathbf{n}} = \nabla \bullet \frac{\nabla \phi}{|\nabla \phi|} = -\frac{\phi_y^2 \phi_{xx} - 2\phi_x \phi_y \phi_{xy} + \phi_x^2 \phi_{yy}}{\left(\phi_x^2 + \phi_y^2\right)^{3/2}} \qquad (14.37)$$

The above equation is discretized using a central difference scheme. For the sake of brevity, details of the performance evaluation for various surface tension models are not provided here. One may refer to Gerlach et al. (2006) and Abadie et al. (2015) for further details.

14.4 Modelling of Boiling Flow Problems

For modelling boiling flows with accurate capturing of interface evolution, the governing equations should be modified for the interface cells to account interface mass transfer, surface tension force and interface heat transfer. In the following subsection, these modifications along with the related calculations are discussed.

14.4.1 Heat and Mass Transfer Across Moving Interface

To account the mass transfer across the interface, a source terms is added in the continuity equation. Here, the model of Welch and Wilson (2000) is presented. As shown in Fig. 14.20, heat and mass transfer occurs simultaneously across the interface in the normal direction.

Let S_i be the surface area of the interface between the liquid and vapour phases and $\boldsymbol{u}_i$ be the velocity of the interface, $\dot{m}_i$ is the mass transfer rate across the interface and $\hat{\boldsymbol{n}}$ is the unit normal vector of the interface points towards the liquid region.

For the vapour phase region of the interface cell, the equation of mass conservation is written as

$$\frac{d}{dt} \int_{V_g(t)} \rho \, dV + \overbrace{\int_{S_g(t)} \rho \, \boldsymbol{u} \cdot \hat{\boldsymbol{n}} dS + \int_{S_i(t)} \rho \, (\boldsymbol{u} - \boldsymbol{u}_i) \cdot \hat{\boldsymbol{n}} dS} = 0 \qquad (14.38)$$

where, V_g is the volume of the vapour phase and S_g is the surface area of the vapour phase in the interface cell, respectively. The continuity equation of the liquid phase region of the interface cell is written as

Fig. 14.20 Interfacial heat and mass transfer

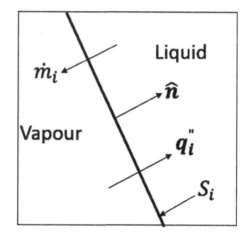

$$\frac{d}{dt}\int_{V_l(t)}\overset{\rho u\cdot\widehat{n}dS-\int_{S_l(t)}\frac{\rho}{}(u-u_i)\cdot\widehat{n}dS}{\rho\,dV+\int_{S_l(t)}} = 0 \qquad (14.39)$$

where, V_l is the volume of the liquid phase and S_l is the surface area of the liquid phase in the interface cell. The mass conservation equation for the interface cell is obtained by combining Eqs. 14.38 and 14.39. Summing up Eqs. 14.38 and 14.39 and considering incompressibility, the modified mass conservation equation for the cell with the interface is obtained as

$$\int_{S_c}\mathbf{u}\cdot\widehat{\mathbf{n}}\,dS+\int_{S_i(t)}\|(\mathbf{u}-\mathbf{u_i})\|\cdot\widehat{\mathbf{n}}\,dS = 0 \qquad (14.40)$$

In the above equation, the symbol $\|\;\|$ represents the jump in a quantity across the interface, and S_c is the surface area of the cell. The mass and energy jump conditions across the interface are written as

$$\|\rho\,(\mathbf{u}-\mathbf{u_i})\|\cdot\widehat{\mathbf{n}} = 0 \qquad (14.41)$$

$$\|\rho\,h(\mathbf{u}-\mathbf{u_i})\|\cdot\widehat{\mathbf{n}} = -\,\|\mathbf{q_i}''\|\cdot\widehat{\mathbf{n}} \qquad (14.42)$$

In the above equation, h represents the enthalpy and $\|\mathbf{q_i}''\|$ represents the jump in the interfacial heat flux $\mathbf{q_i}''$.

Using the interface jump conditions, the jump term, $\|(\mathbf{u}-\mathbf{u_i})\|\cdot\widehat{\mathbf{n}}$ in the modified mass conservation equation (Eq. 14.40) is written as

$$\|(\mathbf{u}-\mathbf{u_i})\|\cdot\widehat{\mathbf{n}} = \left(\frac{1}{\rho_l}-\frac{1}{\rho_g}\right)\frac{\|\mathbf{q_i}''\|\cdot\widehat{\mathbf{n}}}{h_{lg}} \qquad (14.43)$$

where h_{lg} is the latent heat of vaporization. Using Eq. 14.43, the modified continuity equation is expressed as

$$\int_{S_c}\mathbf{u}\cdot\widehat{\mathbf{n}}dS+\int_{S_l(t)}\left(\frac{1}{\rho_l}-\frac{1}{\rho_g}\right)\frac{\|\mathbf{q_i}''\|\cdot\widehat{\mathbf{n}}}{h_{lg}}\,dS = 0 \qquad (14.44)$$

14.4.2 The VOF Advection Equation

To account the interface mass transfer in boiling flow problems, the VOF advection equation has to be modified. The modified VOF advection equation for the boiling flows is written as

$$\frac{\partial F}{\partial t}+\mathbf{u}\bullet\nabla F = F_{\mathrm{mt}} \qquad (14.45)$$

where, F_{mt} is computed as $F_{mt} = u_i \left[\frac{S_i}{V_\Omega} \right]$. Here, u_i is the interface velocity due to mass transfer and V_Ω is the cell velocity.

14.4.3 The Level Set Advection Equation

The level set equation has to be modified as follows to account interfacial mass transfer

$$\frac{\partial \phi}{\partial t} + (\mathbf{u_{int}} \cdot \nabla)\phi = 0 \qquad (14.46)$$

where, $\mathbf{u_{int}} = \mathbf{u_{cell}} + \mathbf{u_i}$ is the sum of the cell velocity of fluid and the velocity of interface due to mass transfer. The 5th order accurate WENO scheme is used to discretize the convective terms of Eq. 14.46.

14.4.4 Momentum and Energy Equations

For two-phase flows including boiling, the momentum conservation equation is written as

$$\rho \left(\frac{\partial \mathbf{u}}{\partial t} + \mathbf{u} \bullet \nabla \mathbf{u} \right) = -\nabla p + \nabla \bullet \mu \left[\nabla \mathbf{u} + (\nabla \mathbf{u})^T \right] + \rho \mathbf{g} + F_{st} \qquad (14.47)$$

where ρ is density and μ is viscosity of fluid. In the above equation, the superscript T indicates the transpose operation of the diffusion terms. F_{st} is the interfacial surface tension force, calculated as $\sigma \kappa \hat{\mathbf{n}} \delta$.

In the case of saturated film boiling, the liquid is maintained at saturation temperature. Hence, only the vapour region is solved to find out the temperature distribution. To obtain the temperature distribution in the vapour region, the following energy equation is solved:

$$\rho C_p \left(\frac{\partial T}{\partial t} + \mathbf{u} \bullet \nabla T \right) = \nabla \bullet (k \nabla T) \qquad (14.48)$$

where C_p is the specific heat at constant pressure and k is the thermal conductivity. For the interface cells, the thermos-physical properties are specified as discussed in Sect. 14.5.1.

At each time step, the interfacial temperature is obtained through extrapolation of temperature in the normal direction to the interface from the neighbouring superheated vapour cells.

14.5   Numerical Simulation of Film and Nucleate Boiling

14.5.1   Saturated Film Boiling over a Horizontal Flat Surface

In the film boiling regime, the heater surface is completely covered by vapour and the movement of liquid surrounding to the vapour region occurs due to the release of vapour from the vapour layer. The modelling of film boiling is easier compared to that of nucleate boiling. Due to the simple flow features of film boiling and no complexities such as microlayer and moving contact lines, film boiling regime is usually considered for the validation of boiling flow models.

Even though film boiling flow over a flat surface positioned horizontally is three-dimensional, Son and Dhir (1997) modelled the film boiling of water at near critical condition using an axisymmetric model using the level set method. Welch and Wilson (2000) used a 2D geometry for the modelling of film boiling at near critical conditions using the VOF method. In their study, they selected a computational domain in such a way that only one vapour bubble releases periodically. Subsequently, many researchers considered the 2D geometry used by Welch and Wilson (2000) for validating their numerical models. Following this model, Ningegowda and Premachandran (2014) simulated film boiling of water at $T_{\text{sat}} = 646.15$ K and $P_{sat} = 21.9$ MPa to validate their numerical model. The scaling parameters associated with the film boiling are the characteristic length, $\lambda = \sqrt{\frac{\sigma}{g(\rho_l - \rho_g)}}$, characteristic time, $t_o = \sqrt{\lambda/g}$ and the characteristic velocity, $u_o = \frac{\lambda}{t_o} = \frac{\lambda}{\sqrt{\lambda g}}$.

The most dangerous Taylor hydrodynamic instability for the 2D film boiling is

$$\lambda_D = 2\pi \sqrt{\frac{3\sigma}{g\left(\rho_l - \rho_g\right)}} \tag{14.49}$$

The computational domain of size $\lambda_D \times \lambda_D$ is used for the modelling film boiling as the bubbles are expected to depart alternately from locations spaced at a distance $\lambda_D/2$.

The periodic interface evolution obtained in the case of saturated film boiling of water at $T_{sat} = 646.15$ K and $P_{sat} = 21.9$ MPa for ΔT_{sup} of 5 K by Ningegowda and Premachandran (2014) from their numerical model is presented in Fig. 14.21a. For this case, the numerical simulation was carried out with 300^2 cells and the CFL number of 0.05. Figure 14.21b shows that the space averaged Nusselt number obtained from the numerical study compares well with the average Nusselt number obtained from the correlations of Berenson (1961) and Klimenko (1981).

Esmaeeli and Tryggvason (2004b) investigated film boiling of water over a flat surface at 169 bar for different wall superheats using the numerical model presented in Esmaeeli and Tryggvason (2004a). In their study, a large heated flat surface was considered for the parametric study to understand the effect of wall superheat and the Grashof number. Most of the simulations were carried out with a 2D geometry as the simulations with the 3D geometry were computationally expensive. Figure 14.22

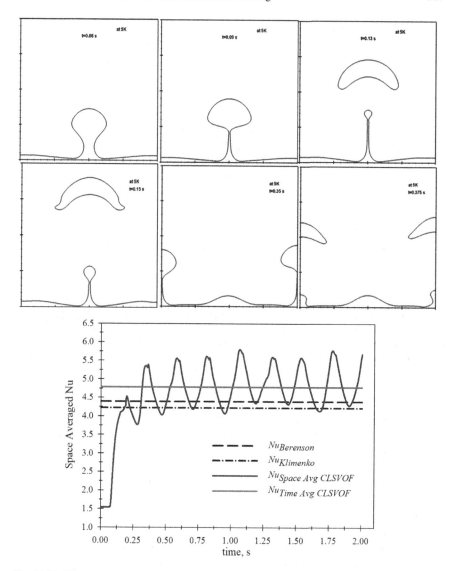

Fig. 14.21 Film boiling of water over a horizontal flat surface simulation **a** periodic vapour bubble release pattern **b** comparison of Nusselt numbers obtained from the numerical study with the correlations of Berenson (1961) and Klimenko (1981) (*Courtesy:* B. M. Ningegowda, Ph.D. thesis, under the supervision of Prof. B. Premachandran, 2014, IIT Delhi)

(a)

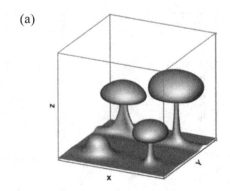

(b)

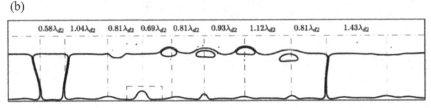

Fig. 14.22 Snapshots of film boiling of water obtained at $Ja_v = 0.064$ by Esmaeeli and Tryggvason (2004b) **a** liquid–vapour interface obtained using 3D model and **b** liquid–vapour interface obtained using the 2D model (*Courtesy:* "Computations of film boiling. Part II: multi-mode film boiling" by Asghar Esmaeeli and Grtar Tryggvason, Int. J Heat Mass Trans., with permission from Elsevier)

shows the interface between liquid water and water vapour obtained from 3D and 2D simulations for $Ja_v = 0.064$.

In order to obtain the results which are independent of the computational domain size used in the numerical simulations, the width of the computational domain was varied from λ_D to $20\lambda_D$ and the average Nusselt number was calculated after the quasi-steady state was reached. No significant effect on the average Nusselt number was observed as the width of the computational domain was increased. At lower computational width, periodic release of vapour bubble was observed. However, as the computational domain increases, the vapour release from the vapour layer become non-periodic. According to the vapour release pattern, the variation of the space averaged local Nusselt number changes from periodic to non-periodic as the computational domain size increased. The vapour Jacob number, Ja_v was varied in the parametric study to understand the effect of wall superheat on interface dynamics and heat transfer. As the vapour Jacob number increased, the distance between the bubble releasing locations in the vapour layer did not change. However, the pattern of vapour release from the vapour layer changed from vapour bubble release to vapour column formation. The average Nusselt numbers obtained for various wall superheats from the 2D simulations compare very well with the prediction of their modified form of Berenson's correlation.

Film boiling over cylinders encounters in a wide range of applications. Son and Dhir (2008a) used the level set method to investigate saturated film boiling of water

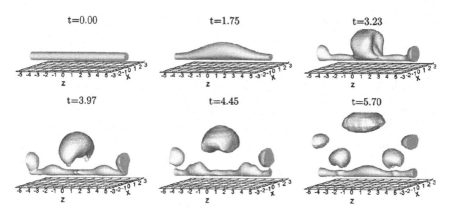

Fig. 14.23 Film boiling of water over a circular cylinder of non-dimensional diameter, $\widehat{D} = 0.5$ at 1 atm pressure. Wall superheat, $T_w - T_{sat} = 200\,°C$ (Son and Dhir (2008a)) (*Courtesy:* "Three-dimensional simulation of saturated film boiling on a horizontal cylinder" by Gihun Son and Vijay K. Dhir, Int. J Heat Mass Trans., with permission from Elsevier)

over a circular cylinder at 1 atmospheric pressure. The computations were performed to understand the effect of non-dimensional diameter of the heated circular cylinder, $\widehat{D} = D/l_o$, where D is the diameter of the circular cylinder and I_o is the reference length defined as $l_o = \sqrt{\sigma/g(\rho_l - \rho_v)}$. They carried out numerical simulations by varying the non-dimensional diameter, $\widehat{D}$ ranging from 0.05 to 5. Figure 14.23 shows the liquid and vapour interface evolution obtained over a cylinder of non-dimensional diameter, $\widehat{D} = 0.5$. The numerical results were compared with the experimental results of Bromley (1950), Breen and Westwater (1962), and Sakurai et al. (1990). The pattern of bubble departure from the vapour layer over the cylinder agrees well with that observed in the experiments of Breen and Westwater (1962). The Nusselt numbers obtained from the numerical simulations of Son and Dhir (2008a) compare well within ±10% of the values predicted from the correlation of Sakurai et al. (1990).

14.5.2 Nucleate Boiling

During the nucleate boiling, the formation of a microlayer of liquid below the vapour bubble and the formation of dynamic contact angle at the trijunction of the dry patch, micro layer and heated solid wall are observed as shown in Fig. 14.24a. The thickness of the microlayer in the nucleate boiling varies from 0.2 μm to 20 μm. Modelling both the micro-region and macro-region in boiling as shown in Fig. 14.24b is challenging computationally due to the cell size requirement in the micro-region. Instead of resolving the microscale region as in the modelling of macro-region, theory of lubrication is used to obtain heat and mass transfer and an apparent contact angle

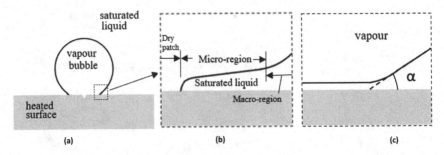

Fig. 14.24 Regions observed in nucleate boiling and the modelling approach used **a** vapour bubble during nucleate boiling **b** microscale region and **c** apparent contact line model

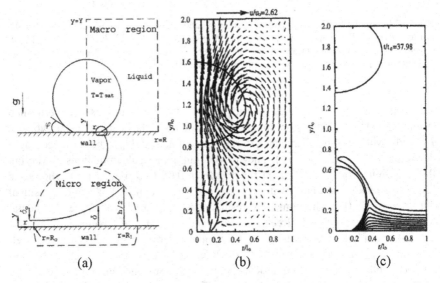

Fig. 14.25 Numerical study of nucleate boiling (Son et al. (1999)) **a** Computational domain **b** liquid–vapour interface and flow pattern **c** temperature distribution (*Courtesy:* "Dynamics and Heat Transfer Associated With a Single Bubble During Nucleate Boiling on a Horizontal Surface" by G. Son, V. K. Dhir and N. Ramanujapu, J. Heat Transfer, with permission from The American Society of Mechanical Engineers)

is used as shown in Fig. 14.24c instead of considering the true contact angle (Refer Fig. 14.24a).

Son et al. (1999) developed a numerical model for the nucleate boiling of water over an isolated nucleation site without accounting microcavity of the nucleation site to study bubble growth and departure. For interface tracking, the level set method was used. An axisymmetric computational domain was used (Fig. 14.25a) to reduce the computational time. In the microlayer region, simplified governing equations are solved considering the lubrication theory. Instead of accounting true contact angle, the apparent contact angle shown in Fig. 14.24c was used in the numerical simulations. Figure 14.25b, c show the flow pattern and the temperature contours, respectively, along with the liquid–vapour interface obtained from their study. Son et al. (2002) extended the above mentioned work to investigate the bubble growth rate, departure frequency and merger of vapour bubbles in detail. Subsequently, many numerical studies on nucleate boiling were carried out using similar simplified models for microlayers. Even though the predictions of bubble growth rate and lift of frequency obtained with the microlayer model agree well with the experimental data, more details on the heat transfer and flow characteristics in the microlayer were not obtained.

Son and Dhir (2008b) carried out a numerical study on nucleate boiling at high critical heat flux using 2D and 3D models. They used the 2D model for all the simulations excluding one 3D simulation which was carried out to compare the results of the 2D and the 3D models. For this study, the microlayer model of Son et al. (1999) was used. Sato and Niceno (2017) introduced a new microlayer model in which the initial thickness of the microlayer was assumed to increase linearly from the nucleation site. In the microlayer region, they assumed that the liquid is stagnant. Due to a very fast growth in computational capabilities recently, Sato and Niceno (2018) could carry out numerical simulations from nucleate boiling regime to film boiling regime. For the modelling of nucleate boiling, they used their microlayer model mentioned above. This work has demonstrated that all the regimes of pool boiling can be effectively investigated through direct numerical simulations. The direct numerical simulations of flow boiling has also been extensively studied using interface capturing and front tracking methods. The review article by Kharangate and Mudawar (2017) highlights the advances, current capabilities and development needs of direct numerical simulation of boiling and condensation.

14.6 Concluding Remarks

In this chapter, various methods used for the interfacial flows were highlighted. The VOF, the LS and the CLSVOF methods widely used for the modelling of two-phase flows were discussed in detail. For the modelling of interfacial flows with surface tension at the interface, the widely used continuum surface force (CSF) method was presented. Subsequently, the modifications required in the governing equations for the modelling of boiling flows were discussed. The results of direct numerical simulations of boiling flows were presented to demonstrate the capabilities of the models used for the boiling flows. The methods described in this chapter are not limited to boiling flows and can be used for modelling of condensation with minor modifications.

References

Abadie T, Aubin J, Legendre D (2015) On the combined effects of surface tension force calculation and interface advection on spurious currents within volume of fluid and level set frameworks. J Comput Phys 297:611–636

Albadawi A, Donoghue DB, Robinson AJ, Murray DB, Delauré YMC (2013) Influence of surface tension implementation in volume of fluid and coupled volume of fluid with level set methods for bubble growth and detachment. Int J Multiph Flow 53:11–28

Balcázar N, Lehmkuhl O, Jofre L, Rigola J, Oliva A (2016) A coupled volume-of-fluid/level-set method for simulation of two-phase flows on unstructured meshes. Comput Fluids 124:12–29

Balsara DS, Shu CW (2000) Monotonicity preserving weighted essentially non-oscillatory schemes with increasingly high order of accuracy. J Comput Phys 160:405–452

Berenson PJ (1961) Film-boiling heat transfer from a horizontal surface. J Heat Transf 83:351–356

Bourlioux, A (1995) A coupled level-set volume-of-fluid algorithm for tracking material interfaces. In: Proceedings of the 6th international symposium on computational fluid dynamics, Lake Tahoe, CA, vol 15

Brackbill JU, Kothe DB, Zemach C (1992) A continuum method for modeling surface tension. J Comput Phys 100:335–354

Breen BP, Westwater JW (1962) Effect of diameter of horizontal tubes on film boiling heat transfer. Chem Eng Prog 58:67–72

Bromley LA (1950) Heat transfer in stable film boiling. Chem Eng Prog 46:221–227

Darwish M, Moukalled F (2006) Convective schemes for capturing interfaces of free-surface flows on unstructured grids. Numer Heat Transf Part B: Fundam 49:19–42

Diwakar SV, Das SK, Sundararajan T (2009) A quadratic spline based interface (QUASI) reconstruction algorithm for accurate tracking of two-phase flows. J Comput Phys 228(24):9107–9130

Esmaeeli A, Tryggvason G (2004) Computations of film boiling Part I: numerical method. Int J Heat Mass Transf 47:5451–5461

Esmaeeli A, Tryggvason G (2004) Computations of film boiling part II: multi-mode film boiling,. Int J Heat Mass Transf 47:5463–5476

Ferrari A, Magnini M, Thome JR (2017) A flexible coupled level set and volume of fluid (flexCLV) method to simulate microscale two-phase flow in non-uniform and unstructured meshes. Int J Multiph Flow 91:276–295

Fischer S, Gambaryan-Roisman T, Stephan P (2015) On the development of a thin evaporating liquid film at a receding liquid/vapour-interface. Int J Heat Mass Transf 88:346–356

Gada VH, Sharma A (2011) On a novel dual-grid level-set method for two-phase flow simulation. Numer Heat Transf Part B: Fundam 59:26–57

Gerlach D, Tomar G, Biswas G, Durst F (2006) Comparison of volume-of-fluid methods for surface tension-dominant two-phase flows. Int J Heat Mass Transf 49:740–754

Hernández J, López J, Gómez P, Zanzi C, Faura F (2008) A new volume of fluid method in three dimensions–part I: multidimensional advection method with face-matched flux polyhedral. Int J Numer Meth Fluids 58:897–921

Jofre L, Lehmkuhl O, Castro J, Oliva A (2014) A 3-D Volume-of-fluid advection method based on cell-vertex velocities for unstructured meshes. Comput Fluids 94:14–29

Juric D, Tryggvason G (1998) Computations of boiling flows. Int J Multiph Flow 24:387–410

Kharangate CR, Mudawar I (2017) Review of computational studies on boiling and condensation. Int J Heat Mass Transf 108:1164–1196

Klimenko VV (1981) Film boiling on a horizontal plate-new correlation. Int J Heat Mass Transf 24:69–79

Kurioka S, Dowling DR (2009) Numerical simulation of free surface flows with the level set method using an extremely high-order accuracy WENO advection scheme. Int J Comput Fluid Dyn 23:233–243

López J, Hernández J, Gómez P, Faura F (2004) A volume of fluid method based on multidimensional advection and spline interface reconstruction. J Comput Phys 195:718–742

Malik M, Fan ESC, Bussmann M (2007) Adaptive VOF with curvature-based refinement. Int J Numer Meth Fluids 55:693–712

Muzaferija S, Peric M, Sames P, Schellin T (1999) A two-fluid Navier-Stokes solver to simulate water entry. In: Proceedings of twenty-second symposium on naval hydrodynamics. The National Academies Press, Washington, DC

Ningegowda BM, Premachandran B (2014) A coupled level set and volume of fluid method with multi-directional advection algorithms for two-phase flows with and without phase change. Int J Heat Mass Transf 79:532–550

Noh WF, Woodward P (1976) SLIC (simple line interface calculation) In: Proceedings of the fifth international conference on numerical methods in fluid dynamics June 28–July 2, 1976, Twenty University, Enschede. Springer, Berlin, Heidelberg, pp 330–340

Nourgaliev RR, Wiri S, Dinh NT, Theofanous TG (2005) On improving mass conservation of level set by reducing spatial discretization errors. Int J Multiph Flow 31:1329–1336

Osher S, Sethian JA (1988) Fronts propagating with curvature-dependent speed: algorithms based on Hamilton-Jacobi formulations. J Comput Phys 79:12–49

Popinet S (2009) An accurate adaptive solver for surface-tension-driven interfacial flows. J Comput Phys 228:5838–5866

Puckett EG, Almgren AS, Bell JB, Marcus DL, Rider WJ (1997) A high-order projection method for tracking fluid interfaces in variable density incompressible flows. J Comput Phys 130:269–282

Renardy Y, Renardy M (2002) PROST: a parabolic reconstruction of surface tension for the volume-of-fluid method. J Comput Phys 183:400–421

Rider WJ, Kothe DB (1998) Reconstructing volume tracking. J Comput Phys 141:112–152

Sakurai A, Shiotsu M, Hata K (1990) A general correlation for pool film boiling heat transfer from a horizontal cylinder to subcooled liquid. Part 2: experimental data for various liquids and its correlation. J Heat Transf 112:441–450

Sato Y, Niceno B (2017) A depletable micro-layer for nucleate pool boiling. J Comput Phys 300:20–52

Sato Y, Niceno B (2018) Pool boiling simulation using an interface tracking method: from nucleate boiling to film boiling regime through critical heat flux. Int J Heat Mass Transf 125:876–890

Scardovelli R, Zaleski S (2000) Analytical relations connecting linear interfaces and volume fractions in rectangular grids. J Comput Phys 164:228–237

Scardovelli R, Zaleski S (2003) Interface reconstruction with least-square fit and split Eulerian-Lagrangian advection. Int J Numer Meth Fluids 41:251–274

Singh NK, Premachandran B (2018) A coupled level set and volume of fluid method on unstructured grids for the direct numerical simulations of two-phase flows including phase change. Int J Heat Mass Transf 122:182–203

Son G (2001) A numerical method for bubble motion with phase change. Numer Heat Transf: Part B: Fundam 39:509–523

Son G, Dhir VK, Ramanujapu N (1999) Dynamics and heat transfer assoicated with a single bubble during nucleate boiling on a horizontal surface. J Heat Transf 121:623–631

Son G, Dhir VK (1997) Numerical simulation of saturation film boiling on a horizontal surface. J Heat Transf 119:525–533

Son G, Dhir VK (2008) Three-dimensional simulation of saturated film boiling on a horizontal cylinder. Int J Heat Mass Transf 51:1156–1167

Son G, Dhir VK (2008) Numerical simulation of nucleate boiling on a horizontal surface at high heat fluxes. Int J Heat Mass Transf 51:2566–2582

Son G, Hur N (2002) A coupled level set and volume-of-fluid method for the buoyancy-driven motion of fluid particles. Numer Heat Transf: Part B - Fundam 42:523–542

Son G, Ramanujapu N, Dhir VK (2002) Numerical simulation of bubble merger process on a nucleation site during pool nucleate boiling. J Heat Transf 124:51–62

Sussman M, Puckett EG (2000) A coupled level set and volume-of-fluid method for computing 3D and axisymmetric incompressible two-phase flows. J Comput Phys 162:301–337

Sussman M, Smereka P, Osher S (1994) A level set approach for computing solutions to incompressible two-phase flow. J Comput Phys 114:146–159

Tomar G, Biswas G, Sharma A, Agrawal A (2005) Numerical simulation of bubble growth in film boiling using a coupled level-set and volume-of-fluid method. Phys Fluids 17(112103):1–13

Tryggvason G, Esmaeeli A, Al-Rawahi N (2005) Direct numerical simulations of flows with phase change. Comput Struct 83:445–453

Tryggvason G, Bunner B, Esmaeeli A, Juric D, Al-Rawahi N, Tauber W, Han J, Nas S, Jan YJ (2001) A front-tracking method for the computations of multiphase Flow. J Comput Phys 169:708–759

Ubbink O, Issa RI (1999) A method for capturing sharp fluid interfaces on arbitrary meshes. J Comput Phys 153(1):26–50

Walters DK, Wolgemuth NM (2009) A new interface-capturing discretization scheme for numerical solution of the volume fraction equation in two-phase flows. Int J Numer Meth Fluids 60:893–918

Welch SW, Wilson J (2000) A volume of fluid based method for fluid flows with phase change. J Comput Phys 160:662–682

Youngs DL (1982) Time-dependent multi-material flow with large fluid distortion. In: Morton W, Baines J (eds) Numerical methods for fluid dynamics. Academic, New York, pp 273–285

Zalesak S (1979) Fully multidimensional flux-corrected transport algorithms for fluids. J Comput Phys 31(3):335–362

Chapter 15
Equipment for Boiling, Evaporation and Condensation

Boiling and evaporation play major roles in power industry as well as chemical process industries such as concentration, crystallization, process heating and cooling, chemical reaction control and even desalination involving the generation of vapour from liquid. This may be carried out by different processes such as evaporation, nucleate boiling, flow boiling or even special techniques such as flashing and direct contact heating. Depending on the application as well as the convention, the equipment for generation of vapour are given different names such as evaporator, reboiler, steam generator, vaporizer or simply, boiler. The following section deals with the classification of vapour generating equipment. It may be mentioned here that the range of vapor generating equipment is very wide and it is difficult to provide a detailed analysis and design for all the types of equipment. Hence, a general description of the major types of evaporation and boiling apparatus is presented and the design procedures for a few extensively used ones are described.

Condensers are also an important class of equipment in chemical process industry. It has got equally important role in power generation. The second law of thermodynamics prohibits a cycle operating with only one energy reservoir. Condensers act as a medium to dump the amount of un-utilized heat to atmosphere which act as the second reservoir in both power generation and refrigeration. In process industry, condensation is used to bring the outlet vapour of distillation column, boiling the liquid in a reboiler, removal of liquid from gas phase reactors, separation and process fluid heating applications. Industries such as petrochemicals, food and beverages, biochemical and industrial chemicals use condensers extensively. In this chapter, a brief introduction of these types of condensers will be made.

S. K. Das and D. Chatterjee, *Vapor Liquid Two Phase Flow and Phase Change*, https://doi.org/10.1007/978-3-031-20924-6_15

15.1   Classification of Vapour Generating Equipment

Vapour generating equipment can broadly be classified into two categories, boiling equipment and evaporating equipment. The major difference between them is in the mechanism of vapour formation. Evaporators usually operate with vapour formation from liquid surface, while boiling equipment have vapour generation through bubble formation at solid surface. However, in flow boiling, both the mechanisms become important. In general, vapour generating heat exchangers can be classified as given in Fig. 15.1.

This chart gives an idea about how many different types of heat exchangers fall under the vapour generating category. There is considerable amount of ambiguity as well, in the classification and nomenclature of these equipment. For example, a plate type equipment is called 'plate evaporator' even though the vapour generation process in it is boiling rather than evaporation. Often, by convention, the name of the equipment is chosen according to their usage; for example

- Where generation of steam for power production or process utilization is the purpose, it is called a boiler or a steam generator. By convention, the equipment which produces steam in a nuclear power plant, is called a 'steam generator', and not a

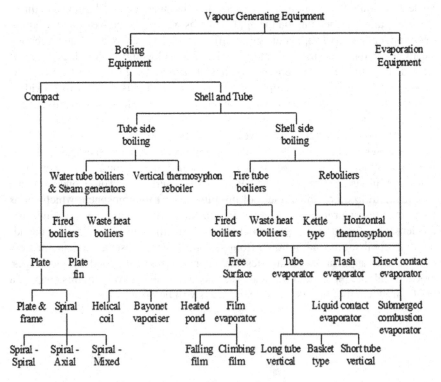

Fig. 15.1 Classification of heat exchangers with vapour formation

'boiler'; while the equipment which generates steam from waste heat is called a 'waste heat boiler' and not a 'steam generator'.

- The equipment that involves vapour generation at the bottom of a distillation column (inside or outside the column) is called reboiler.
- Usually, when concentrating, desalinating, crystallising, refrigerating and other evaporating duties (i.e., vapour generation without nucleation) is carried out in equipment, they are called evaporators. However, in many of the evaporators nucleate boiling also takes place.
- Where revaporisation of liquefied gas or vaporisation of the heat transfer fluid is done in equipment, it is called a evaporator.

15.2 Different Types of Vapour Generators

The large number of vapour generators, given in Fig. 15.1 is difficult to describe, hence only the broader categories are discussed in this section.

15.2.1 Steam Generator or Boilers

The class of equipment, which generate steam from water are called boilers or steam generators. They can be of two types.

15.2.1.1 Fired Steam Generators or Boilers

Boilers or steam generators are used to produce steam. The major use of steam is for power generation, process heating and process requirement. A wide variety of fossil fuel fired boilers and steam generators are available. They can be fire tube or water tube boilers depending on whether the tubes carry the water and steam or the combustion products; this is depicted in Fig. 15.2a and b. Large boilers are usually more common to thermal power plants.

However, in a number of chemical process plants, it is more economical to generate power and process steam simultaneously rather than process steam alone. Such a system is called cogeneration. The generation of steam in a nuclear power plant is different. Usually for heavy water reactors (PHWR), the steam generator uses pressurized heavy water, which carries the heat generated by nuclear fission to the steam generator, which in turn generates steam by utilizing this heat. Figure 15.3a shows a steam generator for a PHWR plant. The steam generator for a liquid metal fast breeder reactor (LMFBR) is more critical, as it employs hot liquid sodium to generate steam; the schematic is shown in Fig. 15.3b. In boiling water reactors (BWR), steam is generated at the reactor core.

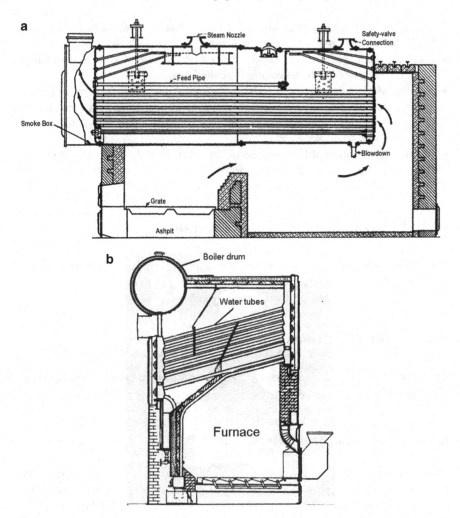

Fig. 15.2 a Fire tube boiler and **b** Water tube boiler

15.2.1.2 Waste Heat Boilers

Utilization of waste heat for process requirement is the main aim to build waste heat boilers. Often it is done with the additional advantage of quenching of hot gases and cooling of effluents, which can be the environmental requirement. Waste heat boilers are widely used in chemical process industries. Generally, waste heat boilers are represented by a schematic shown in Fig. 15.4. Waste heat boilers can be heated by two methods:

(a) Flue gas heated
(b) Process gas heated

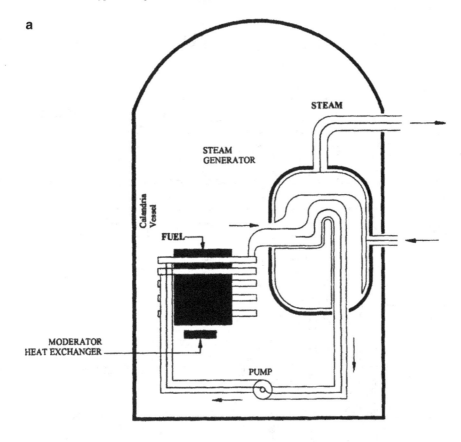

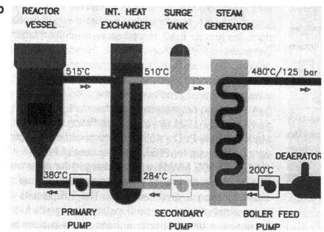

Fig. 15.3 Steam generator for **a** PHWR and **b** LMFBR

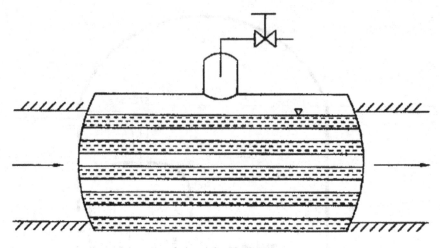

Fig. 15.4 Schematic representation of a waste heat boiler

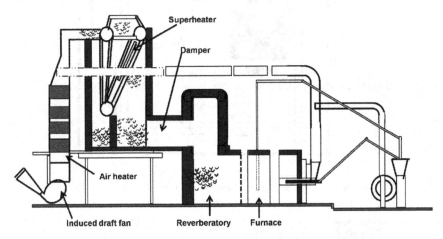

Fig. 15.5 A flue gas heated boiler

Flue gases from furnaces are used in flue gas heated waste heat boilers. Figure 15.5 shows a flue gas heated boiler. They can be of different configurations similar to fired boilers such as horizontal or vertical waste heat boiler. Process gas heated boilers are more typical. They can be vertically mounted such as the vertical calandria boiler (refer Fig. 15.6). In this type of boiler, very hot gases pass through the vertical rod bundle known as calandria. In the passage between tubes, liquid and steam flow, which gives the possibility of flooding.

The other type of waste heat boiler extensively used in industry is the Bayonet tube boiler. In this type of boiler, the outgoing steam–water mixture itself heats the incoming liquid, and the source heat comes from a hot process gas such as nitrogen-hydrogen-ammonia, which in turn needs to be quenched.

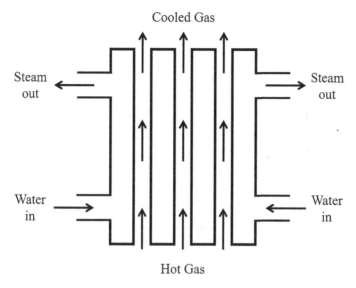

Fig. 15.6 Vertical calandria boiler

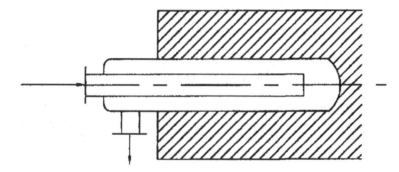

Fig. 15.7 Bayonet boiler in symbol

Figure 15.7 shows a symbolic representation of the Bayonet boiler. There are many other waste heat boilers such as the horizontal fire tube, cross flow boiler or the helical fire tube boiler which are very typical in design for specific applications. However, thermal design of the waste heat boiler is critical. On one hand, it has to utilize the waste heat, and on the other, it has to make sure that the cooled flue or process gas does not condense any of its corrosive components, which in turn may cause damage to process equipment, pipe line and stack.

The cogeneration concept is very important in process plant. Figure 15.8 shows a power and process steam generation circuit. A cogeneration boiler can utilize upto 80% of the energy, compared to 40% in a utility (power generating) boiler.

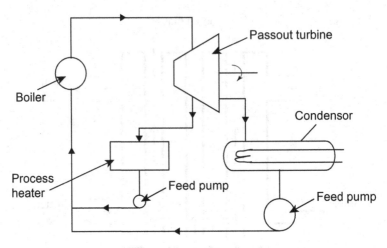

Fig. 15.8 A cogeneration steam circuit

15.2.2   Evaporators

Evaporators have a wide range of applications. It has already been stated that even though the term evaporators usually gives an impression of equipment with vapour generation from the liquid vapour surface without nucleation, the cases where nucleation take place are also called evaporators. In general, any vapour generation (both boiling and evaporation equipment of Fig. 15.1) are called by the generic name 'evaporator' except in special cases of a water–steam system (which is called boiler or steam generator) or a reboiler (evaporation or boiling of liquid from the bottom of distillation column). A number of types of evaporators are available depending on usage. Some major types are discussed below.

15.2.2.1   Tube Evaporators

Tube evaporators are the ones in which the boiling or pre-boiling takes place inside tubes while steam (or other heating fluid) is supplied outside the tube. For a short vertical tube evaporator, the steam enters through the periphery of the vertical shell, heats the tubes and then comes down through a central downcomer as shown in Fig. 15.9.

The weak liquor is fed from the bottom, and often at the bottom, an axial flow impeller is used to assist the liquid circulation. If the device is used for concentration or crystallization of the liquid, boiling inside the tubes is to be avoided as fouling or clogging would have set in inside the tubes. The boiling is carried out in the pool above the tubes in these cases. There is another type evaporator known as the 'basket type' evaporator shown in Fig. 15.10.

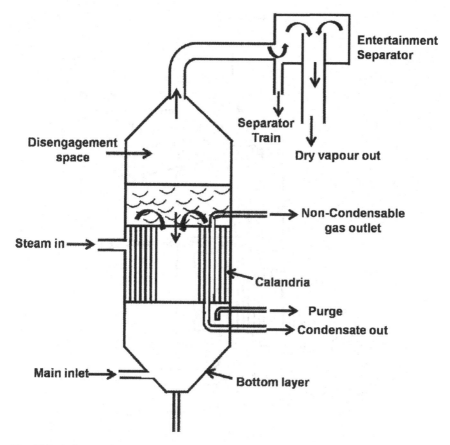

Fig. 15.9 A short vertical tube evaporator

In basket type evaporators, natural circulation takes place. The steam from the bottom enters the space between the tubes, which is baffled and the condensate and non condensibles come out together. The process fluid circulates downward in the annular space between calandria (tube bundle) and the shell. Here, the removal of the calandria from the shell for the purpose of maintenance is easier. This has an elutriator at the bottom, which recirculates the liquor as well as small size crystals (for crystal growth operations) and only the concentrated solution or large size crystals (according to applications) come out. Long tube evaporators have some distinctly different features. They have a fewer number (less than 6) of long tubes, and the downcomer is located outside the shell. Usually, due to a large pressure drop it cannot be driven by a thermosyphon alone and a forced draught system is to be used. Due to high pressure at the bottom, there may be large sensible heat transfer before boiling starts. The long tube evaporator is a once through type, and in contrast to

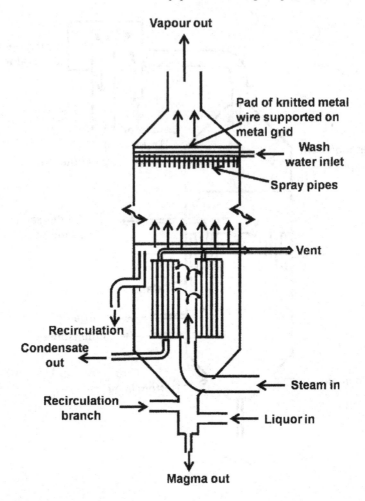

Fig. 15.10 Basket type evaporator

the short type design, they do not contain liquid surface. Figure 15.11 shows a long vertical evaporator.

In all these types of evaporators, a liquid separator is used to bring back the entrained liquid droplets. In the short tube variety, enough space is kept above the liquid level for the liquid droplets to fall back; often a wire mesh is also used at the top. In general, for evaporators with liquid flow upwards, the boiling point decreases in the upward direction due to pressure drop. Thus, usually the bottom part of evaporator carries out sensible heating and the top part boiling. However, in some applications such as concentrators, this is advantageous because of increased concentration.

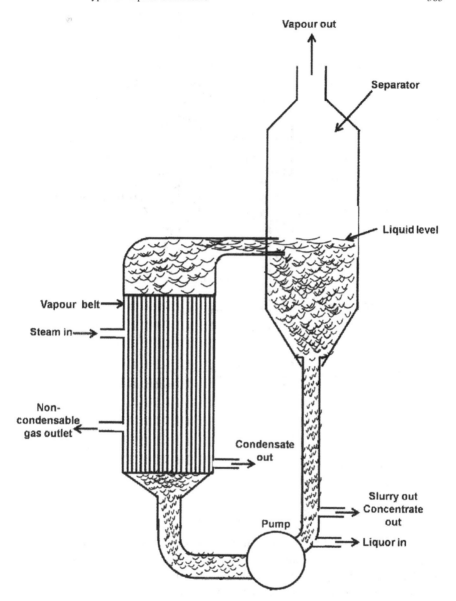

Fig. 15.11 Long tube vertical evaporator

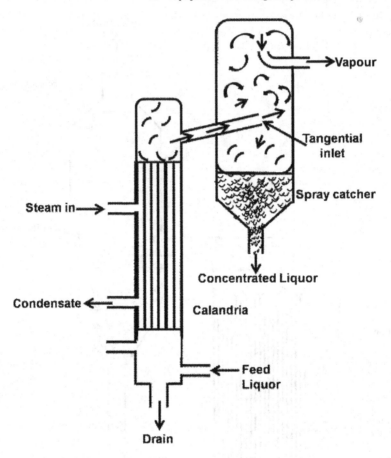

Fig. 15.12 Climbing tube evaporator

15.2.2.2 Film Evaporators

Film evaporators use evaporation from liquid vapour interface without nucleation. They are of two types—falling film and climbing film. The climbing tube design shown in Fig. 15.12 is somewhat similar to long tube vertical evaporator but less common in application. The difference between this type and the long tube type is that it operates under natural circulation and has no external down comer. Here the pressure of liquid at the bottom of the tube is kept low, and hence it starts boiling early utilizing large length of the tube for boiling. As a result, the dryness fraction is smaller and circulation rate is less. This type of evaporator is prone to fouling. Figure 15.12 shows a climbing tube evaporator.

The falling film evaporator is very common. Here the liquid is fed from the top and it falls under gravity, evaporating at the interface. The concentrated solution is collected at the bottom. The vapour usually flows downwards. However, in the case

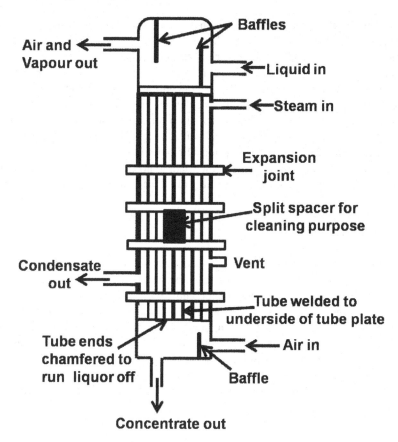

Fig. 15.13 A falling film evaporator

that it is desired to lower the boiling point, either the pressure should be reduced by creating a partial vacuum, or a gas may be introduced from the bottom to decrease the partial pressure of the vapour. Figure 15.13 shows the diagram of a typical falling film evaporator. The counter current arrangement always brings out the limitation in the form of flooding.

15.2.2.3   Compact Evaporators

In recent times, evaporators have been used in compact configurations for enhancement. Plate evaporators are fast gaining importance and plate fin type exchangers are also used for evaporation duties in special applications. However, application of compact heat exchangers in evaporation is far less than that in condensation. One important reason is that the processes such as concentration and crystallization cannot be carried out in narrow passages because they may clog the passage completely.

A narrow passage is a characteristic of both plate and plate fin heat exchangers. Due to this reason, in chemical processes, where a very clean fluid is used (like cryogenics and refrigeration), and thickening or crystallization is not sought, compact heat exchangers can be used for evaporation duty. These compact exchangers are more or less similar to those in their single-phase applications. Some modifications such as port design in plate heat exchangers and header design in plate fin heat exchangers are to be incorporated. The tube fin configuration can also be suggested for condensation on the finned tube and evaporation inside the tube. In such cases, low finned tubes are preferred for the condensing side. Often, more surface area is provided by helical tubes for boiling. However, it is important to note that only clean liquids can be sent through helical coils because cleaning of surfaces may be extremely difficult.

15.2.2.4   Flash Evaporators

Evaporation can also be carried out by sudden reduction of pressure of a liquid which lowers the boiling point. In this case, the latent heat required by the liquid to evaporate is supplied by the loss of the enthalpy of the liquid. As a result, the liquid temperature falls. For liquids such as water, which have a large latent heats of vaporisation, a decrease of 50–60 °C in the temperature is required for evaporation of only 10% of the liquid. The entire process of heat transfer here is in the sensible heat zone. A subsequent pressure increase and sudden flashing is done. The entire process is shown in Fig. 15.14, which shows the circuit for a flash evaporator. A flash evaporator can also be a multi-stage type. This is particularly important in desalination applications.

15.2.2.5   Direct Contact Evaporators

Evaporation can also be carried out by directly mixing the hot gas or the hot liquid with the liquid to be evaporated. Usually, a submerged combustion type evaporator (refer Fig. 15.15) is popular where a chamber is immersed inside the evaporating liquid and combustion is carried out in the chamber by supplying fuel from the top. Hot gases come out from the bottom of the chamber which in turn mixes with the liquid raising the temperature of the liquid beyond the saturation temperature, thereby generating vapour. This gives a problem of contamination which can be avoided by using a helical coil around the submerged chamber for evaporation; but then, the evaporator is a helical coil evaporator and not a direct contact evaporator.

15.2.2.6   Cascading of Evaporators

In order to reduce the steam consumption, evaporators can be cascaded. This is called multiple effect evaporation. In this mode, as shown in Fig. 15.16, the steam generated in the first effect is used for heating the second effect, and steam generated in the

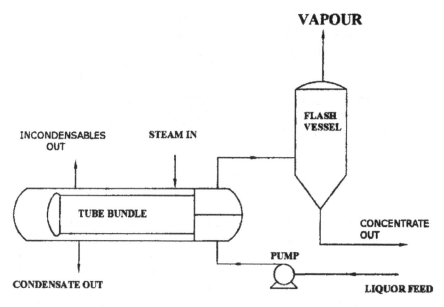

Fig. 15.14 A flash evaporator

Fig. 15.15 A submerged combustion type helical coil evaporator

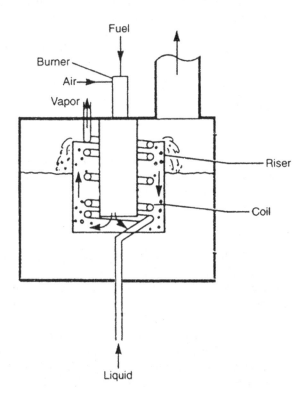

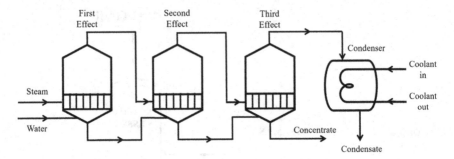

Fig. 15.16 A multiple effect evaporator for concentration process

second effect is used for heating in the third effect. This type of evaporation is ideally suited for a concentration process. The process can be forward or backward feed.

Ideally, the steam requirement is reduced by the number of effects. However, in practice, many other factors come into play. For example, in order to maintain a temperature difference, the temperature should reduce in the upstream direction indicating a reduction of pressure. As a consequence, the temperature difference available decreases progressively, thereby increasing the surface area. Also, since the latent heat increases with decreasing pressure, the steam generation rate also decreases with each effect. In a multiple effect, forward feed is preferred when the feed is hot, and backward feed is preferred when the feed is cold.

15.2.3 Reboilers

Reboilers are the vapour generators used to produce the vapour to be supplied to the lower section of a distillation column. This vapour is generated in the reboiler using the liquid available at the lowest section of the distillation column. Reboilers can be divided into two main categories:

(a) Internal reboilers, and
(b) External reboilers

Internal reboilers are the ones which are located inside the distillation column as shown in Fig. 15.17.

In fact, reboilers can rather be seen as simple heating tube bundles immersed in the bottom liquid sump of a distillation column. The heating fluid for reboilers is steam in most of the cases. However, hot gases and process fluids are also used. The external reboilers can be of a variety of types like thermosyphon, forced flow reboilers or falling film evaporator. Figure 15.18 gives a classification of reboilers.

Usually, all reboilers can also be divided into cross flow and axial flow categories. Kettle, internal and horizontal thermosyphon reboilers fall in the cross flow type, while the vertical thermosyphon, vertical shell side and vertical forced flow type

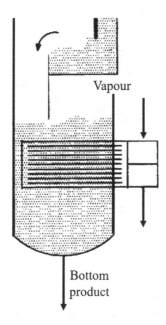

Fig. 15.17 An internal boiler

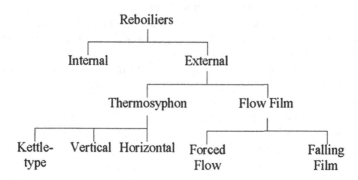

Fig. 15.18 Classification of reboilers

fall in the axial flow category. One more type of evaporator is the 'flooded evaporator', which is very close to Kettle type reboilers except for the enlargement of the shell. They are used for refrigeration and hence not included in the list of reboilers. Following are some reboilers and their relative advantages and limitations.

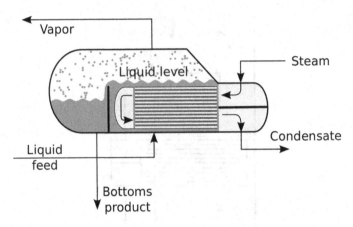

Fig. 15.19 Kettle type reboilers

15.2.3.1 Kettle Type Reboilers

Kettle type reboilers are probably the most widely used reboilers. It has a very unique shell design shown in Fig. 15.19, which is called the K-shell in Tubular Exchangers Manufacturing Association (TEMA) terminology.

The liquid in the shell side is kept just over the tube bundle by means of a vertical baffle. This ensures that the tube bundle never gets overheated. Whenever the liquid is more, it falls over the baffle and is taken through the drain as the bottom product of the distillation column. A level control is employed on the drainage side to avoid dry out. Usually, U-tubes with two passes are used in Kettle type reboilers. Since the liquid level around the tube bundle is kept constant, Kettle type reboilers are less sensitive to liquid and vapour flow characteristics giving a very stable operation. This makes its design considerably simple. Flow re-circulation and hydrodynamics have relatively less influence on the heat transfer coefficient except at very high vapour fraction at the exit or very low temperature difference which are two extreme cases and do not occur during usual operation. Since the liquid and vapour separation takes place, the complex calculation of two phase pressure drop need not be carried out for piping design. However, proper shell design is important to avoid high entrainment. Both at high vacuum and high pressure (near critical point), Kettle type reboilers perform better than thermosyphon reboilers. A major problem of Kettle type reboiler is in duties with heavy fouling. In these duties, this is the worst type of reboiler because of the pool type boiling. Non-volatiles, particularly the ones that lead to fouling and corrosion tend to accumulate in the shell side. The other important effect which causes under performance of the Kettle type reboiler is the liquid re-circulation around tube bundle. Figure 15.20 shows the liquid re-circulation around a tube bundle.

As the liquid flows through the bundle, it becomes more concentrated with the less volatile component and its boiling point increases. Therefore, at the top of the bundle, the liquid remains below its saturation temperature and flows around the bundle and

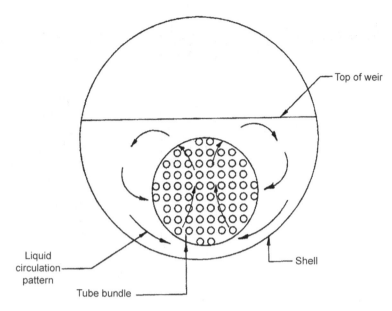

Fig. 15.20 Liquid re-circulation around the tube bundle in a Kettle type reboiler

mixes with the incoming fluid. This increases the heat transfer coefficient in the shell side; but for wide boiling range mixtures, the rise in boiling point decreases the effective mean temperature difference (MTD) on the higher tubes. This will deteriorate the performance of the reboiler. Also, this liquid re-circulation may result in a plume in the vapour zone; to avoid this, sufficient height should be provided in the vapour zone, which in turn will increase the cost. Hence, Kettle type reboilers are preferred in applications where relatively cleaner fluids with narrow boiling range are involved.

15.2.3.2 Vertical Tube Side Thermosyphon Reboilers

The vertical thermosyphon reboilers with tube side evaporation works with flow boiling where the density difference between the two phase mixture and liquid at the bottom drives the flow. Figure 15.21 shows a vertical tube side thermosyphon reboiler along with its connection to the distillation column. For steady operation, the driving static pressure head is kept constant by a vertical wall inside the distillation column. Usually, the vapour exit velocity and the two phase mixture velocity inside the tube and vapour piping are kept low by increasing the tube diameters. Usually, the tube length is kept limited to 4 m but design upto 6 m is available. However, for higher tube length, a thorough check is required so that the dry out does not occur inside the tubes. The recommended exit vapour weight fraction is 0.1–0.35 for hydrocarbons and 0.03–0.1 for narrow boiling range acqueous solutions.

Fig. 15.21 A vertical tube
side thermosyphon reboilers

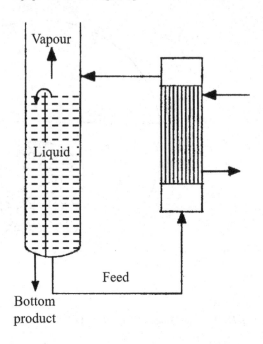

The main advantage of a vertical tubeside thermosyphon reboiler is the high
vapour velocity which reduces fouling. It is also cheaper compared to the Kettle
type due to short piping. Due to pure countercurrent operation, it can give maximum
effectiveness. However, it is sensitive to operating conditions and for high pressure
and high vacuum, instabilities in two phase flow may affect the performance considerably. For viscous wide boiling range liquids, this is unsuitable. For low ΔT
operation, the static head provided by the density difference may not be sufficient
to drive the fluid. The possibility of attaining dry out is another disadvantage for
it. Usually, vertical tube side thermosyphon reboilers are designed with high liquid
re-circulation designated by re-circulation ratio.

$$R = \frac{\text{liquid re} - \text{circulation}}{\text{vapour generation}} = \frac{1 - x_o}{x_o} \qquad (15.1)$$

where x_o is the quality of the two phase mixture leaving the reboiler.

The value of R is in the range 3–6, but in vacuum applications, it can be as high
as 15–18. Usually, the upper tube sheet of vertical tube side thermosyphon reboilers
is kept at the level of the liquid maintained at the distillation column with the help
of the vertical baffle.

Fig. 15.22 A vertical shell
side thermosyphon reboilers

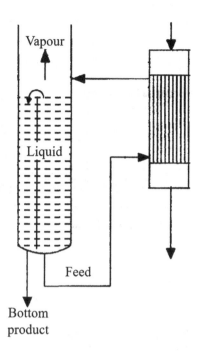

15.2.3.3   Vertical Shell Side Thermosyphon Reboiler

In this type of reboiler, the vaporisation takes place on the shell side. Usually, this is
avoided due to vapour blanketing and dead zone, particularly at the baffle roots. To
avoid this, rod baffles are preferred in this design. This type of reboilers is to be used
where the heating fluid is corrosive which may cause the damage of the costly shell.
In this design, local overheating and accumulation of vapour is the major problem.
Figure 15.22 shows a vertical shell side thermosyphon reboiler.

15.2.3.4   Horizontal Thermosyphon Reboilers

For horizontal thermosyphon reboilers, the usual design is with shell side boiling. For
proper vapour distribution, a cross flow arrangement is preferred. It can be TEMA
X, G, H, J or the simplest E shell. Usually, the pure cross flow or the split flow
models are popular. Similar to the vertical thermosyphon reboiler, the hydrostatic
head is provided by a liquid head inside the distillation column maintained by vertical
baffle. This type of reboilers have a number of advantage such as—good mixing and
turbulence promotion due to cross flow over the tubes and the horizontal baffles giving
high heat transfer coefficient, improvement of the effective temperature difference
particularly in the wide range boiling mixtures, less amount of subcooling due to

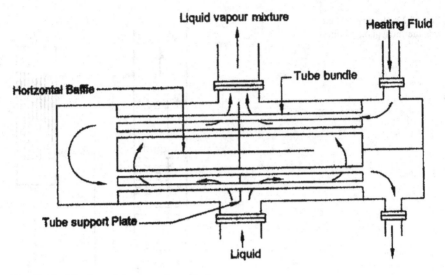

Fig. 15.23 A horizontal thermosyphon reboilers

less amount of static head available at the entry. Figure 15.23 shows a horizontal thermosyphon reboiler.

These reboilers are also susceptible to fouling and localized over heating. Distributing the vapour–liquid mixture along the shell length is a problem. Due to the tubes, the horizontal baffles and the supports, it is possible to have dead zones. The exit pipe design is important so that choking due to high velocity or phase separation due to low velocity does not take place. This type of reboiler can operate with a relatively low temperature difference and low available static head.

15.2.3.5 Forced Convection Reboilers

The reboilers with forced flow (vertical or horizontal) are also available. Here the boiling can be limited to a small fraction when required by using high pressure generated by the pump. For high fouling liquids, often forced convection is used. For low ΔT and viscous fluids, forced flow is necessary. Figure 15.24 shows a forced convection reboiler.

15.3 Analysis and Design of Vapour Generating Equipment

Design of equipment with boiling or evaporation is a difficult task. The rise of interest in nuclear reactors in the 60 and 70s gave a boost in the research in boiling, but many of these works are specific to the nuclear industry and difficult to apply to

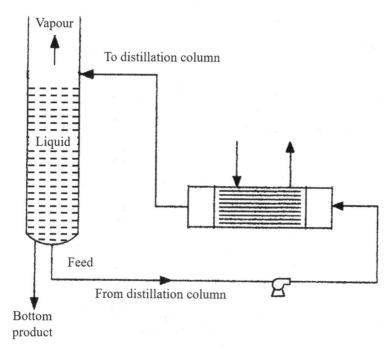

Fig. 15.24 A forced convection horizontal reboilers

the equipment used in the chemical process industry. The chemical process industry often uses chemicals of high viscosity, mixture of liquids, gas–liquid mixtures and non-Newtonian fluids, the requirement for which is very much different and typical. However, without going into much detail, the following section aims at the presentation of the design and analysis of three major categories of vapour generators.

15.3.1 Steam Generator Calculations

Design of steam generators involves a number of issues. In industrial power generators, super heating of the produced steam is a major consideration; however, for process steam, the saturated or slightly superheated state is sufficient. For a natural circulation (thermosyphon type), steam generator as shown in Fig. 15.25, the available pressure head is given by

$$\Delta p = gH(\rho_D - \rho_m) \tag{15.2}$$

where H is drum height, ρ_D is density of water in downcomer, and ρ_m is mean density of vapour water mixture.

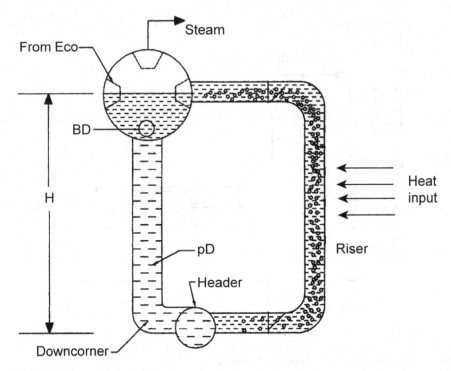

Fig. 15.25 A natural circulation steam generators

In simplified calculations, the mean density can be taken as the mean between the top and bottom density of the riser tube

$$\rho_m = \frac{\rho_{top} + \rho_{bottom}}{2} \tag{15.3}$$

Now, usually the downcomer is treated as adiabatic, and hence the riser bottom density is given by ρ_D.

Hence, the required head may be given by

$$H = \frac{2\Delta p}{g(\rho_D - \rho_{top})} \tag{15.4}$$

where ρ_D usually saturated condition

$$\rho_{top} = \frac{1}{v_f + x_{top} v_{fg}} \tag{15.5}$$

Thus, the height required strongly depends on the steam quality required at the top of the riser and the riser pressure drop. Now, the density difference between the

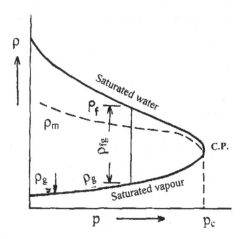

Fig. 15.26 Density variation with pressure

pure vapour and the saturated liquid decreases with the increase of pressure (refer Fig. 15.26); this makes the operation of natural circulation steam generator difficult at high pressure.

For steam generators, the circulation ratio is the amount of saturated water to be circulated through the downcomer for unit steam release from the drum. Downcomers are made large in diameter to reduce pressure drop, but risers are smaller in diameter to increase heating surface area. From Chap. 1, we know that void fraction is given by

$$\varepsilon = \frac{A_G}{A_L + A_G}$$

and slip in the riser tube is given by

$$S = \frac{x^*}{1 - x^*} \times \frac{1 - \varepsilon}{\varepsilon} \times \frac{v_g}{v_f} \tag{15.6}$$

This gives

$$\varepsilon = \frac{1}{1 + \frac{1-x^*}{x^*}\varphi} \tag{15.7}$$

where $\varphi = \frac{v_f}{v_g} S$.

The average density in the riser tube can be accurately calculated by integrating

$$\rho_m = \frac{1}{H} \int_o^H \rho(z) dz \tag{15.8}$$

where

$$\rho(z) = (1 - \varepsilon)\rho_L + \varepsilon \rho_G$$

This integration for uniform heat flux gives

$$\rho_m = \rho_L - \frac{\rho_L - \rho_G}{1 - \varphi}\left\{1 - \left[\frac{1}{\varepsilon_e(1 - \varphi)} - 1\right]\ln\frac{1}{1 - \varepsilon_e(1 - \varphi)}\right\} \qquad (15.9)$$

where ε_e is the exit void fraction. The total pressure drop in the riser tube can be estimated as

$$\Delta p = gH(\rho_D - \rho_m) = \Delta p_{downcomer} + \Delta p_{riser} + \Delta p_{bends} + \Delta p_{header} \qquad (15.10)$$

15.3.2 Evaporator Design

The various types of evaporators described in Sect. 15.2.2 have got typical design procedures of their own. In general, for shell and tube evaporators, the procedure used for single-phase shell and tube heat exchangers can be used with the heat transfer coefficient calculated according to expressions given in Chap. 9 for the evaporating side. This can be done for both shell side and tube side evaporation. Usually, shell side evaporation follows pool type boiling with correction for liquid re-circulation, while the tube side boiling follows the flow boiling characteristics. In these cases, the design of evaporators does not differ significantly from that of reboilers. However, in short tube evaporators, boiling often takes place in the upper part of the tubes and the pool above it, which has typical design procedure. Thus, the evaporators which generally operate similar to boilers in principle are not dealt with here. The typical type of evaporator, which is widely used, is the falling film evaporator which is discussed in detail here.

Falling film evaporators are used in applications where the solvent is to be extracted by evaporation. The concentration of beverages, recovery of solvents in various chemical processes and desalination of sea water are some such applications. The pressure drop in such evaporators is small, and hence the evaporation takes place over a wide length with very little temperature variation. The heat transfer coefficient for falling film evaporators is large, ranging from 1–4 kW/m^{2}K. A parallel flow is chosen with both vapour and liquid film flowing down the wall, usually inside a tube. As the liquid evaporates, the film becomes thinner and the heat transfer increases. Also, the vapour velocity (in the case of tube) increases and this exerts shear on the film and the film becomes thin, this also enhances heat transfer. The process of falling film evaporation can be seen as the inverse case of film condensation. Figure 15.27 shows falling film evaporation. In film condensation, laminar film becomes turbulent and it moves downward. In case of evaporation with large flow rate at the top of the tube (or plate), the film is thick and often turbulent. In the lower part due to thinning of the film, flow may turn into laminar. This is more true for processes such as concentration, where due to the increase of concentration of the solute, the viscosity increases with evaporation, decreasing the Reynolds number further and bringing about the laminarisation of flow. For falling film evaporators,

Fig. 15.27 Falling film
evaporation. (*Courtesy*:
"Heat Transfer in
Condensation and Boiling"
by Karl Stephan, with
permission from Springer)

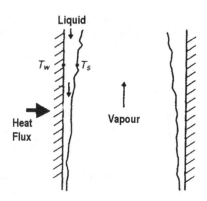

usually the wall superheat is kept low so that the evaporation takes place at the liquid
surface and not at the wall through nucleation.

Falling film evaporation can be treated in a way exactly identical to film condensation. For a laminar film, the heat transfer coefficient can be given by the same
expression as Nusselt's theory of laminar flow condensation as

$$h_{Nu} = \frac{k}{\delta} \tag{15.11}$$

where δ is the local film thickness given by

$$\delta = \left[\frac{3\mu_L \Gamma}{\rho_L (\rho_L - \rho_G) g} \right]^{1/3} \tag{15.12}$$

Obviously, this expression is based on the assumptions of the Nusselt's theory of
film condensation, which includes pure conduction heat transfer in the film, stagnant
vapour and absence of interfacial shear. This equation can be changed to a Nusselt
number type correlation

$$Nu_{Nu} = \frac{h_{Nu} l}{k_L} = \left(\frac{4}{3} \right)^{1/3} Re^{-1/3} \tag{15.13}$$

where, Reynolds number

$$Re = \frac{\bar{u} D_h}{\nu_L} = \frac{4 \bar{u} \delta}{\nu_L} = \frac{4 \Gamma}{\gamma_L}$$

$\bar{u}$ average local film velocity and

$$l = \left[\frac{\mu_L^2}{\rho_L (\rho_L - \rho_G) g} \right]^{1/3}$$

D_h = hydraulic diameter.

For the average heat transfer coefficient upto the length at which the wall dries out

$$\bar{N}u = \frac{\bar{h}_{Nu}l}{k_L} = \left(\frac{4}{3}\right)^{4/3} (\text{Re})_{x=0}^{-1/3} \qquad (15.14)$$

In many cases, the film becomes wavy (Chun and Seban 1971), this takes place when

$$\text{Re} \geq 0.6075 \, \text{Ka}^{0.09}$$

where Ka is Kapitza number = $\frac{\rho_L \sigma^3}{g\mu_L^4}$. In case of wavy film

$$Nu = \frac{hl}{k_L} = 0.756 \, \text{Re}^{-0.22} \qquad (15.15)$$

The average film coefficient being

$$\bar{N}u = 0.922 \, \text{Re}_0^{0.22} \qquad (15.16)$$

Chun and Seban (1971) indicates the criterion for transition from laminar to turbulent flow as

$$\text{Re} > 1460 \, \text{Pr}^{1.05} \qquad (15.17)$$

For turbulent film evaporation

$$Nu = \frac{hl}{k_L} = 6.616 \times 10^{-3} \text{Re}^{0.4} \, \text{Pr}^{0.65} \qquad (15.18)$$

the mean value being

$$\bar{N}u = (6.616 \times 10^{-3})^{4/3} \text{Re}_0^{0.4} \, \text{Pr}^{0.65} \qquad (15.19)$$

In case of all the above situations, where the falling film does not evaporate completely, in the expression for average Nusselt number, Re_0^n should be replaced by

$$\Delta \text{Re}^n = \frac{\text{Re}_o - \text{Re}_L}{\text{Re}_o^{1+n} - \text{Re}_L^{1+n}} \qquad (15.20)$$

where Re_o and Re_L are film Reynolds number at the top and bottom of the tube. These expressions are valid if the entire film is in one regime (laminar, laminar wavy or turbulent). In case of transition from one to the other, piecewise evaluation is required. Figure 15.28 shows the Nu−Re relationship over the entire range of Reynolds number.

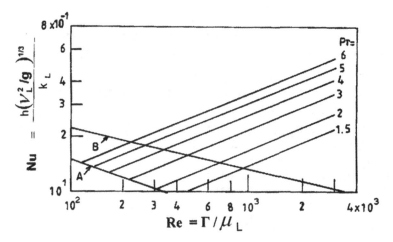

Fig. 15.28 Falling film evaporator's performance. (*Courtesy:* "Heat Transfer in Condensation and Boiling" by Karl Stephan, with permission from Springer)

The effect of vapour flow can also be incorporated, which is normally not very significant. One important check that has to be made in falling film evaporator is the prevention of total dry out. Palen (1988) gives a criterion for avoiding dry out as

$$\frac{\Gamma k_L}{qL} > 1.5 \tag{15.21}$$

where q = heat flux at wall.

15.4 Classification of Condensers

Condensers can be primarily classified as surface condensers and direct contact condensers depending on whether intermixing of coolant and condensing fluid is permitted. Among surface condensers, shell and tube type is more popular, but the compact types as well as plate type condensers are also making good impact in process industry. Air cooled condensers are important in certain applications and are very common for refrigeration. Since in condensers, vapour with large specific volume (dependent on temperature) flows, hence effect of buoyancy is considerable. Thus, due to buoyancy and venting of non-condensibles, orientation (i.e., vertical or horizontal) of the condenser is important. Also, in vertical (to some extent also in horizontal condensers because it is always slightly tilted) condensers, the flow direction of the vapour is important because under upflow (reflux) condition 'flooding' is an important consideration. It will be discussed in greater details at appropriate place. The general classification of condensers is given by Fig. 15.29.

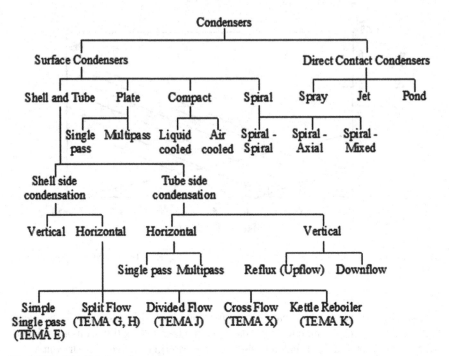

Fig. 15.29 Classification of condensers

It is difficult to fix one criterion to select a particular condenser for a given duty from the large number of condensers given here. Generally, the following requirements are to be kept in mind:

- **Condensation Process**: This is the most important requirement. Whether it is single or multi component is the first thing to be seen. In case of multi component, checking whether condensation is total or partial and whether there are non-condensibles are important considerations.
- **Coolant**: Coolants can be water, air or cooler process steam or other cooler process liquid. The condition of the coolant is a major factor in design.
- **Pressure Level and Pressure Drop**: High pressure condensation is carried out inside tubes, while low pressure in plates or shell sides. Restricting pressure drop is a major requirement of a condenser which operates under vacuum.
- **Fouling and Corrosion**: Corrosive and fouling vapours are usually put in the tube side because they can harm the costly shell. They are not used in the plate HX because they attack the gaskets.
- **Condensate Control**: The condensate and vapour flow should be controlled. It is more important in reflux condenser where increase of vapour velocity may cause flooding. Better control can only be achieved in tube side condensation due to close contact with the cooling surface.

- **Freezing and Crystallization**: Freezing of the condensate is often a problem. In case of such a possibility, the freezing prone fluid is to be put in the shell side to avoid total plugging of the passage. In cold countries, often air is preheated in air condenser in winter to prevent freezing.
- **Venting**: Venting of non condensible is an integral part of condenser. It can be done by suction fans or steam jet air ejector (SJAE).

Many of the above requirements are conflicting. Hence, a compromise of these requirements may be necessary.

15.5 Various Types of Condensers and Their Applications

Before going to their performance analyses and general design procedures, it is first important to know the advantages, disadvantages and applications of different types of condensers. A short review of this is given below:

(i) **Shell and Tube Condensers**; This is the most popular type of heat exchanger which can be utilized for nearly all applications. They can be used for single or multi-component vapour, high pressure to high vacuum, and wide range of cooling liquids. However, they are not used for air cooled condensers. The design procedure for these condensers are essentially same as the famous Bell-Delaware method in Chap. 6 but some special considerations are to be taken into account. A wide range of condensers fall under this name:

(a) *Vertical in tube down flow condenser*: This exchanger is shown in Fig. 15.30. Sometimes a separator is fixed at its head at the top. Venting is provided with the tube sheet and draining of condensate from the bottom header. The bottom header is larger and acts as the hot well for the condensate. Due to close contact of cold wall and condensate, as well as vapour, this type of condenser is preferable for non azeotropic mixtures and high pressure where condensation takes place over a wide range of temperature. In this type, the liquid completely blankets the inner tube surface and thus avoids contact with vapour. For fluids which are more corrosive in vapour state, this type of condenser is preferable. Since the vapour flow path is well known, subcooling effects can be incorporated accurately. However, it has disadvantages like shell side fouling with coolant, larger vapour pressure drop requiring larger diameter tubes (and hence less surface area) and loss of velocity due to condensation which may make it difficult to push the vapour downward against buoyancy.

(b) *Vertical intube upflow or reflux condenser*: The reflux condensers are used to condense the boiled off vapour coming from an exothermic chemical reactor. This is because the reflux condenser can act as a controller for the reactor by changing its coolant flow rate. They are also used when small amount of low boiling component is to be removed from a mixture without condensing

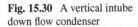

Fig. 15.30 A vertical intube
down flow condenser

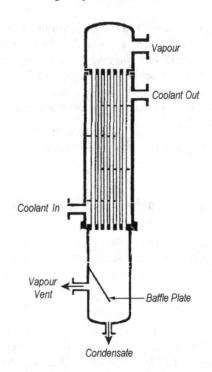

it. The main advantage of reflux condenser lies in its easy venting from the
top because the vapour cannot reach the top under complete condensation
condition. In reflux condenser, the tubes extend beyond bottom tube sheet
and they are chamfered to aid dripping of condensate. One important limitation of reflux condenser is 'flooding'. Flooding occurs at higher velocity
because vapour and liquid condensate flow in opposite direction and the
vapour entrains the liquid and the condenser is flooded. The chamfering of
the tube (as shown in Fig. 15.31) can increase the flooding velocity (and thus
decrease the possibility of flooding) by upto 30%.

(c) *Horizontal in tube condenser*: For high pressure high velocity vapour often
horizontal inclination is preferred. This arrangement is usually single pass
or *U* tube two pass as shown in Fig. 15.32. This condenser is not completely
horizontal but slightly tilted downwards to aid drainage of condensate. Flow
'uphill' should be avoided to prevent flooding. At very high flow rate, the
vapour shear can form waves which entrain mist and form slugs which
can give intermittent or oscillatory flow. Multipassing of tubes in intube
horizontal condensers gives condensation separation and as a result flow
maldistribution.

The *U* tube condenser with entrance and exit at the same end has a better
flow distribution. In *U* tube condenser, the outer tubes are larger than the

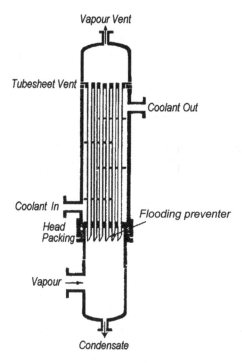

Fig. 15.31   A reflux condenser

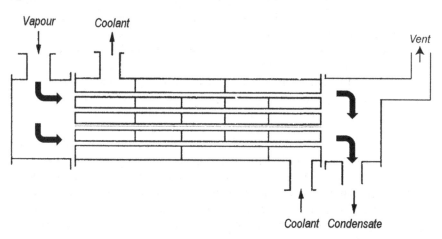

Fig. 15.32   A horizontal tube condenser

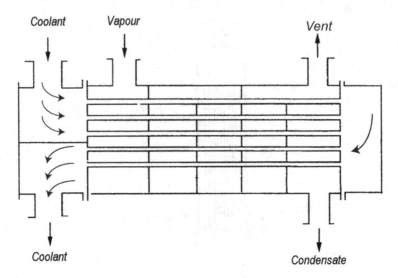

Fig. 15.33 Horizontal shell side condenser (E shell)

inner tubes. This unequal tube length gives a condensation less than equal
tube lengths.

(d) *Horizontal shell side condenser*: The horizontal shell side condensers
can be of different TEMA shell types, e.g., E Shell (Fig. 15.33), X Shell
(Fig. 15.34a), J Shell (Fig. 15.34b) and so on. Figure 15.33 shows such a
heat exchanger where an impingent plate is fixed at the vapour entry to
prevent erosion and vibration of the tubes which are major sources of heat
exchanger mechanical failures.

The TEMA E shell is extensively used in petrochemical industry. The baffles
help the cross flow of vapour over tubes but obstruct the condensate flow.
Hence, vertically cut baffles are usually used to allow the condensate to flow.
However, in this type of condenser, the local equilibrium does not exist; and
sometimes, to enhance mixing of the fluid, horizontally cut baffles are used.
The cross flow design is particularly popular in power plants due to low pressure drop which is a requirement for high vacuum operation. Special venting
system is required for poor distribution of vapour. Split flow (J shell) configuration has got a better vapour distribution and venting system. It is widely
used for condensation of hydrocarbons. There are also other designs such as
rod baffled condenser. This has an advantage of better drainage of condensate. This design has the additional advantage of low pressure drop and less
flow induced vibration and yet preserving the countercurrent flow direction.

(e) *Vertical shell side condenser*: Condensation on the shell side is usually carried out when the prime function is not condensing but heating the coolant
in applications such as evaporator or vertical thermosyphon reboiler where
the liquid from distillation column is vapourized inside tubes. Most often,

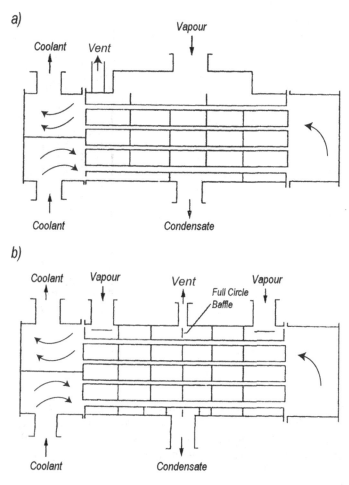

Fig. 15.34 Shell side condenser. **a** X shell and **b** J shell

the shell side is unbaffled and only support plates are provided as shown in Fig. 15.35. This type of condensation is nearest to Nusselt theory of film condensation and hence called a falling film condenser. Venting is somewhat problematic in such condensers. Usually, vent is provided at the lower half of the shell. Sometimes notches are provided (if support plates are used) to drain condensate. This has low pressure drop but high temperature rise of the coolant. The water side fouling can be minimized by online cleaning. However, often liquid distribution is a problem.

(ii) **Compact Condensers**: In recent times, a wide variety of compact heat exchangers are used for condensation duty. The special types are described here.

Fig. 15.35 A falling film condenser

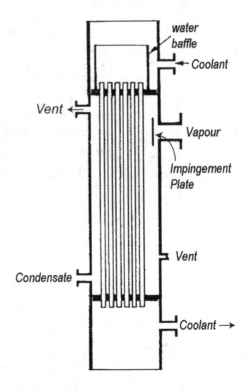

(a) *Plate condenser*: Usually, plate heat exchangers are not considered for condensing duties due to factors such as narrow corrugated passages giving high pressure drop, maldistribution of vapour and liquid inside the channel, condensing fluid may attack gasket material, limitation of pressure and temperature and difficulty to prevent leakage through gasket at high vacuum. However, in applications such as Pasteurization of milk, this type of heat exchanger is extensively used due to the ease of cleaning the plates. For plate condenser, the vapour carrying port is usually larger in size compared to liquid ports and they are often non circular (See Fig. 15.36), OTEC (Ocean Thermal Energy Conversion) plants use plate condensers.

(b) *Plate fin condenser*: For cleaner fluids, aluminium brazed plate fin condenser can be used. Usually, for condensation duties, low finned tubes are used. The main advantage of plate fin heat exchangers for condensation duties lies in their ability to work with small temperature difference. Cryogenics is an area where plate fin condensers are extensively used. The main problem associated with plate fin condensers are flow maldistribution and instability.

(c) *Spiral heat exchangers*: The advantage of spiral heat exchanger lies in their use as internal column reflux condensers at the upper part of a distillation column. This internal positioning of the heat exchanger makes the column operation more stable and less complicated in piping. There are three types

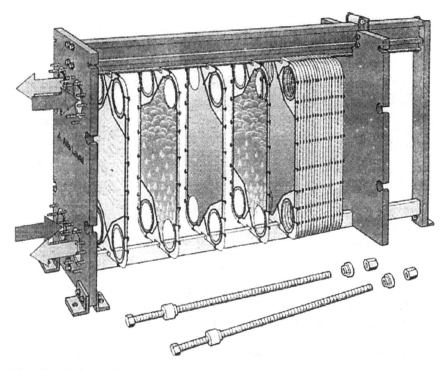

Fig. 15.36 A plate condenser

of spiral condensers: spiral–spiral, spiral–axial, spiral–mixed as shown in Fig. 15.37.

(iii) **Direct Contact Condensers**:   Direct contact condensers have high cooling efficiency but contamination of the fluids is the major concern. Due to cooling efficiency, often even immiscible coolant is used as spray and subsequently separated from the condensate.

  (a) *Jet condenser*: These are more often used in power plants where large volume of vapour is to be condensed.
  (b) *Spray condenser*: This is the most popular type of direct contact condenser. In such condensers, cleaning of jets is a major problem if vapour contains impurity. It is practically infinite surface area condensation.
  (c) *Baffle columns*: Baffle columns can also be called as condensers with near countercurrent operation.
  (d) *Pool condenser*: This is a simple condenser where vapour is bubbled in a pool which can also provide safety against sudden pressure rise. Specially designed vapour ejectors are often used to get desired rise of vapour bubbles.

(iv) **Air cooled Condensers**:   The air cooled condensers are usually compact condensers (usually tube fin). However, the additional feature is the use of air as

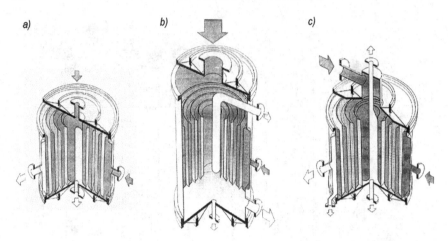

Fig. 15.37 Spiral heat exchangers. **a** Spiral–Spiral, **b** Spiral–Axial and **c** Spiral–Mixed

a cooling fluid. This is important where enough cooling water is not available. The increased cost due to draught fans may be compensated by the elimination of cooling tower which finally takes up heat from water in most water cooled system. Air is to be forced or induced (induced or forced draught) over finned tubes to condense the vapour inside tubes. Refrigeration industry uses air cooled condenser extensively. In domestic refrigerator, even air with natural convection is used for condensing.

15.6   Shell and Tube Condensers: Analysis and Design

Generally, shell and tube condensers can be analyzed and designed in the same manner as shell and tube heat exchangers. However, some special considerations are to be taken into account. Whether it is tube side or shell side condensation, if the condensation is done only for a single component saturated vapour (also for azeotropes) without condensate subcooling, the LMTD can be given by

$$\Delta T_{LM} = \frac{T_{c,out} - T_{c,in}}{\ln \frac{T_{sat} - T_{c,in}}{T_{sat} - T_{c,out}}} \tag{15.22}$$

However, the ideal condition of overall heat transfer coefficient remaining constant is rarely applicable in condensers for a variety of reasons. Firstly, for flow quality varying along the condensation path, the heat transfer coefficient varies significantly. With pressure drop, condensing temperature varies which also brings a significant change in heat transfer coefficient. Finally, blanketing effect of non- condensibles is always there.

15.6.1 Effect of Desuperheating and Subcooling

Often, the condensing vapour enters the condenser in a superheated state and leaves at a subcooled state. Hence, the heat transfer in the part of desuperheating and sub-cooling should be calculated separately.

Desuperheating can be of two types—drywall desuperheating and wet wall desu-perheating. When the superheated vapour enters the condenser with a cold wall temperature higher than the saturation temperature, the flow of vapour can be treated like that of a pure gas flow and the heat transfer can be estimated from

$$q = h_{sv}(T_v - T_w) \qquad (15.23)$$

where T_v = superheated vapour temperature. h_{sv} = heat transfer coefficient for single-phase vapour.

This is called dry wall desuperheating. The heat transfer can also be estimated for single-phase vapour from entry point to the point where vapour gets saturated

$$Q_{\text{sup}} = \dot{m}_G C_G (T_{G,in} - T_{sat}) \qquad (15.24)$$

For superheated vapour entering at a plane where wall temperature is lower than saturated temperature, wet wall desuperheating takes places. Here a film is formed at the wall even though the vapour is superheated and then desuperheating takes place with condensation at the liquid–vapour interface. Heat transfer takes place through the liquid film from the interface to wall.

$$q = h_{sf}(T_{sat} - T_w) \qquad (15.25)$$

The heat transfer coefficient h_{sf} can be taken as the condensate film coefficient.

Bell (1986) showed that for pure as well as narrow range condensing mixtures, the maximum heat flux of the two conditions (dry or wet wall) should be used for calculation. An energy balance gives wall temperature estimation which lies between the local superheated vapour and coolant temperature

$$T_w - \frac{T_c + (h_{sv}/h_c)T_v}{1 + (h_{sv}/h_c)} \qquad (15.26)$$

where, h_c and T_c are the heat transfer coefficient and temperature for the coolant.

Subcooling of condensate is more common as well as more significant for heat transfer considerations. Subcooling is preferable because without subcooling liquid can flush into vapour in valves and lines by partial throttling or cause cavitation in pumps. In the sub-cooled section, heat transfer coefficient is small due to low condensate velocity. Where more subcooling is to be provided, vertical orientation is preferred. The subcooling heat transfer coefficient is more predictable in vertical configuration. In laminar flow, it is given by

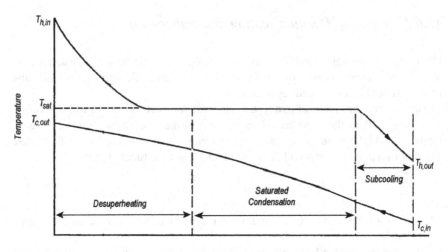

Fig. 15.38 Vapour and coolant temperature of a desuperheating and subcooled condenser

$$h_{sc} = 0.78\,k_L \left[\frac{C_{pL}\mu_L}{k_L L}\right]^{1/3} \left[\frac{\mu_L^2}{g\rho_L^2}\right]^{-2/9} \left[\frac{4\Gamma}{\mu_L}\right]^{1/9} \quad (15.27)$$

where Γ is the mass flow rate per unit periphery. In laminar–turbulent transition region

$$h_{sc} = 0.032\,k_L \mathrm{Re}_L^{0.2}\,\mathop{\mathrm{Pr}}_L^{0.34} \left(\frac{\mu_L^2}{\rho_L^2\,g}\right)^{-1/3} \quad (15.28)$$

for $\mathrm{Re}_L > 2460\,Pr^{-0.65}$. In turbulent region it is given by

$$h_{sc} = 0.0057\,k_L \mathrm{Re}_L^{0.4}\,\mathop{\mathrm{Pr}}_L^{0.34} \left(\frac{\mu_L^2}{g\rho_L^2}\right)^{-1/3} \quad (15.29)$$

A typical temperature plot for a desuperheating and subcooled condensation of a pure (or azeotropic) fluid is given in Fig. 15.38.

15.6.1.1   Special Consideration in Reflux Condenser Flooding

In process industry, reflux condensers are widely used inside distillation column or to control reaction inside a chemical reactor. The main advantage of reflux condenser is very stable operation with pressure and temperature, but it suffers from a serious limitation of flooding. If the vapour velocity increases beyond a point, the interfacial shear creates a wave which entrains a large amount of liquid and reverses the liquid flow flooding the entire condenser.

With increase of vapour flow first, the pressure drop increases with vapour velocity smoothly. At the flooding point, the pressure rises sharply taking maximum liquid upward. This region is of partial reflux and partial carry over. After reaching maximum pressure, the pressure comes down and no liquid comes down, a total carry over takes place. Certain features of flooding are important:

- It is still not clear whether flooding has any effect on heat transfer. Flooding in adiabatic tubes (with air water) and condensing vapour does not differ significantly.
- Flooding depends on tube length. Shorter tubes give lower flooding velocity. This is why the tubes in a reflux condenser are projected at the bottom beyond the tube sheet
- Flooding takes place at the lower end of the tube. The flooding velocity can be significantly increased by angled (bevel cut) tubes at the lower end.
- Since flooding limits the velocity hence mixtures with large amount of non-condensibles is not used in reflux condensers.

A number of studies have been conducted on flooding and a number of correlations are available for flooding velocity. Wallis (1961) presented correlations as

$$
\begin{aligned}
V_L^* &= \frac{V_L \rho_L^{1/2}}{[gd \, (\rho_L - \rho_G)]^{1/2}} \\
V_G^* &= \frac{V_g \rho_g^{1/2}}{[gd \, (\rho_L - \rho_G)]^{1/2}} \\
V_L^{*1/2} + V_G^{*1/2} &= 0.8
\end{aligned} \tag{15.30}
$$

Better correlation is obtained in terms of Kutateladze number (K_G) which includes surface tension.

$$
K_G = 0.286 \, B_0^{0.26} Fr^{-0.22} \left[1 + \frac{\mu_L}{\mu_{water}}\right]^{-0.18} \tag{15.31}
$$

$$
\begin{aligned}
K_G &= \frac{V_G \rho_G^{1/2}}{[g\sigma \, (\rho_L - \rho_G)]^{1/4}} \\
B_o &= \frac{d^2 \, g \, (\rho_L - \rho_G)}{\sigma} \\
Fr &= V_L \frac{\pi}{4} d \left[\frac{g(\rho_L - \rho_G)^3}{\sigma^3}\right]^{1/4}
\end{aligned} \tag{15.32}
$$

The other correlation often used is

$$
\dot{m} = 80 \frac{D^{0.3} \rho_L^{0.46} \sigma^{0.09} \rho_G^{0.5}}{\mu^{0.14} (\cos\theta)^{0.32} (m_L/m_G)^{0.07}} \tag{15.33}
$$

Recently Zapke and Kroger (1961) suggested a correlation for superficial vapour Froude number in terms of a new parameter ZK (Zapke and Kroger 1997), as

$$Fr_{HSV} = K_{fl} \exp\left(-n_{fl} \times Fr_{dsl}^{0.6}/ZK_{d_H}^{0.2}\right) \qquad (15.34)$$

where

$$ZK = \frac{\sqrt{\rho_L d_H \sigma}}{\mu_L}$$

and n_{fl}, K_{fl} = constants depending on angle of inclination.

15.6.2   Compact Condensers

Use of compact heat exchanger for condensing (also evaporation duty) is a matter of increasing interest and intense research activity at present. A stream of such studies have been presented in a series of conferences dedicated to 'compact heat exchangers for process industry' (1997 Snowbird USA, 1999 Banff Canada, 2001 Davos Switzerland). Though air cooled heat exchangers are also compact ones, they are treated separately in the next section. Usually plate and plate fin type of condensers are becoming popular. They are discussed below.

15.6.2.1   Plate Condensers

Plate heat exchangers are catching up fast in condensation duties. Due to their relative light weight, high efficiency and low space requirement they are now even considered for hydrocarbon duties. For non hydrocarbon duties of refineries and petrochemicals, usual gasketted model can be used but for hydrocarbon duties the welded type of plate exchangers with allowable temperature upto 570 °C and pressure upto 70 bar (gauge) should be used.

The flow arrangement used in plate exchangers for condensation duty is little different from that in single-phase applications. The reflux condition is avoided and hence the vapour side is always single pass with downward flow. This is because the flow channel is small and as a result the large vapour velocity will give rise to flooding. Also, the port arrangement for reflux condition will be inconvenient. However, the coolant can follow multi pass arrangement. These single and multi pass coolant arrangements are shown in Fig. 15.39a and b respectively. However it is known from works of Bassiouny and Martin (1984) as well as the present author (Rao 2001) that flow can be severely maldistributed from port to the channels in PHEs. With multi pass flow this maldistribution is likely to be even more acute. Hence, an entry from both sides of the port as shown in Fig. 15.39c is often preferred.

For computation of heat transfer in plate heat exchangers two different regimes can be considered - gravity controlled (falling film) regime and shear controlled regime.

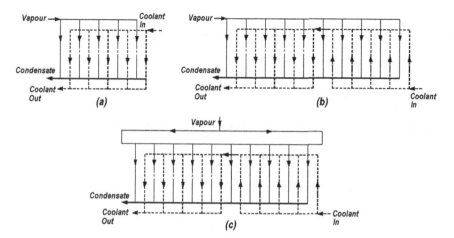

Fig. 15.39 Plate condenser. **a** Single pass, **b** Multi pass, and **c** Double entry

For the falling film regime the heat transfer coefficient for vertical plate following Nusselt's theory is given by,

$$h = C_o \, 1.47 \, \mathrm{Re}_L^{1/3} \qquad (15.35)$$

where C_o = physical property number.

15.6.2.2 Plate Fin Condensers

Plate fin heat exchangers are less common in condensation duties except cryogenic applications other than the air cooled ones. It is generally accepted that the Nusselt theory holds good for plate fin configuration if the interfacial shear is not significant. However, for shear driven flow and mixture of vapours it is suggested that the same method as plate heat exchanger be used but the condensation heat transfer coefficient with fin configuration be used.

15.7 Air cooled Condensers

In many cases it is more convenient to reject heat directly to atmosphere rather than cooling water or other process fluid. This is because other cooling systems involve additional piping and equipment such as cooling tower. Air cooled condensers are extensively used in

1. Condensation vapour from distillation column
2. Refrigeration applications

Air cooled condensers are always finned on the air side to compensate the poor conductance of the air side. They have got major advantages of

1. Cost reduction due to less piping
2. Cost reduction due to elimination of additional devices
3. No need of water treatment
4. Cleaner operation
5. Can operate as leak proof sealed unit where environmentally hazardous fluids are (such as refrigerants) to be handled
6. Water cost and availability do not become a constraint
7. Fouling is much less compared to water cooled units
8. Can be used where water cannot be used due to hazard of leakage (e.g., concentrated acid)

They however have got certain disadvantages as well:

1. Possibility of air leakage is more, which in the form of non condensable can harm the performance
2. Start up takes time
3. Maintenance of the fans is more expensive
4. Performance vary with ambient condition varies
5. Noise associated with draught fans is more

Air cooled units are usually 'A' frame mounted as shown in Fig. 15.40. It has got the additional advantage of occupying less ground area. Usually multi tube rows are avoided due to excess air side pressure drop.

Different types of flow arrangements are possible. Vertical down flow and reflux condensers are possible. In all the units induced draught is preferred because it gives better flow distribution. Figure 15.41 shows a reflux air cooled condenser. Reflux designs are more useful for distillation column where large subcooling is not desirable.

Examples

Problem 1: An air cooled condenser for a package air conditioning unit is to be designed to transfer 22.2 kw of heat from R22 condensing inside 1.27 cm Outer Dia. and 1.12 cm Inner Dia. Tubes at 50 °C. The mass flow rate of the refrigerant is 0.22 kg/s. The refrigerant-side heat-transfer coefficient is given by,

$$Nu = 0.026 \, (Pr)^{(1/3)} \, (Re_m)^{0.8}$$

Air circulated is 120 cmm (cubic meters per minute) entering at 40 °C the air-side heat-transfer coefficient is approximated by,

$$Nu = 0.193 \, (Pr)^{(1/3)} \, (Re)^{0.618}$$

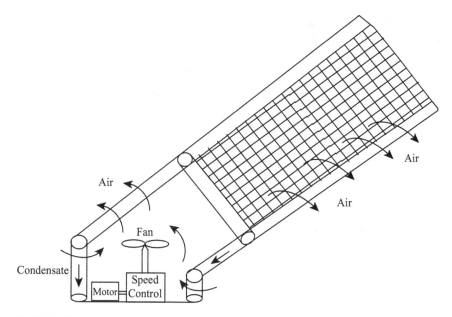

Fig. 15.40 'A' frame air cooled condenser

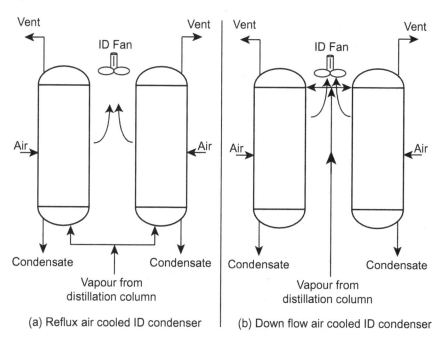

(a) Reflux air cooled ID condenser | (b) Down flow air cooled ID condenser

Fig. 15.41 Reflux, induced draught, air cooled condenser

The Prandtl number of air can be taken as 0.71. The face velocity of air may be taken as 6 m/s. The air side has 6 fins/cm so that finned surface to outside bare tube surface area ratio is 20. Determine the finned surface area of the condenser. Neglect the resistance of metal wall. Assume efficiency of finned surface $\eta_f = 0.9$.

Solution: Refrigerant side heat-transfer coefficient. Inside area of the tube,

$$A_i = \frac{\pi (D_i)^2}{4}$$
$$= \frac{\pi (0.0112)^2}{4}$$
$$= 9.852 \times 10^{-5} \, \mathrm{m}^2$$

Mass velocity of refrigerant,

$$G = \rho u$$

Assume a refrigerant-side temperature drop of 5 °C. So the mean film temperature is

$$T_m = 55 - \frac{5}{2}$$
$$= 52.5 \,°\mathrm{C}$$

At this temperature, R22 liquid film properties are,

$$k_f = 0.08 \, \mathrm{W/mK}$$
$$\rho_f = 1096 \, \mathrm{kg/m}^3$$
$$\mu_f = 1.765 \times 10^{-4} \, \mathrm{kg/ms}$$
$$C_p = 1.36 \, \mathrm{kJ/kgK}$$

Hence,

$$Re_m = \frac{G\,(D_i)}{\mu_f}$$

$$Re_m = \frac{(0.22) \times (0.0112)}{(1.765 \times 10^{-4}) \times (9.852 \times 10^{-5})}$$
$$= 141,700$$
$$Pr_f = \frac{C_p(\mu_f)}{k_f}$$
$$Pr_f = \frac{(1.36 \times 10^3) \times (1.765 \times 10^{-4})}{0.08}$$
$$= 3.0005$$

$$Nu = \frac{h_i D_i}{k_f}$$
$$= 0.026 \, (Pr)^{(1/3)} \, (\text{Re}_m)^{0.8}$$
$$Nu = 0.026 \times (3.0005)^{(1/3)} \times (141,700)^{0.8}$$
$$= 495.604$$
$$h_i = \frac{Nu \, k_f}{D_i}$$
$$= \frac{495.604 \times 0.08}{0.0112}$$
$$= 3540.03 \, \text{W/m}^2 \, \text{K}$$

Air side heat transfer coefficient, face area,

$$FA = \frac{Q_v}{u}$$
$$= \frac{(120/60)}{6}$$
$$= \frac{1}{3} \, m^2$$

Equivalent diameter,

$$D_E = \sqrt{\frac{4 \, FA}{\pi}}$$
$$= \sqrt{\frac{4 \times (1/3)}{\pi}}$$
$$= 0.65 \, m$$

Mean temperature of air,

$$t_a = 40 + \frac{\Delta t_a}{2}$$
$$= 40 + 4.63 = 44.63 \, ^\circ\text{C}$$

At this temperature for air,

$$\mu = 2 \times 10^{-5} \, \text{kg} \, \text{m}^{-1} \, \text{s}^{-1}$$
$$\rho = 1.11$$
$$k = 0.03 \, \text{W} \, \text{m}^{-1} \, \text{K}^{-1}$$
$$Re = \frac{\rho u \, D_E}{\mu}$$

$$= \frac{1.11 \times 6 \times .65}{2 \times 10^{-5}}$$

$$= 216450$$

$$Nu = 0.193 \times (0.71)^{(1/3)} \times (234,000)^{0.618}$$

$$= 341.4$$

$$h_o = \frac{Nu\, k}{D_E}$$

$$= \frac{358 \times 0.03}{(0.65)}$$

$$= 15.8 \ \mathrm{W\, m^{-2}\, K^{-1}}$$

Overall heat transfer coefficient. Neglecting the metal-wall resistance, the overall heat-transfer coefficient based on the fin-side surface area is given by,

$$\frac{1}{U_t\, A_t} = \frac{1}{h_i\, A_i} + \frac{1}{h_o\, A_t\, \eta_f}$$

$$\frac{1}{U_t} = \frac{1}{h_i} \frac{A_t}{A_0} \frac{A_0}{A_i} + \frac{1}{h_o} \frac{1}{\eta_f}$$

$$\frac{1}{U_t} = \frac{1}{3540.03} \times 20 \times \frac{1.27}{1.12} + \frac{1}{15.8} \times (0.9)$$

$$U_t = 13.033 \ \mathrm{W\, m^{-2}\, K^{-1}}$$

Finned surface area. Temperature rise of air

$$\delta t_a = \frac{Q_k}{Q_v} \rho C_p$$

$$= \frac{22.2 \times 60}{120 \times 1.2 \times 1.005}$$

$$= 9.25\,°\mathrm{C}$$

Air leaving temperature $t_{a2} = 40 + 9.25 = 49.25\,°\mathrm{C}$.

Problem 2: The following reading were taken in a test on surface condenser

Mean temperature of condensation	: 35 °C
Temperature of hot well	: 30 °C
Condenser vacuum	: 670 mm of Hg
Barometer reading	: 760 mm of Hg
Condensate per hour	: 930 kg
Inlet temp. of cooling water	: 17 °C
Outlet temp. of cooling water	: 29 °C
Cooling water circulated per hour	: 36500 kg

Calculate:

(a) The weight of air per m^3 in condenser
(b) The condition of steam entering the condenser
(c) The vacuum efficiency

Solution: Condenser pressure

$$(760 - 670) \times 13.6 \times 9.81 = 12\,\text{kPa}$$

Partial pressure of steam corresponding to 30 °C = 5.575 kPa.
So, partial pressure of air = 12 − 5.575 = 6.425 kPa. Now PV = mRT, weight of air per m^3

$$W = \frac{P}{RT}$$
$$= \frac{6.425}{0.287 \times (35 + 273)}$$
$$= 0.0727\,\text{kg}$$

Heat carried away by cooling water = heat lost by steam

$$mC_p \delta T = m(h_f + x(h_{fg}))$$
$$36500 \times 4.2 \times (29 - 17) = 930 \times (146.5 + x(2411.67))$$
$$x = 0.7586$$

Vaccum efficiency = $\dfrac{\text{Vaccum actully obtained}}{\text{Vaccum that would have been obtained if air were absent}}$

$$\eta_v = \frac{670 \times 13.6 \times 9.81}{(101.325 - 5.575) \times 103} = 0.9335\ \text{(or)}\ 93.35\%$$

Problem 3: Design a vertical 4 pass shell and tube heat condenser with 25.2 mm tubes having benzene vapour condensing at a rate 4 t/hr in the shell side while cooling water is flowing in the tube side. The inlet and outlet temperatures of cooling water are 15 °C and 35 °C respectively. The benzene vapour is saturated and condensing at atmospheric pressure.

Fouling factor:
benzene side: 8.59 × 10^{-5};
Water side: 5 × 10^{-4};

424 15 Equipment for Boiling, Evaporation and Condensation

Benzene

Physical Properties at	4 °C	60 °C	80 °C	100 °C
Density (kg/m^3)	858	836	815	793
Dynamic viscosity (Ns/m)	5.2×10^{-4}	4.1×10^{-4}	3.3×10^{-4}	2.5×10^{-4}
Thermal conductivity (W/m.K)	0.143	0.136	0.132	0.125

Heat condensation of benzene at 1 atm pressure = 400.145 KJ/kg.
Basic characteristics of 4-pass shell and tube heat exchanger with tube dia. (25 × 2 mm)

Shell dia. (mm)	No of tubes	Heat exchanger surface with the following tube lengths in mm^2			
		2000 m	2500 m	3000 m	4000 m
400	90		17	21	
600	232		44	53	72
800	446	67	85	102	137
1000	736	100	140		126

The height of condenser should not exceed 3.5 m.

Solution: The condenser temperature of benzene at 1 bar is 80 °C. Calculation of LMTD,

$$\theta_{h1} = 80 °C \quad \theta_{h2} = 80 °C$$
$$\theta_{c1} = 15 °C \quad \theta_{c1} = 35 °C$$

So, $\Delta\theta1 = \theta_{h1} - \theta_{c1} = 65 °C$ and $\Delta\theta2 = \theta_{h2} - \theta_{c2} = 45 °C$

$$LMTD(\Delta\theta_{lm}) = \frac{(\Delta\theta1) - (\Delta\theta2)}{\ln\left(\frac{\Delta\theta1}{\Delta\theta2}\right)}$$

$$\Delta\theta_{lm} = \frac{(65 - 45)}{\ln\left(\frac{65}{45}\right)} = 54.4 °C$$

Calculation of mean temperature of cooling water

$\Delta\theta_{lm} = 54.4 °C$
$\theta_{h1} = 80 °C$
$\theta_{cm} = 25.6 °C$

Calculation of heat load, $q = M_h \times H_{cond}$

$$q = \frac{4000}{3600} \times 400.145 = 444605 \text{ W}$$

Mass flow rate of cooling water

$$M_c = \frac{q}{C_{pc}(\theta_{c2} - \theta_{c1})}$$
$$= \frac{444.605}{4.174 \times (35 - 15)}$$
$$= 5.32 \text{ kg/s}$$

Number of tubes: Let us assume turbulent flow condition for water and $R_e = \frac{\rho u d_i}{\mu}$.

$$M_c = n \times \left(\frac{\pi}{4}\right) \times (d_i)^2 \times u \times \rho$$
$$n = \frac{4 \times M_c}{\pi \times (d_i)^2 \times u \times \rho}$$
$$n = \frac{4 \times M_c}{\pi \times (d_i) \times R_e \times \mu}$$

but according to the data minimum number of tube should be $= 90$. So we adopt $= 90$. There for number of tubes in each pass $= \frac{90}{4} = 22.5$.

Pass	No of tubes
1st	22
2nd	22
3rd	23
4th	23

Calculation of heat transfer coefficient for cooling water

$$R_e = \frac{\rho u d_i}{\mu}$$
$$M_c = \left(\frac{\pi}{4}\right) \times (d_i)^2 \times u \times \rho \times n$$

From these two we get,

$$R_e = \frac{4 \times M_c}{\pi \times (d_i) \times n \times \mu}$$

M_e	d_i	n	μ	Re	Inference
5.32 kg/s	0.021 m	22.5	891×10^{-6}	16089	The flow is turbulent

Nusselt Number:

$$Nu = 0.021 \, (Re_f)^{0.8} (Pr_f)^{0.4} \left(\frac{Pr_f}{Pr_w}\right)^{0.25}$$

For turbulent flow through tubes: $Pr_f = 6.125$ at $25.6\,°C$;

$$\frac{Pr_f}{Pr_w} = 1 \text{ (assumed)}$$

$$Nu = 0.021 \times (16089)^{0.8} \times (6.125)^{0.4} \times (1)^{0.25} = 100$$

Heat transfer coefficient:

$$h_i = \frac{Nu \times k}{d_i} = \frac{100 \times 0.608}{0.021}$$

Heat transfer coefficient for condensing Benzene.

$$h = 3.78 \times k \times \left(\frac{\rho^2 \times n \times d_o}{\mu \times M_h}\right)^{1/3}$$

Overall Heat transfer coefficient (U_0),

$$U_0 = \frac{1}{\left(\frac{d_o}{d_i}\right)\left(\frac{1}{h_i}+\frac{1}{h_{st}}\right)+\left(\frac{r_o}{k_w}\right)\ln\left(\frac{d_o}{d_i}\right)+\left(\frac{1}{h_{so}}+\frac{1}{h_o}\right)}$$
$$U_0 = 417\ W/m^2\,°C$$

Heat transfer surface of the condenser:

$$A_o = \frac{q}{U_o\Delta\theta_{lm}} = \frac{444605}{417 \times 54.4} = 19.5\,m^2$$

Conclusion:

Shell (dia)   : 400 mm
Tube(dia)   : $L25 \times 2$ mm
Number of tubes : 90
Length of tubes : 3 m each.

Problem 4: A condensing plant condenses 12473 kg/hr and leakage of air in steam is 1 kg/2500 kg of steam. The vacuum in the air pump suction is 71.41 cm (barometer) and temperature 32.2 °C. Compute the capacity of air pump which remove both air and water in cm^3/min. taking the volumetric efficiency as 80%. Assume PV = 96T.
Solution: At 32.2 the pressure of steam is 4.826 kPa. The combined pressure of air and steam is = 6.274 kPa. Partial pressure of air = 6.274 − 4.826 = 1.448 kPa. Weight of air per min,

$$W = \frac{12473}{2500 \times 60} = 0.08315 \text{ kg}$$

Volume of air per minute,

$$1448 \, V = 96 \times 0.0831 \times (273 + 32.2)$$
$$V = 1.681 \, \text{m}^3$$

Weight of steam condensed per minute = 12473/60 kg. Volume of condensate

$$\frac{12473}{(60 \times 1000)} = 0.207 \, \text{m}^3$$

Theoretical capacity of air pump = $1.681 + 0.207 = 1.888 \text{m}^3/\text{min}$. Effective capacity

$$\frac{1.888}{0.8} = 2.3611 \, \text{m}^3/\text{min}$$

Problem 5: Find the surface area required in a surface condenser dealing with 30×10^3 kg of steam per hour 90% dry at $0.04 \text{kgf}/\text{cm}^2$ absolute pressure. The cooling water enters the condenser at 15 °C and leaves at 25 °C. Assume the overall heat transfer coefficient 12600 kJ/m^{2}hr °C. If the condenser has two water passes composed of 2 cm outside diameter and 1.2 cm thick with water speed of 1.5 m/s. determine the number of tubes in each pass and length of each tube.

Solution: The saturation temperature of the steam at 0.04 atm is 28.6 °C. Total heat lost by steam per hour = total heat gain by water

$$30 \times 10^3 \times 2440 \times 0.9 = m_w \times 4.2 \times (25 - 15)$$

$$m_w = 1.57 \times 10^6 \text{ kg/hr.}$$

The LMTD can be calculated by using the following expression.

$$\text{LMTD} = \frac{(\theta_i) - (\theta_o)}{\ln\left(\dfrac{\theta_i}{\theta_o}\right)}$$

$$= \frac{(28.6 - 15) - (28.6 - 25)}{\ln\left(\dfrac{28.6 - 15}{28.6 - 25}\right)} = 7.52 \,^\circ\text{C}$$

The surface area required can be calculated using the following equations

$$Q = AU(LMTD)$$
$$6.594 \times 10^7 = A \times 12600 \times 7.52$$
$$A = 696\,\mathrm{m}^2$$

Assume the number of tubes in one pass is n we can write down the following equation,

$$\left(\frac{\pi}{4}\right) \times (d_i)^2 \times V \times \rho \times n \times 3600 = 1.5 \times 10^6$$

$$V = 1.5\ \mathrm{m/s}$$

$$\rho = 998.2\,\mathrm{kg/m^3}\ \text{ at film temperature }\ \left(\frac{15+25}{2}\right) = 20\,°\mathrm{C}$$

$d_i = 20 - 2.4 = 17.6\mathrm{mm}$. Total number of passes in both the tubes $= 2\,n = 2396$. The length of the each tube is

$$\pi \times d_o \times L \times 2n = 696$$

$$\pi \times (0.02) \times L \times 2396 = 696$$

$$L = 4.64\ \mathrm{m}.$$

Problem 6: Air is heated in the shell of an extended surface exchanger. The inner pipe is 38.1 mm IPS schedule 40 pipe carrying 28 longitudinal fins 12.5 mm high and 0.89 mm thick. The shell is 76.2 mm schedule 40 steel pipe. The exposed outside area of the inner pipe is 0.127 m²/m (not covered by the fins); total surface area of the fins and pipe is 0.863 m²/m. Steam condensing at 121 °C inside the inner pipe has a film coefficient 8500 W/m².°C. The thermal conductivity of steel is 45 W/m².°C. The wall thickness of inner pipe is 3.68 mm. The mass velocity of the air is 24400 kg/h.m² and the average air temperature is 54.5 °C, what is the overall heat transfer coefficient based on the inside area of the inner pipe? Neglect fouling factor.

Solution: From tables

ID of shell $= 77.84\,\mathrm{mm}$
OD of inner pipe $= 48.26\,\mathrm{mm}$

The cross sectional area of the shell space is,

$$\left(\frac{\pi \times \left((0.07784)^2 - (0.04826)^2\right)}{4}\right) - (28 \times 0.0127 \times 0.00089)$$
$$= 2.6215 \times 10^{-3}\ \mathrm{m}^2$$

The perimeter of the air space is

$$(\pi \times 0.07784) + 0.863 = 1.10238 \, \text{m}$$

The hydraulic diameter is

$$D_e = \frac{4A}{P} = \frac{4 \times 2.6125 \times 10^{-3}}{1.10238} = 9.48 \times 10^{-3} \text{m}$$

$$Re = \frac{G \, D_e}{\mu} = \frac{24400 \times 9.48 \times 10^{-3}}{0.06879} = 3.37 \times 10^{3}$$

From graph the heat transfer factor is

$$j_h = \left(\frac{h_0}{c_p \times G}\right)\left(\frac{c_p \mu}{k}\right)^{2/3}\left(\frac{\mu}{\mu_w}\right)^{-0.14} = 0.0031$$

The quantities needed to solve for h_0 are

$C_p = 1.05$ KJ/kg K
$k = 0.028$ W/m K

In computing μ_w the resistance of the wall and the steam film are considered negligible so $T_w = 121\,^{\circ}\text{C}$ and $\mu_w = 2.19 \times 10^{-5}\text{kg/ms}$

$$\Phi = \left(\frac{\mu}{\mu_w}\right)^{0.14} = \left(\frac{0.06879}{0.07896}\right)^{0.14} = 0.981$$

$$Pr = \frac{\mu \, c_p}{K} = \frac{1.91 \times 10^{-5} \times 1.05 \times 10^{3}}{0.028} = 0.716$$

$$h_0 = \frac{0.0031 \times 0.981 \times 1.05 \times 10^{3}}{(0.716)^{2/3}} \times \left(\frac{24400}{3600}\right) = 27.05 \, \text{W/m}^2 \, \text{K}$$

For rectangular fins, disregarding the contribution of the ends of the fins to the perimeter, $Lp = 2L$, and $S = Ly_f$ is the fin thickness L is the length of the fin. then

$$a_F \, X_F = X_F \sqrt{\frac{(2 \, L/L \, y_f) \times h_0}{k_m}}$$

$$= X_F \sqrt{\frac{2 \, h_0}{y_f \times k_m}}$$

$$= 0.0127 \sqrt{\frac{2 \times 27.05}{45 \times 0.00089}} = 0.46$$

$\eta_f = 0.933$ (given) inside diameter $D_i = 0.0409$ m

$$D_L = \frac{(0.04826 - 0.0409)}{\ln\left(\frac{0.04826}{0.0409}\right)} = 0.0444\,\text{m}$$

Area inside $= A_i = \pi \times 0.0409 \times 1 = 0.128\,\text{m}^2/\text{m}$

$A_F + A_b = 0.863\,\text{m}^2/\text{m}$
$A_F = 0863 - 0.127 = 0.736\,\text{m}^2/\text{m}$
$L = 0.0444\,\text{m}$

$$U_i = \frac{1}{\frac{0.128}{28\,(0.933 \times 0.763 + 0.127)} + \frac{0.00368 \times 0.0409}{45 \times 0.0444} + \frac{1}{8500}} = 172.08\,\text{W/m}^2\,\text{K}$$

Problem 7: Basic design of a Direct-Expansion Chiller: Design a DX chiller for a refrigeration cycle of R 22 with 20 TR capacity given the following specifications: Effective tube length $= 221.5$ cm, diameter of tubes $= 1.905$ (OD) cm and 1.704 cm ID, 8 refrigerant passes, water enters at $11.1\,°C$ and leaves at $7.2\,°C$. Standard tables for R 22 may be used considering condensing temperature of $2.2\,°C$. Assume a water side heat transfer coefficient of $4650\,\text{W/m}^2\text{K}$, and on R22 side as $h_i = 230\,\Delta TW/\text{m}^2\text{K}$. Find the number of tubes in the last pass.

Solutions: Assuming that there is no fouling (R 22 is a clean fluid) and that conduction term can be neglected,

$$\frac{1}{U_o} = \frac{1}{h_o} + \frac{1}{h_i}\frac{A_o}{A_i}$$

For 20TR, Heat transfer per pass is

$$Q = \frac{20 \times 3.52}{8} = 8.80\,\text{kW}$$

Assuming that enthalpy change is same in each pass,

$$\Delta t_w = \frac{\Delta t_{w,total}}{N} = \frac{11.1 - 7.2}{8} = 0.49\,°C$$

$$\left(\dot{m}\,C_p\right)_w \Delta t_{w,total} = \dot{Q}$$

$$\dot{m}_w = \frac{20 \times 3.52}{4.18 \times (11.1 - 7.2)} = 4.31\,\text{kg/s}$$

For the last pass,

$t_{w,in} = 11.1\,°C$
$t_{w,out} = 11.1 - 0.49 = 10.61\,°C$

$$MTD = \frac{0.49}{\ln\left(\frac{11.1-2.2}{10.61-2.2}\right)} = 8.65\,°C$$

hence $h_i = 230 \times (8.65 - 2.2) = 1484\,W/m^2K$ (R 22 side)

$$\frac{1}{U_o} = \frac{1}{4650} + \left(\frac{1}{1484}\right)\left(\frac{1.905}{1.704}\right)$$

$U_o = 1032.7\,W/m^2K$. Thus,

$$A_o = \frac{\dot{Q}}{U_o\,\Delta t_{MTD}} = \frac{8.8 \times 10^3}{1484 \times 8.65} = 0.6853\,m^2$$

$$A_i = 0.6853 \times \frac{1.704}{1.905} = 0.6129\,m^2$$

Iteration: From here the value of MTD is checked and further iteration is done. Using a scientific calculator, $A_{o,final} = 0.924\,m^2$

$$n_{tubes} = \frac{A_o}{\pi\,D_o\,L}$$

$$= \frac{0.613}{3.14 \times 0.01905 \times 2.215} = 5$$

Number of tubes = 7. Iteratively, are can be adjusted for $n_{tubes} = 5$; otherwise taking $n_{tubes} = 8$ will result in overdesign.

Problem 8: Estimate the capacity of a R22 flooded chiller given the design conditions as $t_{evap} = 2\,°C$, water inlet temperature of $t_{wi} = 11.1\,°C$, $\dot{m}_w = 24.33\,kg/s$. The tube is of OD = 17.04 mm and ID = 13.79 mm with length of 3.69 m and two passes with 14 tubes per pass. No. of fins = 748/m with diameter of over fins, Dfin = 0.01905 m and fin thickness of 0.8 mm.

Solution: Assume a seed value of $\Delta T_o = 5\,°C$ for film temperature and $t_{wo} = 7\,°C$ (water outlet temperature).

Using Rohsenow's correlation for evaporation

$$q = h_o\,\Delta T = \mu_f\,h_{fg}\left[\frac{g\,(\rho_f - \rho_g)}{\sigma}\right]^{\frac{1}{2}}\left[\frac{c_f\,\Delta T}{0.013\,h_{fg}\,Pr_f^{1.7}}\right]^3$$

The standard refrigeration table for R22 is used to substitute values in the above equation.

$$h_o \times 5 = 0.23 \times 10^{-3} \times 203.7 \times 10^3 \times \left[\frac{9.81 \times (1279 - 22.57)}{0.0115}\right]^{\frac{1}{2}}$$

$$\times \left[\frac{1.177 \times 10^3 \times 5}{0.013 \times 203.7 \times 10^3 \times 3.07^{1.7}}\right]^3$$

Hence, $h_o = 350\,\text{W/m}^2$ K. Average temperature of water is

$$t_{wm} = \frac{11.1 + 7}{2} = 9.05\,°\text{C}$$

$$A = 114 \times \pi \times \frac{(0.01379)^2}{4} = 0.0175\,\text{m}^2$$

$$Re = \frac{\rho U D}{\mu} = \frac{\dot{m}}{A} \times \frac{D}{\mu} = \frac{24.33}{0.0175} \times \frac{0.01379}{1.35 \times 10^{-3}} = 14201.5$$

$$Nu = 0.023\,(Re)^{0.8}\,(Pr)^{0.3}\,\text{(Dittus-Boelter equation)}$$

$$h_i = \frac{k}{D_i}$$

$$Nu = \frac{0.579}{0.01379} \times 0.023 \times (14201.5)^{0.8} \times (9.663)^{0.3}$$

$$= 4001.6\,\text{W/m}^2\,\text{K}$$

$$A_i = \pi D_i L = \pi (0.01379) = 0.0433\,\text{m}^2$$

$$A_o = \pi D_o L = \pi (0.01704) = 0.0548\,\text{m}^2$$

$$A_1 = 748\,\pi\,D_{fin}\,b = 748 \times \pi \times 0.01905 \times 0.8 \times 10^{-3} = 0.0358\,\text{m}^2$$

$$A_2 = A_0 - A_1 = 0.0548 - 0.0358 = 0.019\,\text{m}^2$$

$$2\,A_3 = \frac{2\,\pi}{4}\left(D_{fin}{}^2 - D_o{}^2\right) 748$$

$$= \frac{\pi}{2}\left(0.01905^2 - 0.01704^2\right) \times 748$$

$$= 0.0852\,\text{m}^2$$

Fin area

$$A_f = A_1 + 2\,A_3 = 0.0358 + 0.0852 = 0.0121\,\text{m}^2$$

Base area

$$A_{base} = A_2 = 0.019\,\text{m}^2$$

Total effective heat transfer area

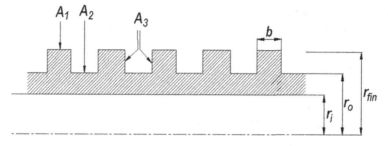

Fig. 15.42 Figure for Problem 8

$$A_t = A_{base} + \eta_f \, A_{fin} = 0.019 + (1)(0.121)0.14 \, \text{m}^2$$

$$\frac{1}{U A} = \frac{1}{h_i A_i} + \frac{1}{h_0 A_t} = \frac{1}{4170 \times 0.0433} + \frac{1}{350 \times 0.14}$$

$$U_t A_t = 38.54$$

$$LMTD = \frac{(\Delta T_{inlet}) - (\Delta T_{outlet})}{\ln\left(\dfrac{\Delta T_{inlet}}{\Delta T_{outlet}}\right)}$$

$$Q = U_t A_t (\delta T_{LM}) F = 38.54 \times 6.85 = 290 \, \text{W/m}$$

(The solution is iterative in nature, only one set of sample calculations is shown above) (Fig. 15.42).

Problem 9: Design the evaporator for a flooded chiller with roughened surfaces using R 22 and following data:

Capacity = 422 kW, $t_{evap} = 2\,°\text{C}$,
$t_{cond} = 42\,°\text{C}$,
$t_{water,in} = 11.1\,°\text{C}$,
$t_{water,out} = 7\,°\text{C}$, 2 passes.

Tube fins as used in example 2 are used.

$$(R_{fouling})_{inside} = 7 \times 10^{-5} \text{m}^2 \, \text{K/W}$$

$$(R_{fouling})_{outside} = 9 \times 10^{-5} \text{m}^2 \, \text{K/W}$$

tube conductivity, $k_{tube} = 386 \, \text{W/mK}$
tube roughness parameter, $R_p = 0.5 \, \mu\text{m}$

Given the correlation,

$$Nu = 230 \, \phi_a^{0.1} (Re_f)^{n_1} \left(\frac{P_h}{Pr_f}\right)^{n_2} \varepsilon^{0.133} \text{ (Slipsevic correlation)}$$

15 Equipment for Boiling, Evaporation and Condensation

Solution: Evaporator Pressure, $P_o = 5.283$ bar at $2\,°$C and critical pressure of R22, $P_c = 49.36$ bar

$$\phi_p = \frac{P_o}{P_c} = 0.107$$

$$\phi_a = \frac{A_t}{A_o} = \frac{0.1399}{0.0548} = 2.5544$$

$$n_1 = 0.75\,\phi_a^{-0.144}\,\phi_p^{0.088}\,\phi_a^{-0.25}$$

$$n_2 = 0.75\,\phi_a^{-0.13}\,\phi_p^{0.28}$$

$$n_1 = 0.75 \times (2.5544)^{-0.144} \times (0.107)^{0.088} \times (2.5544)^{-0.25} = 0.54$$

$$n_2 = 0.75 \times (2.5544)^{-0.13} \times (0.107)^{0.28} = 1.245$$

$$L = \left[\frac{\sigma}{g\,(\rho_f - \rho_g)}\right]^{\frac{1}{2}} = \left[\frac{0.0115}{9.81 \times (1279 - 22.57)}\right]^{\frac{1}{2}} = 9.66 \times 10^{-4}\ \text{m}$$

$$\varepsilon = \frac{R_p}{L} = \frac{0.5 \times 10^{-6}}{9.656 \times 10^{-6}} = 5.178 \times 10^{-4}$$

Assume $N = 228$ tubes, $l = 3.69$ m

$$q_t = \frac{\dot{Q}}{A_t\,N\,L} = \frac{422 \times 10^3}{0.14 \times 228 \times 3.7} = 3583\ \text{Wm}^{-2}$$

$$Re_r = \frac{q_t\,L}{h_{fg}\,\mu_f} = \frac{3583 \times 9.656 \times 10^{-4}}{203.7 \times 10^3 \times 0.231 \times 10^{-3}} = 0.0735$$

$$P_h = \frac{C_f\,T}{h_{fg}} = \frac{1.77 \times (2 + 273)}{203.7} = 1.589$$

$$Pr_f = 2.798$$

$$Nu = \frac{h_o\,L}{k_f} = 230\,\phi_a^{0.1}(Re_f)^{n_1}\left(\frac{P_h}{Pr_f}\right)^{n_2}\varepsilon^{0.133} = 11.2$$

$$h_o = \frac{11.2 \times 0.1}{9.656 \times 10^{-4}} = 1127\ \text{Wm}^{-2}\text{K}^{-1}$$

$$\frac{1}{U_t} = \left(\frac{1}{h_i}\right)\left(\frac{A_t}{A_i}\right) + R_{fi}\left(\frac{A_t}{A_i}\right) + A_t\frac{\ln\left(\frac{D_o}{D_i}\right)}{2\pi k L N} + \frac{1}{h_o\,n_{fin}}$$

$$U_t = 496\ \text{W/m}^{-2}\text{K}$$

$$LMTD = \frac{(11.1 - 2) - (7 - 2)}{\ln\left(\dfrac{9.2}{5}\right)} = 6.89\,°\text{C}$$

Total finned tube surface area required

$$A_t = \frac{\dot{Q}}{U \, \Delta T_m} = 123.6 \, \text{m}^2$$

$$A_i = \frac{A_t}{3.231} = 38.25 \, \text{m}^2$$

Number of tubes = 240.

Exercises

1. In a 112 TR R 22 flooded chiller, the water inlet temperature is 11.67 °C. The mass flow rate of water to be chilled is 18.93 kg/s. The evaporating temperature of the refrigerant is 2.8 °C. The copper tubes used are 1.27 cm OD and 1.12 cm ID. There are two parallel circuits and eight passes. For water the refrigerant-side heat-transfer coefficient is given by Rohsenow's correlation. Calculate the overall heat-transfer surface area of the chiller.

2. In the above problem, investigate the effect of roughening of tubes on the size of flooded evaporator.

3. In the above problem investigate the effect of keeping the number of tubes same as 24 in passes VI to VIII.

4. What is the UA value of a direct expansion finned coil evaporator having the following areas: refrigerant side 15 m^2; air-side prime 13.5 m^2; and air-side extended 144 m^2? The refrigerant-side heat transfer coefficient is 1300 W/m^{2}K, and the air side coefficient is 48 W/m^2K. The fin effectiveness is 0.64.

5. Calculate the mean heat transfer coefficient when refrigerant R 12 evaporates on the outside of the horizontal tubes in a shell and tube condenser. The outside diameter of the tubes is 19 mm, and in vertical rows of tubes there are, respectively two three, four, and two tubes. The refrigerant is evaporating at temperature of 7 °C, and the temperature of the tubes inside is 10 °C.

6. A evaporator manufacturer guarantees the U value under operating conditions to be 990 W/m^{2}K based on the water side area. In order to allow fouling of the tubes, the tubes, what is the value of U required for when the condenser leaves the factory?

7. Compute the fin effectiveness of an aluminium rectangular plate fin of a finned air-cooling evaporator of the fins are 0.18 mm thick and mounted on 16 mm OD tubes. The tube spacing is 40 mm in the direction of airflow and 45 mm vertically. The air side coefficient is 55 W/m^{2}K.

8. A R 22 system having a refrigerating capacity of 55 kW operates with an evaporating temperature of 5 °C and rejects heat to water cooled condenser. The compressor is hermetically sealed. The condensing temperature is 41.2 °C and the heat

transfer coefficient of the condenser is $18\,m^2$ and receives a flow rate of cooling water of $3.2\,kg/s$ at a temperature of $30\,°C$. What is the value of U of evaporator?
9. A shell and tube evaporator has a U of $800\,W/m^2K$ based on the water side area and a water pressure drop of $50\,kPa$. Under this operating condition 40% of heat transfer resistance is on the water side. If the water flow rate is doubled, what will the new pressure drop be?

References

Bassiouny MK, Martin H (1984) Flow distribution and pressure drop in plate heat exchangers-II, Z-type arrangement. Chem Eng Sci 39(4):701–704
Bell KJ (1986) Temperature profiles in pure component condensers with desuperheating and/or subcooling. Course on Process Heat Transfer, London
Chun KR, Seban RA (1971) Heat transfer to evaporating liquid films. Trans. ASME Ser. C. J. Heat Trans. 93:391–396
Nageswara Rao T (2001) Effect of air ingress and spray on performance of a dual condenser. In: Doctoral thesis, Indian institute of technology, Madras
Palen JW, Small WM (1964) A new way to design kettle reboilers. Hyd. Proc. 43:199
Palen JW, Taborek J (1962) Refinery kettle reboilers. Chem. Eng. Prog. 58
Palen JW (1983) Shell-and-tube reboilers. Heat exchanger design handbook, vol 3. Hemisphere Publishing, Sect. 3.6
Palen JW (1988) Falling film evaporation in vertical tubes. In: Chisholm D (ed) Heat exchanger technology. Elsevier Applied Science, London, New York, p 208
Proceedings of 1st international conference on compact heat exchangers and enhancement technology for the process industries (1997). Snowbird
Proceedings of 2nd international conference on compact heat exchangers and enhancement technology for the process industries (1999). Banff
Proceedings of 3rd international conference on compact heat exchangers and enhancement technology for the process industries (2001). Davos
Wallis GB (1961) Flooding velocities for air and water vertical tubes, AEEW-R123
Zapke A, Kroger DG (1961) The influence of fluid properties and inlet geometry on flooding in vertical and inclined tubes. Int J Multiph Flow 22(3):461–472
Zapke A, Kroger DG (1997) Vapor-Condensate Interactions during Counterflow in inclined Reflux Condensers. In: HTD-vol 342, National Heat Transfer Conference, vol 4. ASME

Chapter 16
Boiling in Mini and Microchannels

In this chapter, we look at some special topics which are important during liquid–vapour phase change. These topics are not essential for basic understanding of gas–liquid two-phase flow and heat transfer but important from the viewpoint of applications and emerging research scenario.

The topic of boiling in mini and microchannels has developed in recent years particularly due to the high energy density cooling applications. Cooling of electronic components including computing machinery is the major contributor to this area. However, applications such as cooling of LASER, high power X-Ray, superconducting magnets, etc., are also important in this regard. Micro/mini channel boiling has distinctive features because of very different types of flow regimes compared to that in conventional channels discussed earlier. The confinement of bubbles play a major role here and this in turn affects the pressure drop and heat transfer. The surface tension starts playing a major role as the dimension of channel gets smaller.

16.1 Introduction

Flow boiling heat transfer in microchannels is now the frontline area of research. The need has arisen for cooling of the new generation of computer chips. Heat flux have risen from about 30–100 W/cm² today. The challenge represents a heat flux that is beyond the experience of usual flow boiling. Flow boiling is preferred over single-phase liquid cooling because of high heat transfer coefficient for a given mass flow rate of the coolant. Flow boiling systems can remove a large amount of heat through latent heat of vapourization.

Flow boiling heat transfer is the combination of two different mechanisms, nucleate boiling and convective boiling. In general, nucleate boiling is dominant at high heat flux and low vapour quality, while convective boiling is important at high mass flux and high vapour quality. Developing models for predicting heat transfer coeffi-

S. K. Das and D. Chatterjee, *Vapor Liquid Two Phase Flow and Phase Change*, https://doi.org/10.1007/978-3-031-20924-6_16

cient requires a thorough understanding of the flow patterns existing under different conditions. Fundamental knowledge of boiling mechanisms, the effect of microchannel size, mass flux, vapour quality, heat flux, surface tension, buoyancy on the boiling regimes for such flows remain to be better understood.

From a scientific perspective, flow boiling in microchannels are still works in progress. Their widespread applications are expected in MEMS devices, microscale sensors and actuators, and bio-medical applications.

16.1.1 Macro- to-Microscale Transition in Two-Phase Flow and Heat Transfer

Two-phase flow pattern maps and flow boiling heat transfer prediction methods developed for macroscale applications (typically based on databases with channel sizes ranging from 3 mm up to about 20 mm) are not very accurate or reliable when extrapolated down to channel size below 1–2 mm. Hence, this points specifically to the need to develop macro-to-micro transition criteria to predict the lower boundary of the macroscale and the upper boundary of the microscale, with a transition region in between (mesoscale or miniscale). This means that for the domain of two-phase heat transfer, a well-defined criterion is required for predicting the limiting diameter above which macroscale methods and theory are applicable and backed not only by numerous flow observations but also by experimental results on boiling heat transfer, critical heat flux (CHF), condensation and two-phase pressure drops. Furthermore, methods are also required for predicting thermal performance in channel sizes falling in the mesoscale regime. While flow boiling in microchannels is unexplored, one can expect that a nanoscale heat transfer domain may emerge for smaller-sized channels. As of today, there appears to be no exact definition for definitively distinguishing the transition between macroscale and microscale for two-phase heat transfer.

From a phenomenological viewpoint, some macroscale thermal and fluid phenomena may be suppressed or relegated to secondary importance by the decrease in channel size while others may be enhanced. On the other hand, there does not appear to be a complete disconnect between these two scales, but a gradual change over a range of diameters. Most of the basic premises of two-phase flow and heat transfer are valid in both scales, although extrapolation does not work. From a practical viewpoint, it is necessary to be able to identify the limit to differentiate between macro and microscale flow boiling.

16.1.2 Classification

Kandlikar (2003) more recently recommended the following classifications based on channel size of typical heat exchanger applications:

Conventional channels: $D_h > 3\,\text{mm}$
Minichannels: $3\,\text{mm} > D_h > 200\,\mu\text{m}$
Microchannels: $200\,\mu\text{m} > D_h > 10\,\mu\text{m}$

These dimensional criteria were proposed primarily to differentiate the small diameter channels from that larger channels, for which the experimental trends were observed to be distinctly different. Such fixed dimensional values do not however reflect the influence of the physical mechanisms or the physical properties of the two-phases on the transitions, and hence are not reliable. For the present, the above channel size indicators can be considered to be rule-of-thumb classifications. A more detailed classification is shown below and depicted pictorially in Fig. 16.1. The transition regime is given by

Transitional Channels: $10\,\mu\text{m} > D_h > 0.1\,\mu\text{m}, \approx 100\,\text{nm, or } 1000\,\text{Å}$
Transitional Microchannels: $10\,\mu\text{m} > D_h > 1\,\mu\text{m}$
Transitional Nanochannels: $1\,\mu\text{m} > D_h > 0.1\,\mu\text{m}$
Molecular Nanochannels: $0.1\,\mu\text{m} > D_h$

The above classification is based on air at atmospheric conditions.

However, it serves well for other fluids also. In two-phase applications, as the channel size approaches $200\,\mu\text{m}$, changes in heat transfer characteristics are seen for water. For other liquids, the continuum assumptions are still valid. As the channel hydraulic diameter becomes smaller, the ratio of the heat transfer surface area to the cross-sectional area increases inversely. As the flow velocities are smaller in these channels, laminar flow is generally observed in microchannels. The heat transfer coefficient in single-phase flow then increases inversely with channel diameter as Nusselt number is constant for laminar fully-developed flow.

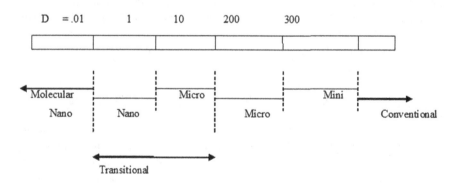

Fig. 16.1 Flow classification length scales (all numbers are in μm)

16.1.3 Difference Between Macro- and Microchannels

Convective flow boiling at low to a moderate wall superheat levels was typically found to consist of a nucleate boiling dominated regime at low quality and regime dominated by annular flow evaporation at moderate to high qualities. Another important feature of internal convective boiling is that often the flow is rapidly accelerating, which induces significant pressure drop along the flow passage. Reducing hydraulic diameter of flow passage impacts all of these important mechanisms, nucleation and bubble growth, annular flow vapourization and pressure loss.

Macrochannel correlations were based on large channels. Secondly, most existing macrochannel correlations were based on turbulent liquid and turbulent vapour flow conditions. However in microchannels flow is laminar and laminar flow correlations need to be applied. The two-phase heat transfer correlations for microchannels are different from conventional channels. Surface tension forces are dominant and gravity forces are insignificant in microchannels. Hence, dedicated studies need to be conducted for microchannels.

16.1.4 Flow Regimes

In the microchannels studied, flow patterns observed via visualizations are classified into five major flow regimes—bubbly, slug, churn, wispy-annular, annular flows and a post-dryout regime of inverted-annular flow. These flow patterns depend on channel sizes, different mass fluxes and flow rates.

Figure 16.2 represents different liquid and gas phases in the flow patterns during adiabatic two-phase flow of air and water mixture. Figure 16.2a shows bubbly flow in which elongated bubbles smaller than the cross section of the channels move with the flow. Bubbles nucleate at the microchannel walls and detach. In bubbly or churn flow, as shown in Fig. 16.2c, bubbles and slugs appear in dispersed form occupying the top portion of the tube, and occur at high liquid and gas velocities. For low liquid and high gas velocities, slug type flow was observed (Fig. 16.2d). Increase in the mass flux leads to an increase in the length of the slugs, leading to merging of the slugs. Thus, the flow is converted to slug-annular flow pattern as shown in Fig. 16.2f. In this pattern the liquid rises in the form of waves while in wavy pattern, gas in the core region. In wavy-annular type flow slugs coalesce with each other and the gas trapped develop a wavy pattern on the surface of the liquid as shown in Fig. 16.2g. The collision of two slugs without merging results in wavy structure giving wavy annular flow pattern. Surface tension leads to the elimination of stratified flow while decrease in tube diameter results in fluid acceleration. The phases do not travel with the same velocity. The coalescence of the slugs occur much faster than in larger tube diameters eliminating wavy annular flow pattern in tube diameters less than 2 mm as shown in Fig. 16.3a. The trailing end of the slug breaks into smaller bubbles. This flow pattern is different from churn flow. Stronger surface tension effects in a

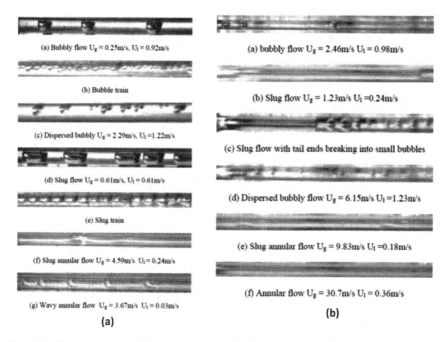

Fig. 16.2 Flow regimes in mini/microchannels. Left side images are for 1.2 mm and right side one is 0.6 mm diameter. (*Courtesy*: M. Venkatesan, Ph.D. Thesis, 2011, IIT Madras)

microchannel allows the liquid film to bridge the gas core more easily than a mini channel. This makes the formation of annular flow less probable. This is visible in Fig. 16.3b of micro channels.

16.1.5 Flow Regime Maps in Micro-Minichannels

There has been lack of agreement on two-phase flow regime maps for microchannels. The shifting of the transition lines with decreasing diameter is unclear.

Two-phase flow patterns in macro/minichannels are presented in Figs. 16.4 and 16.5. The flow regime transition lines obtained Das and Venkatesan with a 3.4 mm diameter tube is compared with that obtained with a 4 mm diameter tube by Damianides and Westwater (1988) and 4 mm diameter tube by Barnea et al. (1983) and are shown in Fig. 16.5. The transition from slug flow to bubbly and churn flow of Damianides and Westwater (1988) agrees well. This is also known as elongated bubble. The term is appropriate since in milli- and micro tubes, the size of the bubble is quite often limited by the diameter of tube. The transition from slug to wavy annular flow agrees well with that of Barnea et al. (1983). The flow pattern observed with 1.2 mm tube diameter were also compared with the flow transitions on a 1.09 mm

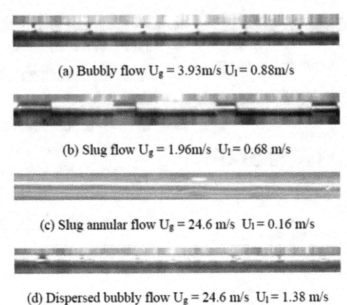

(a) Bubbly flow U_g = 3.93m/s U_l = 0.88m/s

(b) Slug flow U_g = 1.96m/s U_l = 0.68 m/s

(c) Slug annular flow U_g = 24.6 m/s U_l = 0.16 m/s

(d) Dispersed bubbly flow U_g = 24.6 m/s U_l = 1.38 m/s

Fig. 16.3 Flow boiling regimes in macrochannel. Diameter of the channel is 3.4 mm. (*Courtesy:* M. Venkatesan, Ph.D. Thesis, 2011, IIT Madras)

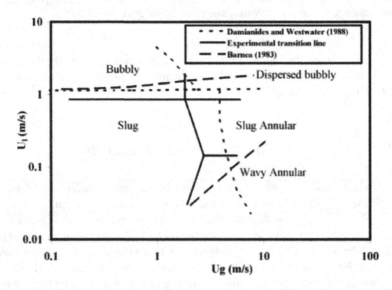

Fig. 16.4 Flow pattern maps with experimental transition lines for 3.4 mm diameter tube. (*Courtesy:* M. Venkatesan, Ph.D. Thesis, 2011, IIT Madras)

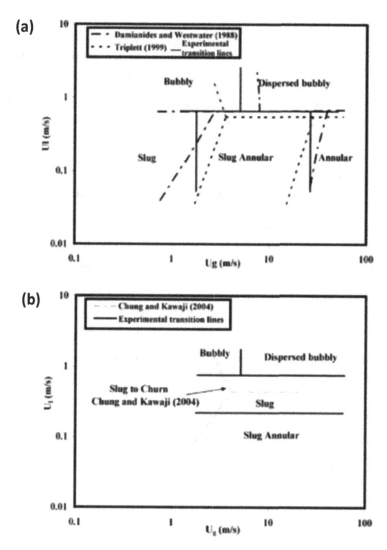

Fig. 16.5 Two-phase flow pattern transition for **a** 1.2 mm and **b** 0.6 mm diameter tubes. (*Courtesy:* M. Venkatesan, Ph.D. Thesis, 2011, IIT Madras)

diameter tube obtained by Triplett et al. (1999) and on a 1 mm tube by Damianides and Westwater (1988) and are shown in Fig. 16.5a. The transition from slug to slug annular flow and from slug annular to annular flow observed by Das and Venkatesan agrees reasonably well with that obtained by Triplett et al. (1999) and Damianides and Westwater (1988). Bubbly flow was observed much earlier when compared with the transition from slug to bubbly flow reported by Triplett et al. (1999). The transition from slug to bubbly flow, slug to slug annular and slug annular to churn flow pattern show agreement with that reported by Damianides and Westwater (1988).

Figure 16.5b shows the flow regimes obtained on a 0.6 mm diameter tube by Das and Venkatesan and compared with a 530 μm diameter tube of Chung and Kawaji (2004). Smaller bubbles and bubble size of the order of the tube diameter were observed in the churn flow regime as mentioned by Chung and Kawaji (2004). Transition from slug to slug annular regime agrees reasonably well. Annular flow was not observed even at the highest gas velocities.

16.1.6   Flow Boiling Heat Transfer in Micro-Minichannels

The rapid miniaturization of the electronics components has resulted in an urgent need to develop efficient heat transfer mechanisms that can effectively remove heat from these components. Since boiling of a liquid involves the latent heat absorption besides the sensible, it offers a great potential for effective cooling of such miniaturized electronic components. Flow boiling in compact heat exchanger passages is relevant to automotive, aerospace, cryogenic and chemical industries also.

In the macroscale we broadly discuss two boiling types, i.e., nucleate boiling which is heat flux governed and the flow boiling which is mass flux governed. In microchannels, we essentially study flow boiling and two important distinctions from the macroscale are

- Due to small diameters, Reynolds number is low and the flow is thus inherently laminar.
- The heat transfer mechanism becomes dominated by nucleate boiling mode.

A number of studies are available on two-phase heat transfer coefficients with water and refrigerant flows. In the case of flow boiling, rapid growth of bubbles, and slug flow have been observed to be most common.

16.1.6.1   Nucleation in Mini/Microchannels

There are two ways in which flow boiling in small diameter channels can occur. They are

(i) two-phase entry after a flashing by throttling
(ii) subcooled liquid entry to the channels followed by evaporation.

The first case takes place in the evaporators used in refrigeration devices. The throttle valve prior to the evaporator is designed to provide a two-phase entry with quality between 0 and 0.1. Although this is a more practical system, the difficulties encountered with proper flow distribution in inlet headers have been a major stable operation.

Subcooled liquid entry is an attractive option, since the higher heat transfer coefficients associated with subcooled flow boiling can be utilized. In either of the systems,

Fig. 16.6 Schematic
representation of bubble
growth in **a** minichannels
and microchannels **b**
conventional large diameter
tube (> 3 mm)

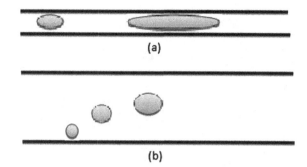

nucleation is an important consideration. Even with a two-phase entry, it is expected that a slug flow pattern will prevail, and nucleation in liquid slugs will be important. With subcooled liquid entry, early nucleation is desirable for preventing the rapid bubble growth.

In micro- and minichannels, similar to the conventional channels, the initial growth of the bubble is inertia controlled. However due to channel diameters of the order of millimetres to micrometres the bubble soon encounters the channel wall before entering the thermally controlled region. The large surface area to fluid volume ratio causes the liquid to heat up rapidly. The bubble continues to grow and spread over the channel wall. The availability of heat from the superheated layer causes a rapid expansion of the bubble, giving long bubble regime. The bubble occupies the entire channel cross section as shown in Fig. 16.6a. This differs from the annular flow pattern mainly in that the liquid near the wall acts similar to the microlayer under a growing vapour bubble in pool boiling. This makes the heat transfer mechanism very similar to the nucleate boiling mechanism. The strong heat flux dependence of the heat transfer coefficient during flow boiling in microchannels is widely accepted, indicating the dominance of nucleate boiling.

In a systematic study by Wambsganss et al. (1993) reported results for R-113 evaporating in a circular channel of 2.92 mm diameter while Tran et al. (1996) reported corresponding results for R-12 evaporating in a 2.46 mm circular channel Fig. 16.7. At wall superheats larger than 2.75 K, they noted that the local heat transfer coefficients did not change with mass velocity (over the range from 50 to 695 kg/m²s) nor with vapour quality (from 20 to 75%) but were dependent on heat flux (from 7.5 to 59.4 kW/m²). Consequently, applying macro-layer logic, they concluded that heat transfer was always nucleate boiling dominated in microchannel evaporation and fitted their data to a nucleate boiling curve of the form.

$$q = a\Delta T^n$$

where ΔT is the wall superheat and q is the heat flux with a being proportionality constant, obtaining an exponent of $n = 2.7$ typical of nucleate pool boiling correlations.

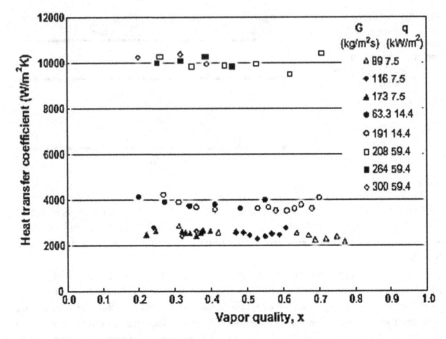

Fig. 16.7 Flow boiling data for R-12 in 2.46 mm tube of Tran et al. (1996). (*Courtesy*: "Small circular- and rectangular-channel boiling with two refrigerants" by T.N. Tran, M.W. Wambsganss and D.M. France, Int. J Multiphase Flow, with permission from Elsevier)

Similar experimental results were obtained by Bao et al. (2000). These results reaffirm the conclusion that microchannel flow boiling does not depend on the mass flux but is a function of heat flux just like nucleate pool boiling.

However, this theory of dominance of nucleate boiling in microchannels has been challenged by many. In other words, some researchers have shown that heat flux is not the sole factor on which microchannel boiling heat transfer depends. Lin et al. (2001) studied evaporation of R-141b in a vertical 1.1 mm tube over a mass velocity range of 300–2000 kg/m² and heat flux range of 18–72 kW/m², although they only presented data at the mass velocity of 510 kg/m²s. Figure 16.8 depicts their results for heat transfer versus vapour quality at these conditions. The outlet pressure of the test section was atmospheric while the inlet pressure varied from 1.35 to 2.20 bar. As opposed to most of the studies, they found a significant influence of vapour quality. The heat transfer model proposed by Jacobi and Thome (2002) also demonstrates that nucleate boiling is not the dominant mechanism. Hwang et al. (2000) also demonstrated significant effects of mass flux. Hence, the complete picture of flux evaporation in micro/mini tubes is yet to be completely confirmed.

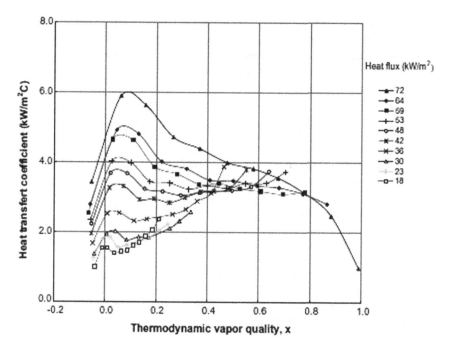

Fig. 16.8 Flow boiling data for R-141b in 1.1 mm tube of Lin et al. (2001). (*Courtesy:* "Flow Boiling of Refrigerant R141B in Small Tubes" by S. Lin, P. A. Kew and K. Cornwell, Chem. Engg. Res Des., with permission from Elsevier)

16.1.6.2   Heat Transfer Correlations

Boiling in mini- and microchannels has fascinated many researchers to carry out studies and experiments. There are numerous heat transfer correlations available in the literature and here we cite a few of them.

Based on the experimental results of boiling heat transfer with two refrigerants in small channels, Tran et al. (1996) correlated boiling heat transfer coefficients as a function of boiling number Bo, liquid Weber number We_l and density ratio $\frac{\rho_l}{\rho_v}$ as

$$h_{two\ phase} = 840000(Bo^2 We_l)^{0.3} \left(\frac{\rho_l}{\rho_v} \right)^{-0.4}$$

where $We_l = \frac{G^2 D}{\rho_l \sigma}$ and $Bo = \frac{q}{G h_{fg}}$. On the similar lines Yu et al. (2002) presented a corrected form of the same equation given by Tran et al. (1996)

$$h_{two\ phase} = 6400000(Bo^2 We_l)^{0.24} \left(\frac{\rho_l}{\rho_v} \right)^{-0.2}$$

It should be noted that these correlations were developed for circular tubes and use of these correlations to rectangular channels can be done using hydraulic diameter in place of diameter. However, the most practical heat sinks, involve heat transfer along only three walls, the top wall being typically adiabatic. To accommodate the difference, a correction factor is introduced.

$$h_{two\ phase} = h_{two\ phase,\ correlation}\left(\frac{Nu_3}{Nu_4}\right)$$

Nu_3 and Nu_4 are the single-phase fully developed Nusselt number for the conditions of three- and four-wall heat transfer, respectively, given as

$$Nu_3 = 8.235(1 - 1.883\beta + 3.767\beta^2 - 5.815\beta^3 + 5.361\beta^4 - 2.0\beta^5)$$

$$Nu_4 = 8.235(1 - 2.204\beta + 3.085\beta^2 - 2.477\beta^3 + 1.058\beta^4 - 0.186\beta^5)$$

where β is the aspect ratio. Also for rectangular channels we use hydraulic diameter in place of tube diameter.

Warrier et al. (2002) suggested another empirical relation for the boiling heat transfer in narrow rectangular channels

$$h_{two\ phase} = \left(\frac{Nu_3}{Nu_4}\right)(Eh_{single\ phase})$$

$$h_{single\ phase} = Nu_4\left(\frac{k_f}{D_h}\right)$$

$$E = 1 + 6Bo^{\frac{1}{16}} + f(Bo)x_e^{0.65}$$

$$f(Bo) = -5.3(1 - 855Bo)$$

Correlations by Yu et al. (2002) and Warrier et al. (2002) employ only the heat flux dependent terms similar to Tran et al. (1996). However, the presence of both nucleate boiling and convective boiling terms to a varying degree is reported by Kandlikar and Balasubramanian (2004). They gave the flow boiling correlation considering both the nucleate boiling dominant (NBD) region and convection boiling dominated (CBD) region. The correlation based on the available data is given as

For $Re_l > 100$:

$$h_{two\ phase} = larger\ of\ h_{two\ phase,\ NBD}, h_{two\ phase,\ CBD}$$
$$h_{two\ phase,\ NBD} = 0.6683Co^{-0.2}(1 - x)^{0.8}h_l + 1058.0Bo^{0.7}(1 - x)0.8F_{FL}h_l$$
$$h_{two\ phase,\ NBD} = 1.136Co^{-0.9}(1 - x)^{0.8}h_l + 667.2Bo^{0.7}(1 - x)0.8F_{FL}h_l$$

where

$$Co = \left(\frac{1-x}{x}\right)^{.8} \frac{\rho_v}{\rho_l}$$

and Bo is the boiling number. The single-phase all-liquid flow heat transfer coefficient h_l is given by

For $10000 \le Re_l \le 5 \times 10^6$

$$h_l = \frac{Re_l \, Pr_l (\frac{f}{2})(\frac{k_l}{D})}{1 + 12.7(Pr_l^{\frac{2}{3}} - 1)(\frac{f}{2})^{0.5}}$$

For $3000 < Re_l < 10000$,

$$h_l = \frac{(Re_l - 1000) \, Pr_l (\frac{f}{2})(\frac{k_l}{D})}{1 + 12.7(Pr_l^{\frac{2}{3}} - 1)(\frac{f}{2})^{0.5}}$$

For $100 < Re_l < 1600$,

$$h_l = \frac{Nu_l k_l}{D}$$

Note that the subscript l refers to the liquid only phase in each term. In the transition region between Reynolds numbers of 1600 and 3000, a linear interpolation is suggested.

For laminar flows, when the Reynolds number is less than 100, nucleate boiling regime is dominating and the effect of convection is negligible, and the following Kandlikar Correlation is proposed,

$$h_{two \; phase, \; NBD} = 0.6683 Co^{-0.2}(1-x)^{0.8} h_l + 1058.0 Bo^{0.7}(1-x)0.8 F_{FL} h_l$$

where

$$h_l = \frac{Nu_l k_l}{D}$$

The fluid surface parameter (F_{FL}) for different fluid surface combinations is given in Table 16.1. These values are for copper or brass surfaces. For stainless steel surfaces, use $= 1.0$ for all fluids. Use of the values listed in Table 16.1 for copper is suggested. Table 16.1 is based on the recommended values of F_{FL} by Kandlikar (1990; 1991).

Flow boiling in small diameter channels is a complex phenomenon. Liquid–vapour interactions, presence of expanding bubbles with thin evaporating film is difficult to analyse. Moreover nucleation in microchannels occurs both in flow as well as thin film. All the factors make it difficult to predict a complete heat transfer mechanism.

Table 16.1 Values of F_{FL}

Fluid	F_{FL}
Water	1.0
R-11	1.3
R-12	1.5
R-22	2.2
R-113	1.3
R-114	1.24
R-134a	1.63
R-152a	1.10
Kerosene	0.488
R-13B1	1.31
R-32/R-132	3.3
R-141b	1.8
R-124	1.0

16.1.7   Critical Heat Flux for Mini/Microchannels

The critical heat flux (CHF) is an important consideration required for the design of cooling using two-phase heat transfer. Before the use of flow boiling in microchannels becomes widely accepted in cooling of electronics and lasers, it is necessary to have a clear understanding of the CHF and estimate its value.

Bergles and Bar-Cohen 1994 carried out extensive experiments with subcooled boiling in tubes as small as 3 mm to ~ 300 μm diameters (Fig. 16.9). They concluded that CHF increased with increasing subcooling and mass flux in a non-linear way. For exit pressures of 0.2–2.2 MPa, CHF has a weak dependence on pressure. Over a diameter range from 0.3 to 3.0 mm, CHF changed by a factor of two; the effect of decreased diameter in increase of CHF is a function of mass flux and subcooling.

Qu and Mudawar (2004) carried out extensive studies on measurement and correlation of critical heat flux in two-phase micro-channel heat sinks using R-113. For these typical microchannel conditions, an unusual vapour backflow from all of the channels into the inlet plenum was observed (Fig. 16.10).

The mild parallel channel instability common to microchannel heat sink gets amplified as heat flux approaches CHF. This causes the vapour to flow backwards into the inlet plenum and increase the liquid temperature to the local saturation temperature as the liquid enters the microchannels. CHF in mini/micro-channel heat sinks increases with increasing mass velocity but, because of the decrease of subcooling due to the backward vapour flow, CHF is independent of inlet temperature (Fig. 16.11). This is a significant departure of mini/micro-channel heat sink behaviour from that of single mini-channels.

They also gave the correlation for the CHF calculation as below

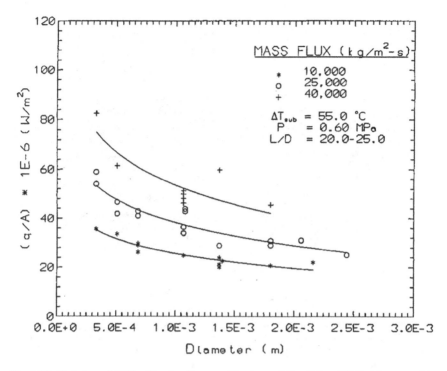

Fig. 16.9 Variation of CHF with tube diameter (Bergles and Bar-Cohen 1994). (*Courtesy:* "An experimental study of critical heat flux in very high heat flux subcooled boiling" by C.L. Vandervort, A.E. Bergles and M.K. Jensen, Int. J Heat Mass Trans., with permission from Elsevier)

$$\frac{q_{p,m}}{Gh_{fg}} = 33.43 \left(\frac{\rho_v}{\rho_l}\right)^{1.11} We^{-0.21} \left(\frac{L}{d_e}\right)^{-0.36}$$

where

$q_{p,m}$ = CHF based on channel heated inside area

We is the Weber number based on the channel heated length $L \dfrac{G^2 L}{\sigma \rho_l}$

d_e is heated equivalent diameter of rectangular channel

This equivalent diameter of rectangular channel is given by

$$d_e = \frac{4(\text{area of cross section of channel})}{\text{heated perimeter of channel}}$$

For circular channels, channel diameter can be used.

The CHF behaviour in multiple parallel microchannels is complex and hence, quite often, determined empirically.

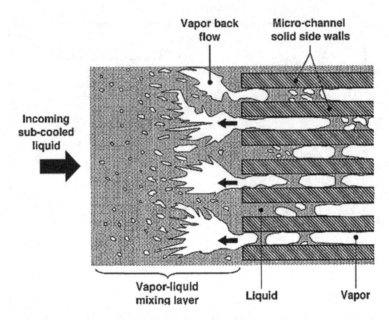

Fig. 16.10 Schematic representation of observed vapour backflow as heat flux approaches CHF (Qu and Mudawar (2004)). (*Courtesy*: "Measurement and correlation of critical heat flux in two-phase micro-channel heat sinks" by Weilin Qu and Issam Mudawar, Int. J Heat Mass Trans., with permission from Elsevier)

16.1.8   Flow Boiling Pressure Drop in Micro/Minichannels

If we consider a practical case of subcooled liquid entering the channel, the total pressure drop in a microchannel can be written as the sum of following pressure drops:

$$\Delta p_{total} = \Delta p_{entry} + \Delta p_{single\ phase,\ frictional} + \Delta p_{2-phase\ frictional}$$
$$+ \Delta p_{acceleration} + \Delta p_{gravity} + \Delta p_{exit}$$

Often, as the fluid enters a microchannel, it encounters a sudden contraction. As in practical systems, the microchannels may be connected to macro reservoirs or pumping systems, and similar loss is there due to expansion at exit. Single-phase entry and exit losses are well known to us. For the two-phase entry and exit losses, Coleman (2003) recommends the following scheme.

$$\Delta p_{entry} = \frac{G^2}{2\rho_l} \left\{ \left(\frac{1}{C_0} - 1 \right)^2 + 1 - \frac{1}{\sigma_c^2} \right\} \psi_h$$

where $G =$ mass flux

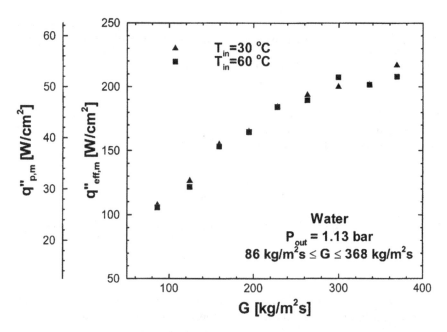

Fig. 16.11 Variation of CHF with mass velocity (Qu and Mudawar 2004). (*Courtesy:* "Measurement and correlation of critical heat flux in two-phase micro-channel heat sinks" by Weilin Qu and Issam Mudawar, Int. J Heat Mass Trans., with permission from Elsevier)

σ_c = contraction area ratio

C_0 = contraction coefficient= $\dfrac{1}{0.639\sqrt{1-\frac{1}{\sigma_c}}-1}$

ψ_h = Two-phase homogeneous multiplier = $\left[1 + x\left(\frac{\rho_l}{\rho_v} - 1\right)\right]$

x = local quality

The exit pressure loss is calculated from the homogeneous model:

$$\Delta p_{exit} = G^2 \sigma_e (1 - \sigma_e)\psi_s$$

σ_e = area expansion ratio

ψ_s = the separated flow multiplier given by

$$\left[1 + \left(\frac{\rho_l}{\rho_v} - 1\right)(0.25x(1 - x) + x^2)\right]$$

The frictional pressure drop in the single-phase region prior to nucleation over length L is calculated from the following equation

$$\Delta p_{single\ phase,\ frictional} = \frac{2f\rho U^2 L}{D}$$

U is the mean flow velocity in the channel; f is the Fanning friction factor, D is the diameter of the channel. For non-circular channels, the hydraulic diameter is given as

$$d_e = \frac{4(\text{area of cross section of channel})}{\text{perimeter wetted by flow}}$$

In the two-phase region, the same equations may be used to calculate the frictional pressure drop as in the case of macrochannels. The local friction pressure gradient at any section is calculated with the following equation:

$$\left(\frac{dp_f}{dz}\right) = \left(\frac{dp_f}{dz}\right)_L \phi_L^2 \qquad (16.1)$$

The two-phase multiplier ϕ_L^2 is given by the following equation by Chisholm (1973):

$$\phi_L^2 = 1 + \frac{C}{X} + \frac{1}{X^2}$$

The value of the constant C depends on whether the individual phases are in the laminar or turbulent region. Chisholm recommended the following values of C:

Both phases turbulent $C = 21$
Laminar liquid, turbulent vapour $C = 12$
Turbulent liquid, laminar vapour $C = 10$
Both phases laminar $C = 5$

The Martinelli parameter X is given by the following equation:

$$X^2 = \frac{\left(\dfrac{dp_f}{dz}\right)_L}{\left(\dfrac{dp_f}{dz}\right)_V}$$

Mishima and Hibiki (1996) found that the constant C depends on the tube diameter, and recommended the following equation for C

$$C = 21(1 - e^{-0.319D_h})$$

where D_h is in mm.
   English and Kandlikar (2005) found that their adiabatic air-water data was over-predicted by using above equation. Thus the following modified equation is recommended for frictional pressure drop calculation in microchannels and minichannels by English and Kandlikar (2005) which showed agreement with their experimental data.

$$\phi^2 = 1 + \frac{C}{x}(1 - e^{-0.319D_h}) + \frac{1}{x^2}$$

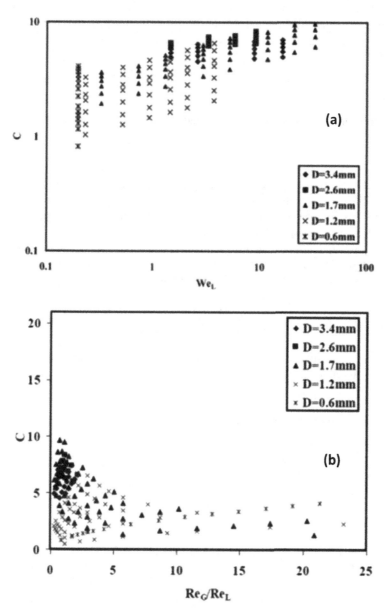

Fig. 16.12 Dependence of Chisholm's parameter (C) on **a** liquid phase Weber number and **b** $\frac{Re_G}{Re_L}$. (*Courtesy*: M. Venkatesan, Ph.D. Thesis, 2011, IIT Madras)

However, to predict the frictional pressure drop for a boiling flow, the quality variation along the flow direction should be taken into account.

$$\Delta p_{boil, friction} = \frac{L}{(x_{out} - x_{in})} \int_{x_{in}}^{x_{out}} \phi_L^2 \left(\frac{dp_f}{dz}\right)_{L, frictional} dx$$

The acceleration pressure drop is calculated from the following equation assuming homogeneous flow

$$\Delta p_{acceleration} = G^2(\vartheta_l - \vartheta_v)x_e$$

where $(\vartheta_l - \vartheta_v)$ is the difference between the specific volumes of vapour and liquid phases. The above equation assumes the inlet flow to be liquid only and exit quality to be x_e. For a two-phase inlet flow, the x_e should be replaced with the change in quality between the exit and inlet sections.

The gravitational forces are rather subdued at the micro level and thus the component of gravitational pressure drop is rather small. The gravitational pressure drop assuming single-phase entry of liquid is given by homogeneous model.

$$\Delta p_{gravitational} = \frac{g \sin \theta L}{(\vartheta_l - \vartheta_v)x_e} \ln \left(1 + \frac{(\vartheta_l - \vartheta_v)x_e}{\vartheta_l}\right)$$

In case of two-phase inlet flow, the difference between the exit and inlet quality should be used in place of x_e.

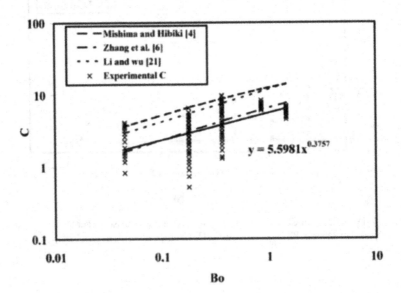

Fig. 16.13 Comparison of correlations with the proposed correlation. (*Courtesy*: M. Venkatesan, Ph.D. Thesis, 2011, IIT Madras)

Figure 16.12a shows the effect of liquid Weber number (We_L) ratio on Chisholm's parameter C. Larger liquid Weber number results in increase in value of parameter C while lower value of C appears to be at lower liquid Weber number. Figure 16.12b shows the effect of $\frac{Re_G}{Re_L}$ ratio on parameter C. For larger tube diameters a higher C value is expected which is shown and for lower tube diameters with the proposed correlation "C" values of up to 0.8 is obtained. Figure 16.13 shows the parameter C value plotted against the Bond number for proposed correlation and the results are compared with the correlations of Mishima and Hibiki (1996). Weber number includes surface tension effects while the inertial and viscous effects are taken care of by the ratio of Reynolds number. Thus the gravitational, inertial, viscous and surface tension forces which are important in two-phase flow in mini channel are included in the proposed correlation.

16.2   Concluding Remarks

Boiling in mini and microchannels is a complex phenomenon and is quite different from what we observe in the macroscale. How a microchannel can be defined is not clear but it can be said that it remains between confined bubble regime and total nucleation suppression regime. All the data on which we can rely is purely experimental and contradictory results have come from various researchers, like in some experiments flow boiling heat transfer coefficients have been shown to be dependent exclusively on heat flux and saturation pressure, almost independent of mass flux, similar to nucleate pool boiling. On the other hand in a new elongated bubble flow heat transfer model of Jacobi and Thome (2002) have shown that transient evaporation of liquid film surrounding elongated bubble is the primary mechanism. Some studies also demonstrate a mass velocity and vapour quality effect. Thus, the two-phase flow and heat transfer in microchannels is still an evolving area in which both experimental and modelling efforts need to be carried out.

Examples

Problem 1: Find flow boiling heat transfer coefficient at quality = 0.3 for water flow rate of 1.88 kg/hr in a channel of 200 μm diameter. The wall temperature is held constant at 140C and inlet temperature of water is given to be 75 °C.

Solution:

Properties of water at 75 °C
Density $\rho = 976.75 \, \text{kg/m}^3$
Dynamic viscosity $\mu = 365.79 \times 10^{-6}$ Pa-s
Thermal conductivity $K_l = 0.6644$ W/m-K
Prandtl number $Pr_l = 2.42$

Surface tension $= 0.0635$ N/m
Specific heat $C_p = 4192$ J/kg-k
$h_f = 419.1$ kJ/kg
$D_h = 200\,\mu$m

Mass Flux

$$G = \frac{\dot{m}}{A_c} = \frac{(5.22 \times (10)^{-4})}{((200 \times (10)^{-6}))^2 \times (3.14/4)} = 16660\,\text{kg/m}^2 - \text{s}$$

$$Re = \frac{GD_h}{\mu} = \frac{(16660 \times 200 \times (10)^{-6})}{(365.79 \times (10)^{-6})} = 9109$$

$$f = 0.0791/(Re)^{-0.25} = 0.0081$$

$$
\begin{aligned}
h_{lo} &= \frac{((Re - 1000)(Pr)_L(f/2)(\frac{k_L}{D_h}))}{(1 + 12.7((Pr)_L^{2/3} - 1)(f/2)^{0.5})}\\
&= \frac{(9109 - 1000) \times 2.42 \times (0.0081/2) \times (0.6644/(200 \times 10^{-6}))}{(1 + 12.7(2.42^{2/3} - 1)(0.0081/2)^{0.5})}\\
&= 168398\,\text{W/m}^2\,\text{C}
\end{aligned}
$$

$$
\begin{aligned}
\Delta T_{sat} &= 413 - 373 = 40\,\text{K}\\
\Delta T_{sub} &= 373 - 348 = 25\,\text{K}\\
q &= h_{lo}(\Delta T_{sat} + \Delta T_{sub})\\
&= 168398(40 + 25) = 10945.87\,\text{kW/m}^2\\
Co &= ((1 - x)/x)^{0.8}(\rho_v/\rho_l)^{0.5}\\
&= ((1 - 0.3)/0.3)^{0.8}(0.5977/957.85)^{0.5} = 0.049\\
Bo &= \frac{q}{Gh_{fg}}\\
&= (10945.87 * 10^3)/(16660 * 2256.9 * 10)^3\\
&= 0.00029\\
h_{tp,NBD} &= (0.6883(Co)^{-0.2}(1 - x)^{0.8}h_{lo}) + (1058(Bo)^{0.7}(1 - x)^{0.8}F_{fl}h_{lo})\\
&= 601.929\,\text{kW/m}^2\,\text{K}\\
h_{tp,CBD} &= (1.136(Co)^{-0.9}(1 - x)^{0.8}h_{lo}) + (667.2(Bo)^{0.7}(1 - x)^{0.8}F_{fl}h_{lo})\\
&= 2452.852\,\text{kW/m}^2\,\text{K}
\end{aligned}
$$

Problem 2: How many microchannels can be designed on a microchip of 20×20cm dimensions for carrying heat flux of $3\,MW/m^2$ and the heat transfer coefficient is limited to $500\,kW/m^2K$. Total channel carries $0.001\,kg/s$ of coolant fluid which is water. Assume an exit enthalpy of 0.35 and fluid entering the channel is saturated. Assume the channels are spaced equally and use the correlation of rectangular narrow channel. Given $\beta = 0.1$. Assume fully developed laminar flow through the channels with three side heating.

Solution: For fully developed laminar flow Nusselt number

$$Nu_3 = 8.235(1 - 1.883\beta + 3.767\beta^2 - 5.815\beta^3 + 5.361\beta^4 - 2.0\beta^5)$$
$$= 8.235(1 - (1.883(0.1)) + 3.767(0.1^2 - 5.815(0.1)^3$$
$$+ 5.361(0.1)^4 - 2.0(0.1)^5)$$
$$= 6.95$$
$$Nu_4 = 8.235(1 - 2.2042\beta + 3.085\beta^2 - 2.477\beta^3 + 1.058\beta^4 - 0.186\beta^5)$$
$$= 8.235(1 - 2.2042(0.1) + 3.085(0.1)^2 - 2.477(0.1)^3$$
$$+ 1.058(0.1)^4 - 0.186(0.1)^5)$$
$$= 6.654$$

$$h_{single\ phase} = (Nu_4)\left(\frac{k_f}{D_h}\right)$$

$$= 8.235 * \left(\frac{0.6804}{\frac{2ab}{(a+b)}}\right)$$

$$Bo = \frac{q}{(Gh_l v)}$$

$$= \frac{(3000 \times 10^3)}{((0.001/nab)2256.9 \times 10^3} = 664.62nab$$

$$f(Bo) = -5.3(1 - 855\,Bo) = -5.3(1 - 855(664.62nab))$$

$$E = 1 + 6Bo^{(1/16)} + f(Bo)x_e^{0.65}$$

$$E = 1 + 6(664.62nab)^{(1/16)} - 5.3(1 - 855(664.62nab)(0.35^{0.65})$$

Here $n = \dfrac{0.2}{2a}$ and $\beta = \dfrac{a}{b}$

$$E = 1 + 6(664.62a)^{(1/16)} - 5.3(1 - (568.250 \times 10^3)(0.35^{0.65})$$

$$h_{two\ phase} = \frac{Nu_3}{Nu_4}(Eh_{single\ phase})$$

$$500 \times 10^3 = \left(\frac{6.95}{6.654}\right) * \left(8.235 * \left(\frac{0.6804}{(2ab/((a+b)}\right) \times 1 + 6(664.62a)^{(1/16)}\right.$$

$$\left. - 5.3(1 - (568.250 * 10^3)(0.35^{0.65})\right)$$

Using this equation we get value

$$a = 84.5\,\mu m$$
$$b = 845\,\mu m$$
$$n = 1183\ microchannels.$$

Problem 3: Determine the Critical heat flux in a rectangular microchannel of dimensions $215 \times 821\,\mu m$ through which de-ionized and deaerated water supplied at mass flux of $100\,kg/m^2s$ is flowing at inlet temperature of $40\,°C$ and outlet pressure of 1.13 bar. The channel is $30\,mm$ long.

Solution: Properties of water at 1.13 bar

Density $\rho_l = 955.56\,kg/m^3$
$\rho_g = 0.6725\,kg/m^3$
Surface tension $= 0.058\,N/m$
$h_{fg} = 2248.2\,kJ/kg$

$$D_h = \frac{(4A_c)}{P_w}$$
$$= \frac{(2 \times 215 \times 10^{-6} \times 821 \times 10^{-6})}{(215 \times 10^{-6} + 821 \times 10^{-6})}$$
$$= 340.76\,\mu m$$

$$q_{crit} = 33.43 \times Gh_{fg} \times \left(\frac{\rho_g}{\rho_l}\right)^{1.11} \times ((G^2L)/(\sigma\rho_f))^{-.21} \times (L/D_h)^{-0.36}$$

$$= 100 \times 2248.2 \times 33.43 \times \left(\frac{0.6725}{955.56}\right)^{1.11} \left(\frac{(100^2 0.04)}{(0.058 * 955.56)}\right)^{-.21}$$

$$\times \left(\frac{0.04}{(340.76 \times 10^{-6})}\right)^{-0.36}$$
$$= 288.12\,kW/m^2$$

Problem 4: Water flows through a circular tube microchannel of 100 μm diameter at temperature of 80 °C at atmospheric conditions. Flow rate is 0.001 kg/s. The channel walls are maintained at uniform heat flux of 500 kW/m². Bubble contact angle is given as 20 °. Calculate the critical radius of bubble at the incipient boiling location.

Solution: Water properties at 80 °C

Density $\rho = 974\,\text{kg/m}^3$
Dynamic viscosity $\mu = 354.536 \times 10^{-6}\,Pas$
Thermal conductivity $K_l = 0.6687\,\text{W/mK}$
Prandtl number $Pr_L = 2.22$

Mass Flux

$$G = \frac{\dot{m}}{A_c} = 0.001/((3.14/4)100 \times 10^{-6})$$
$$= 12.73 \times 10^{-4}\text{kg/m}^2\text{s}$$
$$Re = (GD_h)/\mu$$
$$= (12.79 \times 10^{-4} \times 100 \times (10)^{-6})/(354.536 \times (10)^{-6})$$
$$= 3590$$

For $Re > 3000$ friction factor

$$f = 0.0791/Re^{-0.25} = 0.0018$$

The heat transfer coefficient for $Re > 3000$

$$h_{lo} = \frac{((Re - 1000)Pr_L(f/2)(\frac{k_l}{D_h}))}{(1 + 12.7(Pr_L^{2/3} - 1)(f/2)^{0.5})}$$
$$= \frac{((3590 - 1000) \times 2.22 \times (0.0018/2) \times (0.6687/(100 * 10^{-6})))}{(1 + 12.7(2.22^{2/3} - 1)(0.0018/2)^{0.5})}$$
$$= 27324.66\,\text{W/m}^2\,\text{K}$$

$$\delta_t = \frac{k_L}{h_{lo}}$$
$$= 0.6687/27324.66 = 24.47 \mu\text{m}$$

$$T_{b,z} = T_{b,i} + \frac{(q\,P_w z)}{\dot{m}Cp} = 353 + \frac{(500 * 10^3 \times 3.14 \times 100 \times 10^{-6} \times 0.1)}{(0.001 \times 4195)}$$
$$= 356.74\,\text{K}$$

$$T_{w,z} = T_{b,z} + \frac{q}{h}$$
$$= 356.74 + \frac{(500 \times 10^3)}{27324.66}$$

$$= 375.04 \, \text{K}$$
$$\Delta T_{sat} = 375.04 - 373 = 2.038 \, \text{K}$$
$$\Delta T_{sub} = 373 - 353 = 20 \, \text{K}$$
$$r_{c,crit} = \frac{(\delta_t \sin \theta_r)}{2.2} \frac{\Delta T_{sat}}{(\Delta T_{sat} + \Delta T_{sub})}$$
$$= (24.47 \times 10^{-6} \sin 20^\circ \times 2.038)/(2.2(2.038 + 20)) = 0.35 \, \mu\text{m}$$
$$r_{c,crit} = 0.35 \, \mu\text{m}.$$

Exercises

1. Calculate the liquid sub-layer thickness when water is entering the rectangular microchannel tube of dimensions $120 \times 100 \, \mu\text{m}$ with three side heating at $80 \, ^\circ\text{C}$. Water is flowing at the rate of $0.005 \, \text{kg/s}$. Assume the flow to be laminar fully developed.
2. Water flows through a circular tube microchannel of $200 \, \mu\text{m}$ diameter at temperature of $78 \, ^\circ\text{C}$ at atmospheric conditions. Flow rate is $0.001 \, \text{kg/s}$. The channel walls are maintained at uniform heat flux of $600 \, \text{kW/m}^2$. Bubble contact angle is given as 18^0. Calculate the critical radius of bubble at the incipient boiling location.
3. Circular tube minichannels are arranged in a cooling system to take away heat flux of 1 MW. Water flows through $500 \, \mu\text{m}$ minichannels at $0.001 \, \text{kg/hr}$. The inlet conditions are given as $150 \, ^\circ\text{C}$ and 10 bar. The length of the cooling system is $0.2 \, \text{m}$. Calculate the pressure drop through the mini channel.
4. Heat flux of $500 \, \text{kw/m}^2$ is expected to be carried away by microchannel of 100 /mum diameter. The flow remains single phase throught the channel. What should be the mass flow rate to achieve two-phase flow in atleast 80% of the channel length if the two-phase heat transfer coefficient is limited to $150 \, \text{kW/m}^2$. The inlet temperature is $80 \, ^\circ\text{C}$ and the length of the channel is $500 \, \text{mm}$.
5. A cooling system is to be designed for rejecting $500 \, \text{kW/m}^2$ with mass flow rate of $3.6 \, \text{kg/hr}$ and length of channel $200 \, \text{mm}$ and saturated inlet conditions of 25 kpa. Calculate the heat transfer coefficient inside the mini channel. If the heat transfer coefficient is limited to the same value in a $200 \, \mu\text{m}$ microchannel. What will be the mass flow rate through the microchannel if the exit quality remains the same.
6. A rectangular microchannel heat sinks to be designed to reject heat flux of $80 \, \text{W/cm}^2$. Saturated water enters the channel cut on the chip whose dimensions are given as $10 \times 5 \, \text{cm}$. three sides of the channel wall provide cooling. Calculate the dimensions of rectangular channel and number of channels required if the

heat transfer coefficient is limited $10 \text{W/cm}^2\text{C}$ for mass flow rate of 0.8 kg/hr. The exit quality is given as 0.6 and the channel dimension ratio is 5. Assume fully developed laminar flow.

References

Bao ZY, Fletcher DF, Haynes BS (2000) Flow boiling heat transfer of freon R11 and HCFC123 in narrow passages. Int J Heat Mass Transf 43:3347–3358

Barnea D, Luninski Y, Taitel Y (1983) Flow pattern in horizontal and vertical two phase flow in small diameter pipes. Can J Chem Eng 61(5):617–620

Bergles AE, Bar-Cohen A (1994) Immersion cooling of digital computers. In: Cooling of Electronic Systems. Springer, Dordrecht, 539–621

Chisholm D (1973) Pressure gradients due to friction during the flow of evaporating two-phase mixtures in smooth tubes and channels. Int J Heat Mass Transf 16(2):347–358

Chung PY, Kawaji M (2004) The effect of channel diameter on adiabatic two-phase flow characteristics in microchannels. Int J Multiph Flow 30(7–8):735–761

Coleman JW (2003) An experimentally validated model for two-phase sudden contraction pressure drop in microchannel tube headers. InInternational Conference on Nanochannels, Microchannels, and Minichannels, Vol 36673, pp 241–248

Damianides CA, Westwater JW (1988) Two-phase flow patterns in a compact heat exchanger and in small tubes. In Proc. Second UK National Conf. on Heat Transfer, Glasgow, pp 1257–1268

English NJ, Kandlikar SG (2005) An experimental investigation into the effect of surfactants on air-water two-phase flow in minichannels. In International Conference on Nanochannels, Microchannels, and Minichannels. Vol 41855, pp 615–624

Hwang Y-W, Kim MS, Ro ST (2000) Experimental investigation of evaporative heat transfer characteristics in a small-diameter tube using R134a. In: Proceedings of symposium on energy engineering in the 21st century, Hong Kong, vol 3, pp 965–971

Jacobi AM, Thome JR (2002) Heat transfer model for evaporation of elongated bubble flows in microchannels. J Heat Transf 124:1131–1136

Kandlikar SG (2003) Flow boiling in microchannels: non-dimensional groups and heat transfer mechanisms

Kandlikar SG, Balasubramanian P (2004) An extension of the flow boiling correlation to transition, laminar and deep laminar flows in minichannels and microchannels. Heat Transf Eng 25(3):86–93

Lin S, Kew PA, Cornwell K (2001) Flow boiling of refrigerant R141B in small tubes. Chem Eng Res Des 79(4):417–424

Mishima K, Hibiki T (1996) Some characteristics of air-water two-phase flow in small diameter vertical tubes. Int J Multiph Flow 22(4):703–712

Qu W, Mudawar I (2004) Measurement and correlation of critical heat flux in two-phase microchannel heat sinks. Int J Heat Mass Transf 47:204–2059

Tran TN, Wambsganss MW, France DM (1996) Small circular- and rectangular-channel boiling with two refrigerants. Int J Multiph Flow 22(3):485–498

Triplett KA, Ghiaasiaan SM, Abdel-Khalik SI, Sadowski DL (1999) Gas-liquid two-phase flow in microchannels Part I: two-phase flow patterns. Int J Multiph Flow 25(3):377–394

Wambsganss MW, France DM, Jendrzejczyk JA, Tran TN (1993) Boiling heat transfer in a horizontal small-diameter tube. J Heat Transf 115(4):963–972

Warrier GR, Dhir VK, Momoda LA (2002) Heat transfer and pressure drop in narrow rectangular channels. Exp Therm Fluid Sci 26(1):53–64

Yu W, France DM, Wambsganss MW, Hull JR (2002) Two-phase pressure drop, boiling heat transfer, and critical heat flux to water in a small-diameter horizontal tube. Int J Multiph Flow 28:927–941

Index

Printed in the United States
by Baker & Taylor Publisher Services